UNDERSTANDING BASIC MECHANICS

Text

UNDERSTANDING BASIC MECHANICS

Text

FREDERICK REIF

Carnegie Mellon University

JOHN WILEY & SONS, INC.

New York Chichester Brisbane Toronto Singapore

ACQUISITIONS EDITOR	Cliff Mills
MARKETING MANAGER	Catherine Faduska
PRODUCTION EDITOR	Deborah Herbert
DESIGNER	Lynn Rogan
COVER ART	Ruth Chabay
MANUFACTURING MANAGER	Susan Stetzer

This book was set in 10/12 Times Roman, with illustrations created by the author. The book was printed and bound by Malloy Lithographing, Inc. The cover was printed by Malloy Lithographing, Inc.

Recognizing the importance of preserving what has been written, it is a policy of John Wiley & Sons, Inc. to have books of enduring value published in the United States printed on acid-free paper, and we exert our best efforts to that end.

The paper in this book was manufactured by a mill whose forest management programs include sustained yield harvesting of its timberlands. Sustained yield harvesting principles ensure that the number of trees cut each year does not exceed the amount of new growth.

Library of Congress Cataloging in Publication Data:

Reif, Frederick

Understanding Basic Mechanics – Text / Frederick Reif.

Includes index.

ISBN 0-471-11624-6

Printed in the United States of America

10 9 8 7 6 5 4 3 2

Printed and bound by Malloy Lithographing, Inc.

Preface

Recent investigations (like those cited below[1-3]) reveal that many students emerge from prevailing introductory physics courses with a superficial knowledge of miscellaneous facts and formulas, with many misconceptions and persistent prescientific notions, with poor problem-solving skills, and with little ability to apply what they ostensibly have learned.

The present book has been written as part of an effort to remedy this situation. My aim has been to help students learn some basic physics principles and methods that they can properly interpret and flexibly use — so that they may acquire a sound basis facilitating their subsequent work in various scientific, engineering, or other fields.

As the title of the book indicates, I have sought to implement this aim in the context of a first-semester introductory college-level physics course that deals with mechanics and is addressed to prospective science or engineering students. (The specific topics discussed are indicated in the table of contents.) Some prior student acquaintance with calculus is desirable, but not essential.

Structure of the book

The book consists of two closely coupled parts, a Text and a Workbook addressing complementary functions. The Text is designed to introduce new ideas, and also to facilitate reference and review. The Workbook is designed to get students actively engaged in learning how to interpret and apply these ideas. (It also contains hints and answers to help students study on their own with less need to rely on outside help.)

The "Guide to the Book", which follows this preface, describes in greater detail how the book is structured and how it is to be used.

Pedagogical design

The design of the book is based on the premise that instructional effectiveness can be significantly improved if one gains a better understanding of the thought processes needed for work in physics, and then tries to teach these thought processes more explicitly.

A number of investigations have led me to identify the following cognitive abilities as centrally important for any significant ability to use scientific knowledge. As discussed more fully in a review paper[4] and in the instructor's manual, these abilities include those needed to interpret properly scientific concepts and principles, to describe knowledge in complementary qualitative and quantitative fashions, and to organize scientific knowledge effectively. They also involve more complex problem-solving abilities, including methods to analyze problems, to construct their solutions by judicious decisions, and to check these solutions.

The following paragraphs outline briefly why these abilities are important and how the book attempts to teach them.

Interpretation of concepts and principles. Scientific concepts and principles are necessarily abstract to satisfy the scientific need for generality. But they also need to be unambiguously interpreted in any particular instance. Furthermore, they often require fine discriminations to avoid confusions with other scientific notions or with prior conceptions from daily life.

To achieve unambiguously *accurate* interpretations of scientific concepts or principles, one requires not only explicit descriptive specifications but also procedural specifications (i.e., interpretation methods specifying what one must actually *do* to identify or apply a concept or principle in any specific instance). To achieve *efficient* interpretations, one requires also knowledge about special cases that can be readily recognized and applied in familiar situations.

The design of the book tries to foster these abilities in the following ways: (a) All definitions of important concepts, or statements of important principles, are clearly summarized and highlighted. (b) Many of these are accompanied by explicit interpretation methods which students themselves are asked to apply in diverse specific cases (including those likely to be error-prone). (c) Finally, students are led to summarize their results so that they accumulate useful knowledge about special cases. (For example, Sec. F of Chapter 4 illustrates this approach in dealing with the concept *acceleration*.)

Complementary qualitative and quantitative descriptions. Many scientific tasks are facilitated by using different descriptions in complementary ways. For example, quantitative mathematical descriptions are well suited to achieve precision and to construct long inference chains. On the other hand, qualitative verbal or pictorial descriptions are often better suited to facilitate search processes (like those needed for retrieving appropriate information, for planning, for making decisions, for discovering, for diagnosing difficulties, etc.).

To help students exploit such complementary kinds of description, the book embeds detailed quantitative knowledge within less precise qualitative notions. Thus, new concepts and principles are usually discussed qualitatively before they are formulated more quantitatively. This makes it often easier to motivate the introduction of such concepts or principles. Many potential errors or misconceptions can then also be addressed at a qualitative level before students get submerged in more complex quantitative details.

The following are some examples: (a) The important concept *acceleration* is introduced in Chapter 4 and is there examined qualitatively at some length. It is only later that it is quantitatively elaborated for linear motion in Chapter 5 and for circular motion in Chapter 8. (b) Newton's laws are not simply stated, but emerge as plausible hypotheses generated by trying to discover how the motion of particles is affected by their interactions. Thus the laws of mechanics are introduced as basic relations between motion and interaction, and are later successively elaborated into more quantitative mechanics laws (i.e., Newton's second law, and the laws of momentum, of energy, and of angular momentum).

(c) In this approach the qualitative notion of interaction is gradually elaborated into several quantitative concepts describing different aspects of interaction (i.e., concepts such as *force*, *work*, *potential energy*, and *torque*).

To emphasize the utility of different descriptions, students are asked to solve both quantitative and qualitative problems. Indeed, sometimes the same problem is given in both quantitative and qualitative forms. Furthermore, qualitative considerations are routinely used to check the correctness of quantitative conclusions.

Knowledge organization. To ensure that acquired scientific knowledge is flexibly usable, it is essential that this knowledge be effectively organized so as to facilitate remembering, retrieving desired information, checking consistency, and extending existing knowledge. In particular, hierarchical forms of organization can greatly facilitate these tasks.

To help students acquire well-organized physics knowledge, the book deliberately stresses only a few basic definitions and principles — rather than a large collection of facts and formulas. Thus the entire discussion of motion (through Chapter 8) is centered on the two concepts of velocity and acceleration. After that, everything else deals with the relation between motion and interactions, a relation elaborated into the three basic mechanics laws (of momentum, energy, and angular momentum) which form the backbone of the rest of the book. All discussions revolve around these laws. In particular, the Workbook tries to ensure that students can interpret these laws properly and apply them consistently to deal with all problems.

The book aims to build up students' knowledge gradually in a carefully designed sequence apparent from the table of contents. New knowledge is introduced only when it can be immediately used and practiced. For example, the subtraction of vectors, introduced in Chapter 3, is immediately used to deal with velocity and acceleration. However, components of vectors are not introduced until Chapter 7, where they help deal with projectile motion; and the dot product of vectors is not introduced until Chapter 13, where it helps with calculations of work.

The organization of the knowledge introduced in the book is also made visually apparent by various charts and diagrams (especially in the summary sections and the appendices).

Problem solving. The flexible use of knowledge requires the ability to solve problems (i.e., to devise actions that lead to desired goals). Problem solving is essential to achieve the scientific goals of predicting or explaining many diverse phenomena. Furthermore, problems arise constantly in all scientific work where they may range all the way from simple concept applications to difficult problems involving long inference chains.

Problem solving requires one to deal with the following two fundamental difficulties: (a) How can one describe and analyze a problem effectively so as to facilitate its subsequent solution? (b) How can one construct its solution by making judicious decisions so as to choose, among very many alternative sequences of actions that lead to nowhere, the one (or very few) that lead to the desired goal?

These difficulties are *not* adequately addressed by merely showing students examples of problem solutions and giving them further problems for practice. Hence the book deliberately teaches a simple, but systematic, problem-solving method that explicitly addresses the preceding difficulties. This method specifies (a) how to analyze a problem initially and how to describe it in terms of relevant physics knowledge; (b) how to construct its solution by explicit decisions that exploit a well-organized knowledge base; and (c) how to check the solution so that it can be appropriately revised and improved.

This problem-solving method is first introduced in Chapter 6 to deal with simple kinematics problems; it is then elaborated in Chapter 11 in conjunction with mechanics problems involving Newton's laws; and it is then, with minor modifications, consistently used throughout the rest of the book. The details are more fully spelled out in the cited chapters.

Needed instructional context

Even the best teaching materials cannot guarantee effective learning (no more than perfect pills can cure a disease if patients don't take them according to the recommended regimen). Thus it is important that the book be used in an instructional context or course ensuring that students actually engage in effective learning activities.

Perhaps the most important need is to keep students *actively engaged* in thinking and learning, while they are provided with enough guidance and feedback to ensure that they study properly and acquire good work habits. Thus it helps to minimize lecturing in favor of activities where students are less passively involved; to ask many questions and to give frequent diagnostic tests; and to encourage interaction and collaboration among students.

One must resist the temptation of "covering" too much material. This can easily be counterproductive, and is more likely to create comfortable illusions in teachers than a useful understanding in students. In education, as well as in many other endeavors, less is often more.

Needless to say, the book can usefully be supplemented by many other instructional devices such as demonstrations, special worksheets, computer tutorials, simulations, and hands-on laboratory exercises. On the other hand, these may not necessarily lead to increased student learning. To assess their actual utility, it may be useful to keep in mind the following minimal criterion of instructional effectiveness: If students emerge from a course unable to answer questions like those in the Workbook (except for the more complex questions included in the summary sections), then they have failed to acquire a basic understanding of elementary mechanics.

Acknowledgments

Some of the work leading to this book has been carried out in collaboration with my colleague Jill H. Larkin. Bat-Sheva Eylon collaborated with me in the past and provided some useful suggestions. My research assistants, Lisa Scott and Molly Johnson, have helped substantially with the implementation of my efforts and have become valuable colleagues. I am particularly grateful to Lisa Scott for extensive proofreading and for preparing the index to the book.

The National Science Foundation has partially supported this work through grant #MDR-9150008. The Physics Department at Carnegie Mellon University has permitted me to use its introductory physics course to try this book and some innovative teaching methods. I am grateful to Professor Gregg Franklin who collaborated with me in these trials on several occasions. I also owe thanks to the students who helped me by their willingness to be closely observed while they worked on physics problems.

Finally, I am indebted to some people who have *not* been directly involved in my work (and have sometimes differing points of view), yet indirectly helped by sharing some ideas and making me feel less isolated in pursuing my own endeavors. Among these people are Lillian McDermott, Alan Van Heuvelen, Patricia Heller, Goéry Delacôte, and especially my colleagues Bruce Sherwood and Ruth Chabay. Herbert Simon, even before I knew him personally, stimulated me to take more seriously my own interests in cognitive processes — and has been supportive of efforts to apply these interests to practical education.

I am very grateful to Ruth Chabay for producing the illustrations for the covers of the book. (The rest of the cover design is due to Wiley.)

A book is never finished; it is merely abandoned at some stage of a process of never-ending possible improvements. Some of its deficiencies are always frustratingly apparent to the author. If I abandon the book at this point, it is in the hope that (despite its flaws) it may be useful to some students — and might perhaps suggest to other potential authors ways of writing a better one.

Despite all efforts, there are undoubtedly errors and misprints in the book. I would be grateful if these could be brought to my attention.

Frederick Reif
Carnegie Mellon University
Pittsburgh, PA 15213
November 1994

(1) Halloun, I. A., & Hestenes, D. (1985). The initial knowledge state of college students. American Journal of Physics, 53, 1043-1055.

(2) McDermott, L. C. & Shaffer, P. S. (1992). Research as a guide for curriculum development: An example from introductory electricity. Part I: Investigation of student understanding. American Journal of Physics, 60, 994-1003.

(3) Reif, F. & Allen, S. (1992). Cognition for interpreting scientific concepts: A study of acceleration. Cognition and Instruction, 9, 1-44.

(4) Reif, F. (1995, in press). Millikan Lecture 1994: Understanding and teaching important scientific thought processes. American Journal of Physics.

Guide to the Book

Structure of the book. The book consists of two closely linked parts called *Text* and *Workbook*. The Text is designed to present the basic physics ideas (and strategies for using them), and also to facilitate reference and review. The Workbook aims to ensure that students become actively engaged in learning and understanding what has been presented.

The separation into these complementary functions makes it possible to keep the Text itself more compact. The main ideas can then remain clearly prominent without being obscured by too many details or questions asked of students. In this way the Text becomes easier to use for review or reference. It can also be more readily consulted while working through the Workbook.

The Workbook plays an essential role since it is the primary means facilitating students' learning. After the first introductory chapter, almost every section of the Text is accompanied by a corresponding section of the Workbook. *It is extremely important that students work through the corresponding section of the Workbook* ***immediately after*** *studying each section of the Text.* Otherwise, they will learn very little and will experience difficulties in dealing with later material in the book.

Workbook problems. Each section of the Workbook contains questions or problems designed to assess whether students have understood what they have studied in the Text and can interpret it appropriately. Many of these questions are rather simple and can be answered in a fairly short time. However, it is important that students work through these conscientiously. Indeed, an inability to answer these questions is a clear indication that the material in the corresponding Text section has *not* been adequately understood.

The last section of every chapter contains a summary of the knowledge discussed in it. The corresponding summary section in the Workbook contains some review questions to help students check their basic understanding, and additional practice problems (including some that are more comprehensive or complex). These problems are designed to help students consolidate their knowledge and extend their capabilities beyond the most basic ones.

It is highly recommended that students work through most of the questions accompanying the non-summary sections, and through at least some of the practice problems accompanying the summary sections. Problems marked by a dagger (†) are somewhat more difficult and those marked by a double dagger (‡) are appreciably more difficult.

Organization of the chapters. The chapters in the book are sequentially numbered 1, 2, 3, ... Each chapter is further subdivided into sections labeled A, B, C, ... Equations, statements, and figures are numbered according to the section in which they appear. For example, a reference to (B-7) refers to equation 7 in Section B of the *current* chapter. Similarly, (Text D-2) refers to equation 2 in Section D of the current chapter in the Text. A reference to an item in *another* chapter includes also the chapter number; for example, (4B-7) refers to equation 7 in Section B of Chapter 4.

Appendices. The appendices at the end of the Text contain a review of basic mathematics and charts that summarize concisely the most important knowledge discussed throughout the entire book.

Notational conventions. The right wide margin of every page is reserved for figures and occasional remarks.

Important results are highlighted by being boxed. (The abbreviation *Def.* before such a box indicates an important definition.) Subsidiary material of lesser importance is indented and marked by a vertical line at the left.

Conventional mathematical symbols are used throughout. The symbols $\approx$ ("is approximately equal to") and $\propto$ ("is proportional to") may be slightly less familiar to students.

Hints and answers. Problems in the Workbook are accompanied by numbers (like <*h-5*> or <*a-8*>) which refer, respectively, to hints or answers listed at the back of the Workbook. The purpose of these is to help students study on their own with less need for any outside assistance.

In particular, the hints should reduce the danger that students might get stuck without knowing how to proceed. Similarly, the answers should provide students with useful feedback about whether their work is correct and whether they have understood the requisite knowledge. If a student's answer does *not* agree with that given in the Workbook, this is an indication that something is wrong with the student's work and needs to be corrected. (The available hints may also provide students with clues about the nature of their mistakes.)

Proper use of hints and answers. Hints and answers are helpful only if students use them properly. Thus it is very important that students first try to answer every question *without* referring to the accompanying hints and answers. They should refer to a hint only if this is really necessary, and should refer to an answer only *after* they have tried to answer the question as well as possible. (Hints and answers are listed in random order to help ensure that they are not read without deliberate intent.)

Some students will, undoubtedly, be tempted to use hints and answers irresponsibly by looking at them *before* they have tried to answer questions themselves. I can only urge them to resist this temptation. If they don't, they will remain mostly passive, avoid thinking, and learn very little. They will then fool themselves about what they understand and may soon discover that their seeming knowledge is an illusion hiding many deficiencies.

Unlisted hints and answers. Some hints or answers, indicated in the Workbook by crossed-out numbers like ~~<*h-7*>~~ or ~~<*a-9*>~~, are *not* listed at the back of the Workbook (although they may be available to instructors). Students are thereby given the opportunity to assess their ability to work independently under conditions more closely similar to those encountered in real life or on examinations. Furthermore, they should then have greater incentives to engage in useful discussions with their fellow students or with instructors.

Contents

Motion and Interaction of Systems

UNDERSTANDING BASIC MECHANICS

Text

1 Introduction to the Study of Physics

A. Physics and its relevance

Physics, like any other science, has the following central goal: To discover basic knowledge enabling one to predict and explain many observable phenomena. The main distinguishing characteristic of physics is its generality, i.e., its predominant interest in widely relevant aspects of nature. (For example, physics is centrally interested in the general principles needed to understand all atoms and molecules; but it leaves to chemistry more specific studies of various kinds of molecules.)

The great generality of physics makes it widely applicable and important:

(1) Physics has become a highly successful science capable of explaining or predicting a very large range of diverse phenomena, extending all the way from the behavior of subatomic particles to the structure of the universe.

(2) Physics provides essential knowledge needed for almost all other pure and applied sciences, such as chemistry, biology, geology, astronomy, engineering, metallurgy, and others. (The wide relevance of physics is also evident from the names of many recent scientific fields — such as chemical physics, biophysics, geophysics, astrophysics, etc.)

(3) Some knowledge of physics is needed for the appropriate use of the many electrical, mechanical, optical, and other devices which have come to pervade everyday life and most occupations.

(4) The reasoning and problem-solving methods used in physics are widely applicable in many other domains.

(5) Physics has also philosophical and psychological interest. For example, it raises profound questions about the nature of knowledge and about the symbolic concepts involved in complex thought processes.

(6) Lastly physics, and the many technologies derived from it, have come to permeate our world and have resulted in consequences (beneficial and sometimes detrimental) that profoundly affect all our lives. Thus even people with no particular scientific interests need to know enough about physics to avoid gross misconceptions and, as citizens, to make sensible political decisions about our technological society.

B. Mechanics

This book begins the study of physics with *mechanics*, i.e., the science of motion. The central goal of mechanics is to explain and predict how objects move, and how their motion is influenced by the presence of other objects.

Importance of mechanics

Mechanics is the cornerstone of most of physics and is of central interest for the following reasons:

(1) The world is full of many kinds of motions which one would like to understand and often use for practical purposes (for example, the motions of falling objects, of cars, of boats, of planes, of rolling wheels, of flowing liquids, of moving air masses in meteorology, etc.).

(2) Basic mechanics can be extended in increasingly sophisticated ways to encompass most of physics. For example, it can be extended to "statistical mechanics" which deals with systems consisting of very many atoms, and thus also with heat and temperature; to "relativistic mechanics" which deals with objects moving with very high speeds (close to the speed of light); or to "quantum mechanics" which deals with very small objects like atoms or molecules. Furthermore, mechanics is centrally interested in understanding how the interactions between objects affect their motions. Hence it leads also to the study of various important kinds of interactions (such as gravitational, electric, and magnetic interactions).

(3) Lastly we know that all matter consists of atomic particles. If a science of mechanics leads to an adequate understanding of the motions and interactions of such particles, it then ultimately helps one to understand the properties of all forms of matter (e.g., gases, liquids, solids, biological organisms, ...) and of all complex systems.

Overview of the book

The entire book is directed at one central goal: To understand how the motion of objects is influenced by the interactions between them — and thus to acquire the ability to explain and predict all motions.

In pursuing this goal, we shall follow good scientific practice by examining simple situations before considering progressively more complex ones.

Accordingly, the first part of the book (Chapters 2 through 8) discusses useful ways of describing the motion of particles (i.e., of simple objects small enough that their motions can be adequately specified by those of single points). To do this, these chapters introduce the basic concepts of "position", "velocity", and "acceleration". Then they discuss how these concepts are interrelated and how they can be used to deal with some simple problems.

The second part of the book develops a theory (originally formulated by Newton) which specifies how the motion of particles is related to their interactions — and which is highly successful in predicting the motions of particles. Chapter 9 introduces this theory, Chapters 10 and 12 discuss the properties of some common interactions, and Chapter 11 introduces a problem-solving method needed to apply the theory in various situations. Finally, Chapters 13 and 14 elaborate the theory to derive the widely useful "energy law" and to explore its applications.

The third part of the book extends the Newtonian theory to deal with more complex and interesting systems consisting of any number of particles. Thus Chapter 15 generalizes Newton's law to obtain the "momentum law" applicable to any complex system. Chapter 16 similarly extends the "energy law" to deal with any complex system. Finally, Chapters 17 through 19 discuss the motion of rotating objects and introduce the "angular-momentum law", the third of the fundamental mechanics laws studied in this book.

C. Learning implications

Scientific goal

The central goal of science can be summarized in a deceptively simple way:

> ***Scientific goal:*** To discover knowledge that can be used to predict or explain the maximum number of observable phenomena on the basis of a minimum number of basic concepts and principles. (C-1)

"The aim of science is, on the one hand, a comprehension, as complete as possible, of the connection between the sense experiences in their totality, and, on the other hand, the accomplishment of this aim by the use of a minimum of primary concepts and relations." (Albert Einstein)

A sufficiently well-developed ability to predict also allows one to use scientific knowledge to design the means for achieving many practical goals.

What needs to be learned?

The primary goal of *learning* science is *not* to memorize a large collection of facts and formulas, nor to remember rotely the solutions of some standard problems. Such knowledge is not very useful (and has little to do with science) since it does *not* allow one to deal flexibly with novel or previously unfamiliar situations.

Instead, the scientific goal (C-1) implies the following learning requirements:

(1) One must learn a small number of centrally important scientific concepts and principles which one must be able to interpret properly in any specific instance.

(2) One must learn reasoning and problem-solving methods for applying these concepts and principles so that one can infer their implications in many diverse situations.

(3) One must learn not only to reason accurately and precisely, but also to explore effectively (so that one can discover, plan, design, invent, or diagnose the reasons for encountered difficulties). Hence one must learn both quantitative methods using mathematics, as well as qualitative methods using words or pictures.

By pursuing the preceding learning goals, one can acquire the ability to predict and explain a wide range of phenomena, to deal with unfamiliar situations, and to extend one's knowledge along new directions.

How can it be learned?

The preceding comments imply that scientific work requires one to learn some ways of thinking which are significantly different from those used in everyday life or encountered in earlier schooling. Thus one must also be prepared to transcend familiar notions that may turn out to be naive, misleading, or wrong.

One's efforts must, therefore, be deliberately directed at learning new ways of thinking that are scientifically more effective. This is why the book emphasizes scientifically important reasoning methods and how to apply them in various situations.

However, new ways of thinking cannot be learned without practicing them actively and consistently. One cannot learn to speak French or to play baseball merely by reading books or listening to lectures — or by sporadically practicing now and then. In the same way, one cannot learn physics merely by reading books about physics or listening to lectures — or by feverishly studying just before examinations.

To learn physics, you must must thus actively practice applying the physics principles and methods introduced in the book. You need to solve physics problems, to explain and predict, to raise and answer questions, and to discuss physics with instructors or fellow students. In this way you can gradually gain experience with scientifically useful ways of thinking, improve these in your own ways, and thus acquire intellectual tools that you can use in your future work.

Description of Particle Motions

The central goal of the science of mechanics is to explain and predict how objects move. In order to attain this ambitious goal, one must first address the following simpler question: How can one devise useful ways of *describing* how objects move? The following several chapters address this question and introduce important basic concepts that will be used throughout the study of mechanics.

---

2 Length, Time, and Units

A. Length
B. Time
C. Precision and errors
D. Summary

Length and time are the most fundamental concepts in all the sciences and are essential for describing motion. Although these concepts seem intuitively obvious, they are not easily defined with the precision necessary for scientific purposes. This chapter will indicate briefly how they can be adequately specified so that they can be measured and related to specific observations.

A. Length

Definition of length

Need for operational definitions. Since the goal of science is to explain and predict observations, any scientific concept must be unambiguously related to observations. Hence the definition of any scientific concept must be *operational*, i.e., it must specify what one must actually *do* to identify the concept (or to decide whether any statement about the concept is true or false).

Inadequacy of verbal definitions. The dictionary defines length as "a measure of how long a thing is". Such a definition merely relates one word to other words, but relates none of them to anything observable. It does *not* specify what one means by "long", how one would measure the length of an object, or what one would need to *do* to decide whether some statement about length (e.g., "the length of the table is 5.3 feet") is true or false. The dictionary definition deals purely with words, but is scientifically meaningless. Hence we need define the concept *length* in a scientifically more satisfactory way.

Rough definition of length. Length is a property describing any finite line, whether straight or curved. (For example, Fig. A-1 illustrates such a line whose end points are P and P'.) To relate our intuitive notion of length to observable actions rather than mere words, we can use some rod as a measuring instrument. We can then count the number of times that this rod can be laid end-to-end along the line, starting at one end of the line and without extending beyond the other end (as illustrated in Fig. A-1a). The number N of times that this can be done is a rough measure of what we mean by the *length* of the line.

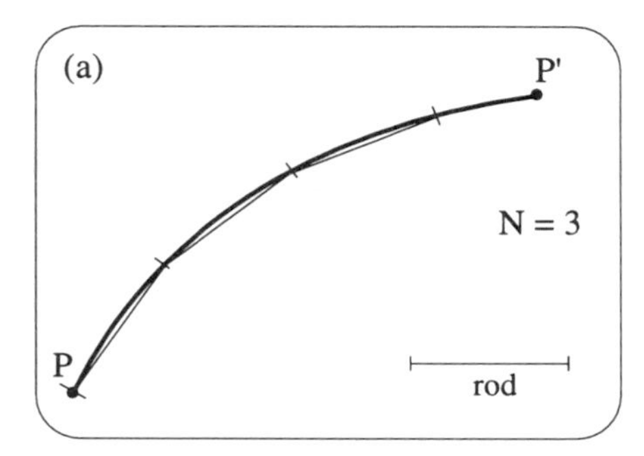

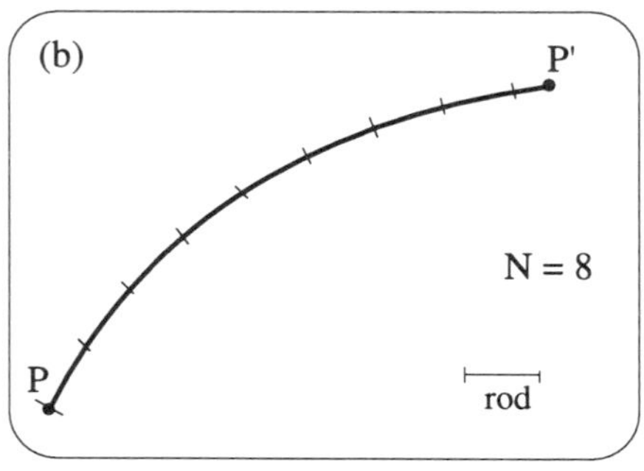

Fig. A-1. Measurement of a line with different rods. (Only the successive positions of the ends of the rod are indicated in the bottom figure.)

The preceding is only a crude specification of length. For example, when the rod is laid along the line the last time, its end point ordinarily does *not* coincide with the end of the line (e.g., with the point P' in Fig. A-1a). Hence the same measured value of N might be obtained even if the line had a somewhat different end point.

Precise comparison of lengths. The precision of the measurement can be improved by using a shorter rod, i.e., a rod which can be laid end-to-end a larger number of times. As indicated in Fig. A-1b, the end points of the line and of the last-laid rod would then coincide more closely.

The preceding remark can be exploited by specifying how to *compare* the lengths of any two lines. To do this, consider any two lines 1 and 2, such as those illustrated in Fig. A-2. Suppose that some rod can be laid end-to-end N_1 times along line 1, and can be laid end-to-end N_2 times along line 2. The ratio L_1/L_2 of the length L_1 of the first line, compared to the length L_2 of the second line, is then *approximately* defined as the ratio of these numbers, i.e.,

$$\frac{L_1}{L_2} \approx \frac{N_1}{N_2}. \tag{A-1}$$

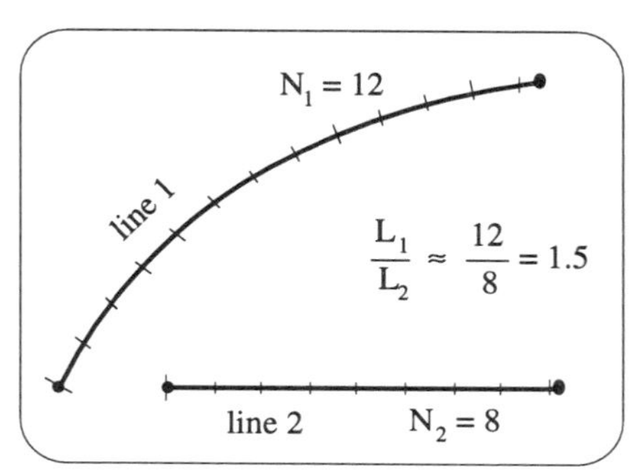

Fig. A-2. Approximate comparison of the lengths of two lines.

For example, if N_1 is twice as large as N_2, one would say that the length of line 1 is approximately twice as large as the length of line 2.

The precision of the comparison can be improved by repeating the preceding measuring process with progressively shorter rods (thereby ensuring closer coincidence between the end of the line and the end of the last-laid rod). The numbers N_1 and N_2 of times that these rods can be laid end-to-end become then progressively larger. However, the *ratio* N_1/N_2 approaches some constant value which remains unchanged when these numbers become large enough (i.e., when a sufficiently short rod is used). This constant *limiting value* provides the following precise definition of the ratio of the lengths of the lines:

Def: ***Length ratio:*** $\frac{L_1}{L_2} = \frac{N_1}{N_2}$ (N_1, N_2 very large). (A-2)

Standards and units

The precise definition (A-2) of length is comparative, i.e., it specifies the *ratio* of any two lengths, but not any one length by itself. Hence it is convenient to specify all lengths by comparison with the length of one particular object chosen as a *standard* of comparison.

Def: **Standard:** A standard for some quantity is a particular *object* with which all such quantities can be compared. (A-3)

The quantity describing the standard can then be denoted by an algebraic symbol called a *unit*.

Def: **Unit:** An algebraic *symbol* denoting the value of the quantity specified by a standard. (A-4)

For example, if an object S is chosen as the standard of length, its length can be denoted by the algebraic symbol L_S. By comparing the length L of any other object with this standard, one then obtains some specific number n for the ratio of their lengths, i.e.,

$$\frac{L}{L_S} = n$$

or

$$L = n\, L_S\,. \tag{A-5}$$

Any length can thus be expressed as a number multiplied by the unit L_S describing the standard.

SI standards and units of length. Standards and units for all the quantities useful in science have been specified, by international convention, in a measurement system called SI (an abbreviation for *Système International*, the French word for *international system*). Until some years ago, the standard of length in this SI system was a particular metal bar, carefully stored near Paris. The length of this bar was denoted by the unit *meter*, conventionally abbreviated by *m*. The length of any other object could then be measured by comparison with this standard.

The unit *meter*, or its abbreviation *m*, is just an algebraic symbol denoting the length of the standard.

Example of length measurement

By comparing the length L of some particular object with the length of the standard, the measurement method (A-2) might yield the following result for the ratio of their lengths

$$\frac{L}{\text{meter}} = 3.7.$$

Hence

$$L = 3.7 \text{ meter} = 3.7 \text{ m}. \tag{A-6}$$

This simply means that the length of this particular object is 3.7 times as large as the length of the standard.

Present standard of length

In recent years, a different standard of length has been adopted by international convention because it is more permanent and allows more precise measurements of length. This new standard is light, and the meter has been redefined to be the distance traversed by light in vacuum during a time of (1/299,792,458) second. (The unit of time *second* is defined in the next section.) This redefinition has been made so that the length of the meter bar in Paris is still, to excellent approximation, one meter long, i.e., so that all previous length measurements are still very closely correct.

Basic and derived units. A *basic* unit is one (like *meter*) which is directly defined in terms of a standard. Many other *secondary* units can then be conveniently defined in terms of such a basic unit.

For example, the unit *kilometer* (abbreviated *km*) is defined in terms of the *meter* by the relation

$$\text{km} \equiv 1000\ \text{m}. \tag{A-7}$$

(Here we used the *identity symbol* ≡ to emphasize that the left side is defined to be exactly equal to the right side.) As another example, the unit *millimeter* (abbreviated *mm*) is defined by the relation

$$\text{mm} \equiv 0.001\ \text{m}. \tag{A-8}$$

Units are algebraic symbols. As pointed out in (A-4), any unit is merely an algebraic symbol, just like x or y. Hence there is *no* need to put units into plural form (e.g., by adding an *s* at the end, as is often done in everyday life).

For example, it is simplest (and in accord with the conventions of the SI system) to say 0.95 meter, 1.05 meter, or 5 meter. There is no more reason to say 5 meters than to say 5xs if one means 5x (i.e., the quantity x multiplied by 5).

A more important implication is that conversions between different units can be accomplished by using simple algebra, i.e., by dealing with units just as with ordinary algebraic symbols.

> ***Example of unit conversion***
>
> To express the length L = 3.7 m in terms of the unit kilometer, one needs merely to use the definition (A-7) to express meter in terms of kilometer. Thus
>
> $$\text{m} = 0.001\ \text{km}.$$
>
> Simple substitution then yields
>
> $$L = 3.7\ \text{m} = 3.7\ (0.001\ \text{km}) = 0.0037\ \text{km}.$$

Unit consistency. Any equation of the form A = B asserts that the quantities A and B are equal. This equality must be true irrespective of what particular standards, and associated units, are used to measure the quantities.

Suppose that both quantities have been expressed in terms of the same set of basic units (e.g., SI units). Then A and B can each be written as some *pure* number (i.e., a number without units) multiplied by some basic units. The equality A = B then implies not only that the pure numbers of both sides of the equation must be equal, but also that the basic units on both sides of the equation must be equal. Thus one arrives at the following conclusion:

Requirement of unit consistency: Both sides of any equation must be expressible in terms of the same basic units.	(A-9)

This requirement provides a very useful way of checking whether any equation is correct. For example, the equation is certainly incorrect if this requirement is *not* satisfied.

➜ ***Go to Sec. 2A of the Workbook.***

B. Time

Definition of time

The dictionary defines time as "the duration in which things are considered as happening in the past, present, or future". Such a definition is scientifically meaningless since it merely relates the word *time* to other words equally unspecified in terms of any observations. Hence we need to provide a scientifically more useful definition of time.

Rough definition of time. Time is a property describing the relation between two events. (For example, one event might be seeing a flash of lightning, and the other event might be hearing a clap of thunder.) Let us then consider any two such events E and E' observed at the same place. In order to relate the *time between these two events* to observable actions one needs a measuring instrument called a *clock.*

Def: **Clock:** A system which repeatedly returns to the same state. (B-1)

For example, a pendulum swinging back and forth is a clock since the pendulum bob returns repeatedly to the same position. A conventional clock with rotating hands is a clock because each of these hands repeatedly returns to the same position (e.g., repeatedly points to the number 12 on the dial). The earth rotating about its axis is also a clock because the sun can be repeatedly observed passing its highest point in the sky on every successive day.

To measure the time between the two events E and E' occurring at the same place, one needs to use a clock located at this place. As schematically indicated in Fig. B-1, one can observe the state of the clock when the first event E occurs. One can then count the number N of repetitions of this state until the occurrence of the second event E'. The number N of repetitions is a rough measure of what we mean by the *time between the two events.*

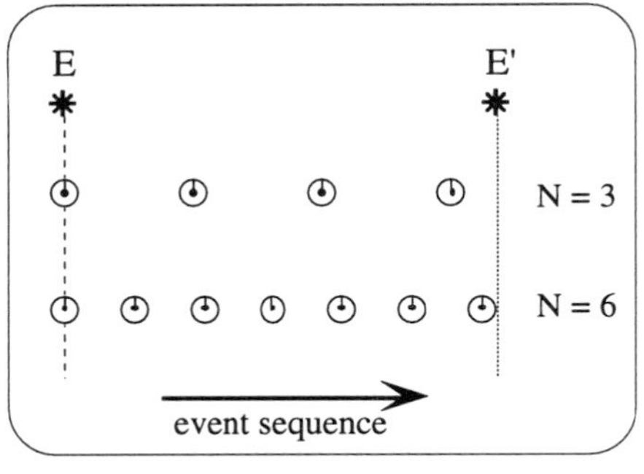

Fig. B-1. Measuring the time between two events E and E' by using two different clocks.

The preceding is only a crude specification of this time. For example, the occurrence of the second event E' ordinarily does *not* exactly coincide with the repetition of the clock . Thus the measured number N of clock repetitions might be the same even if the second event occurred slightly later.

Precise comparison of times. The precision of the measurement can be improved by using a faster clock, i.e., a clock showing a larger number of repetitions. The temporal coincidence between the second event and a clock repetition would then be correspondingly closer (as indicated by the bottom row of clocks in Fig. B-1).

The preceding remark can be exploited by specifying how to *compare* the times between any two pairs of events. To do this, consider a first pair of events E_1 and E_1', and a second pair of events E_2 and E_2' (as illustrated schematically in Fig. B-2). Some clock can then be used to count the number N_1 of clock repetitions between the first pair of events and the number N_2 of clock repetitions between the second pair of events. The ratio T_1/T_2 of the time T_1

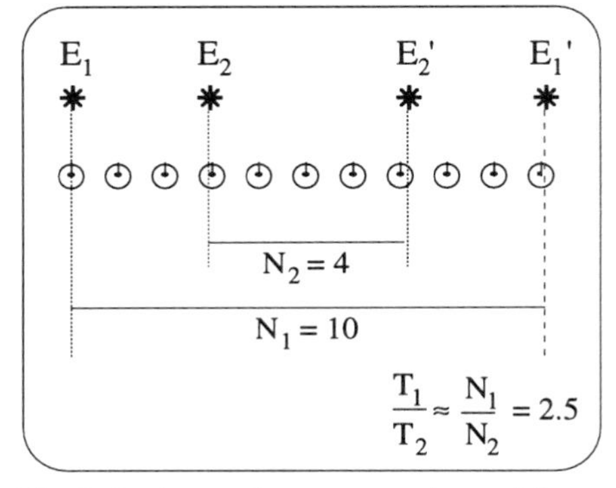

Fig. B-2. Approximate comparison of the times between two pairs of events.

between the first pair of events, compared to the time T_2 between the second pair of events, is then *approximately* defined as the ratio of these numbers of repetitions, i.e.,

$$\frac{T_1}{T_2} \approx \frac{N_1}{N_2}. \qquad \text{(B-2)}$$

For example, if N_1 is twice as large as N_2, one would say that the time between the first pair of events is approximately twice as large as the time between the second pair of events.

The precision of the comparison can be improved by repeating the preceding measuring process with progressively faster clocks (thereby ensuring closer temporal coincidence between the events and the repetitions of the clock). The numbers N_1 and N_2 of clock repetitions become then progressively larger. However, the ratio N_1/N_2 approaches some constant value which remains unchanged when these numbers become larger (i.e., when a faster clock is used). This constant *limiting value* provides the following precise definition of the ratio of the times between the pairs of events:

Def: ***Time ratio:*** $$\frac{T_1}{T_2} = \frac{N_1}{N_2} \quad (N_1, N_2 \text{ very large}). \qquad \text{(B-3)}$$

Standards and units

The precise definition (B-3) of time is comparative, i.e., it specifies only the *ratio* of any two times, but not any one time by itself. It is then useful to specify all times by comparison with one particular clock chosen as a standard of comparison.

SI standard and unit of time. Until some years ago, the clock adopted as the international standard in the SI system was the earth rotating about its axis. The unit *day* was then defined as the time elapsed between successive observations of the sun at its highest point in the sky, and the basic SI unit *second* (abbreviated by *s*) was defined so that

The time between successive observations of the sun at its highest point in the sky is approximately the time required for one revolution of the earth about its axis. (It is not exactly that because the earth also revolves around the sun).

$$\text{day} \equiv 86{,}400 \text{ second}. \qquad \text{(B-4)}$$

Other familiar units of time can all be defined in terms of the second (e.g., minute = 60 second, hour = 60 minute, day = 24 hour). The definition (B-3) allows measurement of any time by comparison with this standard clock, and correspondingly allows one to express any time as a multiple of the basic unit *second* (the algebraic symbol used to describe the time associated with the standard clock).

Choice of a good standard of time

A clock chosen as the standard of time should be permanent, should allow precise measurements, and should lead to a simple description of the world.

To illustrate this last requirement, imagine that one had 100 clocks each of which might be selected as the standard. Suppose that, if the first clock is chosen as standard, all the other 99 clocks run fast compared to this standard, but keep running well-synchronized with respect to each other. In that case, the choice of any of the other 99 clocks would lead to a simpler description of the world. For then all of the clocks would run well-synchronized with this standard, except for the first clock which would run slow compared to this standard.

The preceding situation reflects the actual history of choices of time standards. When highly precise atomic clocks were built, it was found that these run well-synchronized relative to each other, but that the earth slows down relative to them (by about one second per year). Hence the rotating earth is *not* the most desirable standard of time. Accordingly, a cesium atomic clock has more recently been adopted as the SI standard of time and the second has been redefined in terms of this standard. However, this redefinition has been made so that the time between successive passages of the sun at its highest point in the sky is still 86,400 second to excellent approximation (i.e., so that all previous time measurements remain very closely correct).

Inadequacy of the definition of time

Although we may seem to have been pedantically careful in defining the concept of time, we have actually not been careful enough. In particular, we have only defined the time between two events occurring at the *same* place, but have *not* defined the time between two events occurring at different places. Indeed, how would one measure such a time (e.g., the time between the departure of a plane from San Francisco and its arrival in New York)?

At first blush this may seem easy. All one needs is a clock in San Francisco and another one in New York, so that one can compare these clock readings. But how would one properly synchronize these clocks (so that when one of them indicates 3 PM, the other one also indicates exactly the same time)? A good way of doing this might be to send a radio signal from one place to the other. Since such a signal travels with very high speed (about 300,000 km/second), a signal sent from San Francisco would arrive in New York at very nearly the same time, and the clocks could be synchronized accordingly.

For most ordinary purposes, the preceding synchronization would be sufficiently accurate. However, it would *not* be if one is interested in very high precision or in dealing with very fast-moving objects. For then one would need to take into account the time required by the synchronizing radio signal to travel from San Francisco to New York. This could be done if one knew the speed with which the radio signal travels. But, to determine such a speed, one would have to measure the time required for the signal to travel between two different points — and this could only be done if one had two properly synchronized clocks at two different places. Thus one is led into paradoxical circular arguments which are not readily resolved.

A much deeper analysis is required to give a clear self-consistent definition of the time between two events at different places. Such a definition was first provided by Albert Einstein in 1905 and led to his theory of relativity. This theory has many far-reaching consequences, particularly important when one deals with very fast-moving objects or with very precise measurements.

However, in most of this book we shall not deal with such concerns and can proceed without worrying about a more adequate definition of time.

➜ ***Go to Sec. 2B of the Workbook.***

C. Precision and errors

Most quantities in science (like length or time) are measured by comparison methods. Since such comparisons cannot be made with unlimited precision, the value of any such quantity is necessarily somewhat imprecise.

The precision can be specified by the probable error *e* associated with the quantity. Thus one can write Q ± e to specify that a quantity has a value which lies probably in the range between Q − e and Q + e. [For example, one might specify that the length of a desk is (1.524 ± 0.002) m. This would mean that the length is probably between 1.522 m and 1.526 m.] If a quantity is determined more precisely, the error associated with it is correspondingly smaller.

Significant figures. If the precision of a value is not explicitly specified by a stated error, it is implicitly specified by the number of digits used to state the numerical value of the quantity. (The number of such digits, other than the digit 0 used to specify location of the decimal point, is called the number of *significant figures*.) It is then implied that the probable error in the last digit (or last non-zero digit if there is no decimal point) is at least ±1, but unlikely to be ten times larger than that.

> ***Examples***
>
> A length of 1.524 m is specified by 4 significant figures and implies an error of roughly ±0.001 m. A length of 1.5240 m is specified by 5 significant figures and implies an error of roughly ±0.0001 m. A length of 30 m is specified by only 1 significant figure (since the 0 merely specifies location of the decimal point) and implies an error of roughly ±10 m.
>
> Suppose that one wants to indicate that the last zero digit is not merely used to indicate the decimal point, but is really significant (i.e., that the error is only about ±1). Then one can do this either by writing 30. m (with an explicitly indicated decimal point at the end) or by using the scientific notation 3.0×10^1 m.

To avoid conveying misleading impressions about precision, one must be careful to display the numerical values of quantities with an appropriate number of significant figures. For example, many electronic calculators display results with 10 digits. But, if one starts a calculation with an imprecise value specified by two significant figures, the calculator cannot magically yield a precise result specified by 10 significant figures, even if it displays a result with 10 digits.

➜ ***Go to Sec. 2C of the Workbook.***

D. Summary

Length

Determined by comparing two lines with a rod (laid end-to-end N times).

Length ratio: $\frac{L_1}{L_2} = \frac{N_1}{N_2}$ $(N_1, N_2$ very large)

Time

Determined by comparing two pairs of events with a clock (N repetitions).

Time ratio: $\frac{T_1}{T_2} = \frac{N_1}{N_2}$ $(N_1, N_2$ very large)

SI system of standards and units

Basic unit of length = meter.

Basic unit of time = second.

New abilities

You should now be able to do the following:

(1) Convert between values expressed in terms of different units.
(2) Check equations by assessing the consistency of units.
(3) Express numerical values with appropriate numbers of significant figures.

→ ***Go to Sec. 2D of the Workbook.***

3 Displacements and Vectors

To describe the motion of an object, one must be able to specify changes of its position, i.e., its displacements from one position to another. Hence we shall use this chapter to address the following questions: How can one devise convenient ways of describing simple displacements? And how can one combine these to describe more complex displacements? By trying to answer these questions, we shall obtain generally useful methods of describing spatial relationships. These methods are often far simpler than those of traditional geometry, will be constantly used throughout the rest of the book, and are widely useful in mathematics and in many sciences.

A. Position and displacement

Reference frame. Roughly speaking, the *position* of a point is a specification of the relationship between this point and some other objects. These other objects are said to constitute a *reference frame* according to the following definition:

Def: **_Reference frame:_** A chosen set of objects relative to which one specifies the positions of various points of interest. (A-1)

For example, the positions of various points in a room might be described relative to a reference frame consisting of the walls of the room. As another example, the positions of the earth and the planets might be specified relative to a reference frame consisting of the sun and some distant stars.

Coordinate system. The position of a point relative to some reference frame can be most conveniently specified by choosing a particular *coordinate system* in this frame. Such a coordinate system, illustrated in Fig. A-1, is defined as follows:

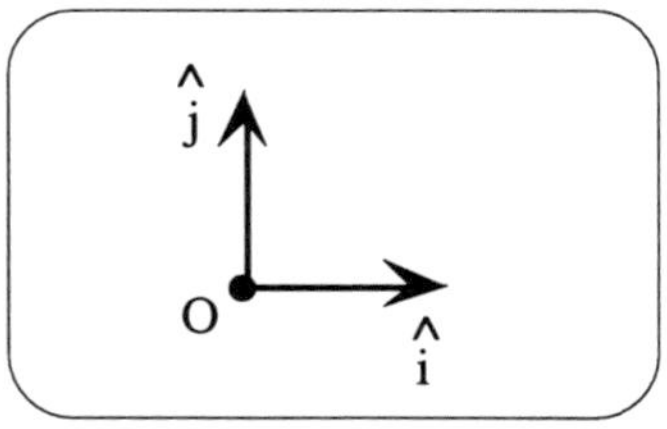

Def: **_Coordinate system:_** A particular point (called *origin*) and a set of mutually perpendicular reference directions specified relative to some reference frame. (A-2)

Fig. A-1. A coordinate system with origin O and directions specified by $\hat{i}$ and $\hat{j}$.

Many different coordinate systems may be chosen within the same reference frame. However, some of these may be more convenient than others for describing the positions of the points of interest.

Specification of position. The position of a point P can be precisely specified by the following information: (a) the *distance* of P from some specified point O in the reference frame (i.e., the length L of the straight line from O to P) and (b) the *direction* of this line relative to some particular directions in the reference frame (e.g., the angle θ relative to the $\hat{i}$ direction in Fig. A-2).

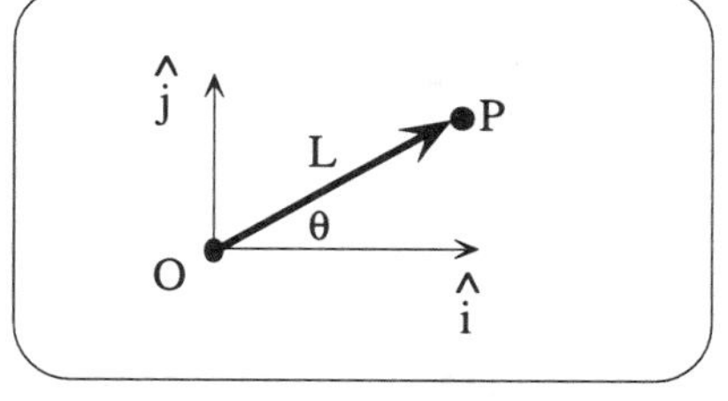

Fig. A-2. Specification of the position of a point P.

The preceding specification of position can be expressed more compactly by introducing the word *displacement* to denote jointly the length and direction of the straight line from one point to another.

Def: | ***Displacement:*** A quantity specifying jointly the length of a straight line and its direction relative to some reference frame. | (A-3)

The definition of position can then be stated as follows:

Def: | ***Position:*** The position of a point is a specification of the displacement of this point from a particular point in a reference frame. | (A-4)

Example of position specification

The position of the point P in Fig. A-3 can be specified by saying that its displacement from the particular point O is 2.3 km at an angle of 30° north of east.

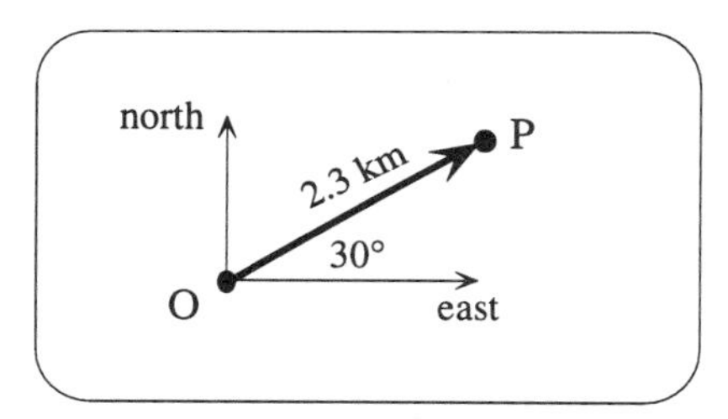

Fig. A-3. Example of position specification.

Note that position is only defined relative to some reference frame. The notion of position is, therefore, meaningless without the specification of such a reference frame.

Particle. If the position of an object can be adequately described by the position of a single point, the object is called a *particle*.

Def: | ***Particle:*** An object whose position can be adequately specified by that of a single point. | (A-5)

Any object can be considered a particle if it is sufficiently small compared to other lengths of interest. The notion of a particle is thus a relative one. (For example, the planet Mars may be considered a particle by someone on the earth. But Mars could not be considered a particle by an astronaut standing on its surface.)

→ ***Go to Sec. 3A of the Workbook.***

B. Vectors

By generalizing the notion of a displacement, one can define a *vector*, a basic concept very useful for describing spatial relationships.

Def:

Vector: A quantity specified jointly by a magnitude (i.e., a positive number, including units) and a direction relative to some reference frame.	(B-1)

According to this definition, a displacement is a special kind of vector whose magnitude is some length. However, in general, the magnitude of a vector is *not* necessarily a length. (For example, it might be a pure number, i.e., a number without any units.)

Symbols denoting a vector. A vector is conventionally denoted by a letter ornamented with an arrow on top (e.g., by $\vec{A}$) or by the letter printed in boldface type (e.g., by **A**). The *magnitude* of such a vector, without a specification of direction, is denoted by $|\vec{A}|$ or even more simply by the letter A unornamented by an arrow or by boldface.

Hence it is very important not to omit the arrow over a letter denoting a vector. Otherwise, the letter would be interpreted as a *scalar*, i.e., as a number without any associated direction.

Arrow representation of a vector. A vector can be pictorially represented by an arrow whose properties correspond to those of the vector. (Note that an arrow is *not* a vector, but *represents* a vector. In the same way a photograph of a person is *not* the person, but merely *represents* the person.)

The correspondence between the vector and its arrow representation is as follows: (a) The *length* of the arrow is chosen proportional to the magnitude of the vector. (b) The *direction* of the arrow is the same as that of the vector (relative to the specified reference frame). (c) The *position* of the arrow on a page does, however, *not* correspond to any property of a vector (since a vector is characterized only by a magnitude and direction, but by no other properties). Hence the same vector can be represented by many possible arrows at different positions, as long as they have the same length and direction. (For example, see Fig. B-1.)

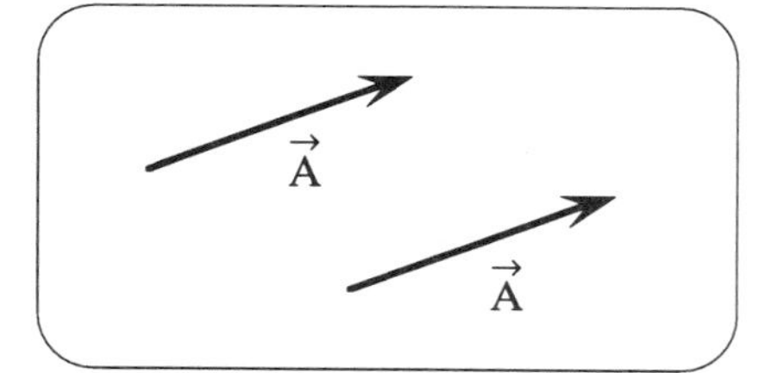

Fig. B-1. Two arrows representing the same vector $\vec{A}$.

A *zero vector* (i.e., one which has magnitude zero) has no associated direction. It can be denoted by $\vec{0}$, or simply by 0, and can be represented by a point.

Equality of vectors. Since a vector is specified jointly by a magnitude and a direction, two vectors are considered equal only if they have the same magnitudes and the same directions.

Def:

Equality of vectors: Two vectors are equal if their magnitudes are equal and their directions are the same.	(B-2)

One denotes the equality of two vectors $\vec{A}$ and $\vec{B}$ by writing

$$\vec{A} = \vec{B}.$$

Angle between two vectors. .i.Angle between vectors,;The angle between any two vectors is defined as illustrated in Fig. B-2:

Def:

Angle between two vectors: The angle (maximum 180°) between their representing arrows drawn from the same point.	(B-3)

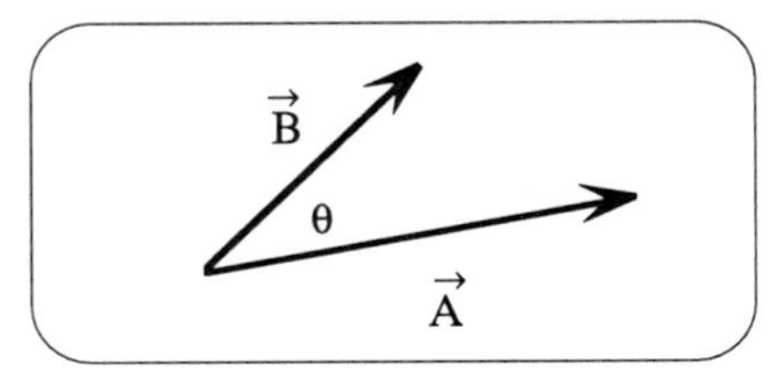

Fig. B-2. Angle between two vectors.

To determine the angle between two vectors, one must thus do the following: (a) Draw their representing arrows so that they start from the same point. (b) Then determine the angle between these.

➜ ***Go to Sec. 3B of the Workbook.***

C. Multiple of a vector

One can associate with every vector some related vectors defined in the following ways, as illustrated in Fig. C-1.

Def:

Negative of a vector: The vector $-\vec{A}$ denotes the vector having the same magnitude as $\vec{A}$, but the opposite direction.	(C-1)

The definition (C-1) is a special case of the following general definition:

Def:

Multiplication of a vector by a number: The vector $m\vec{A}$, where m is any number, denotes the vector having a magnitude $\lvert m \rvert$ times as large as that of $\vec{A}$. The direction of this vector is the same as that of $\vec{A}$ if m is positive, and is opposite to that of $\vec{A}$ if m is negative.	(C-2)

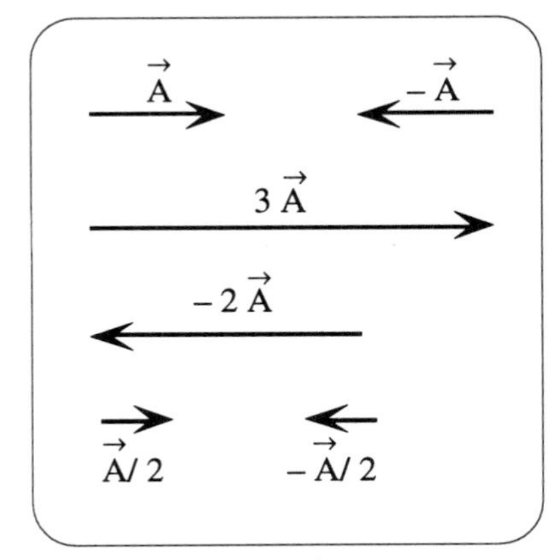

Fig. C-1. Multiples of a vector.

Here $\lvert m \rvert$ denotes the *magnitude* of the number m, i.e., the positive number without attention to its sign. Thus the definition (C-1) corresponds merely to the special case where $m = -1$.

Dividing by a number is the same as multiplying by its reciprocal. Hence the division of a vector by a number is defined in the following obvious way:

Def:

Division of a vector by a number: The vector $\vec{A}/m = (1/m)\,\vec{A}$.	(C-3)

In other words, $\vec{A}/m$ denotes the vector having a magnitude equal to the magnitude of $\vec{A}$ divided by the magnitude of m. Its direction is the same as that of $\vec{A}$ if m is positive, and is opposite to that of $\vec{A}$ if m is negative.

Unit vector. A *unit vector* is a vector whose magnitude is simply 1 (i.e., the pure number 1 without any units).

Def: **Unit vector:** A vector whose magnitude is 1. (C-4)

A unit vector is denoted by a letter with a caret on top of it, e.g., by $\hat{A}$.

A unit vector is particularly useful to indicate a direction since a magnitude specification is then irrelevant. (For example, the mutually perpendicular directions of the coordinate system in Fig. A-1 are indicated by the unit vectors $\hat{i}$ and $\hat{j}$.)

→ ***Go to Sec. 3C of the Workbook.***

D. Sum of vectors

Two vectors can be combined in useful ways to yield other vectors. The following example illustrates this in the case of successive displacements.

Suppose that a person walks along a straight line from a point P_1 to another point P_2, and then along another straight line from P_2 to another point P_3. These two successive displacements are indicated by $\vec{A}$ and $\vec{B}$ in Fig. D-1. The person's resultant displacement is then specified by the straight line from P_1 to P_3, as denoted by $\vec{S}$ in Fig. D-1. (Note that this *displacement* is not the same as the *distance* traveled by the person. Indeed, this distance is a number and not a vector, i.e., it is just the sum $|\vec{A}| + |\vec{B}|$ of the *magnitudes* of the person's successive displacements.)

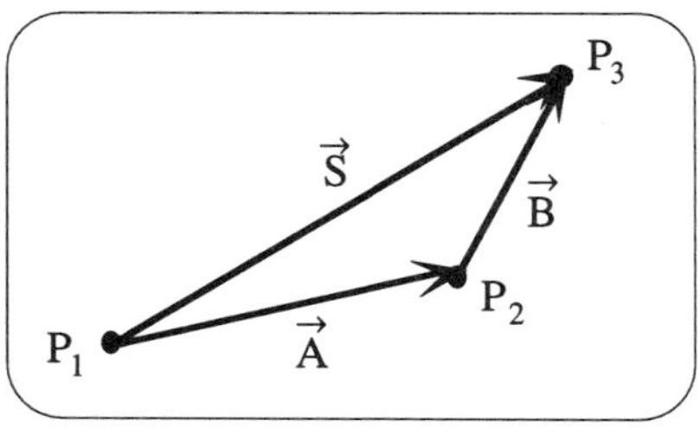

Fig. D-1. Two successive displacements.

Addition of vectors

Definition of vector sum. The preceding example suggests that *any* two vectors may be combined like two successive displacements. Thus one can introduce the following definition specifying how to *add* any two vectors so as to find another vector called their *sum*:

Def: ***Sum of vectors:*** The sum $\vec{A} + \vec{B}$ of two vectors is the vector represented by the arrow $\vec{S}$ in Fig. D-2. (D-1)

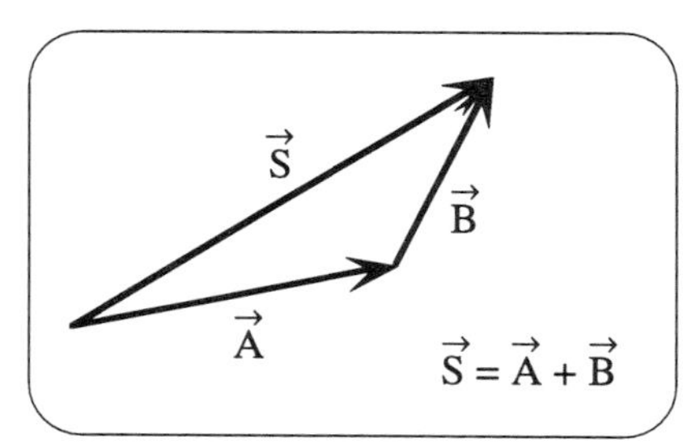

Fig. D-2. Sum of two vectors.

Defining method. The following method specifies *how* one can actually find the sum of two vectors:

Finding the vector sum $\vec{A} + \vec{B}$ (D-2)

(1) Draw the representing arrows tip-to-tail (i.e., so that the tail of the second starts from the tip of the first, as in Fig. D-2).

(2) Draw the arrow from the tail of the first arrow to the tip of the second. (This arrow represents the vector sum $\vec{A} + \vec{B}$.)

Properties of the sum of vectors

Order of addition. Does it matter in which order vectors are added? For example, suppose that one adds $\vec{B}$ to $\vec{A}$, or that one adds $\vec{A}$ to $\vec{B}$. Is the result the same? As indicated in Fig. D-3, the answer is yes, i.e.,

$$\vec{A} + \vec{B} = \vec{B} + \vec{A} . \quad \text{(D-3)}$$

Thus the order of addition is irrelevant (i.e., vector addition is *commutative).*

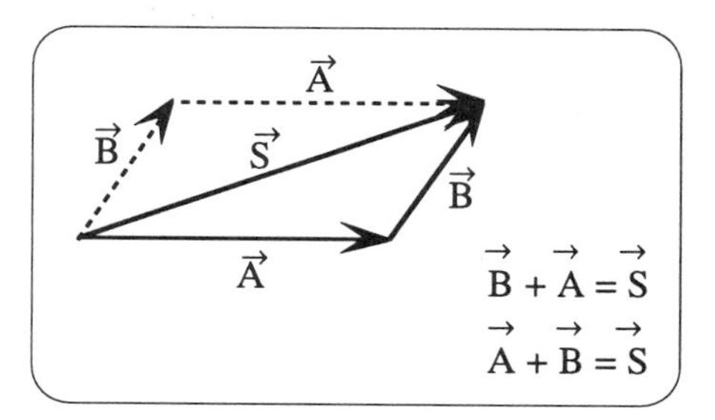

Fig. D-3. Reversing order of addition.

Grouping in addition. If more than two vectors are added, does it matter how they are grouped? For example, suppose that one adds $\vec{B}$ to $\vec{A}$, and then adds $\vec{C}$ to their sum. Alternatively, suppose that one starts with $\vec{A}$ and then adds to it the sum obtained by adding $\vec{C}$ to $\vec{B}$. Is the result the same? As indicated in Fig. D-4, the answer is yes, i.e.,

$$(\vec{A} + \vec{B}) + \vec{C} = \vec{A} + (\vec{B} + \vec{C}) . \quad \text{(D-4)}$$

Thus it does not matter how one groups the vectors when adding them (i.e., vector addition is *associative*). Hence the parentheses, used to indicate different groupings, are really unnecessary and (D-4) can simply be written $\vec{A} + \vec{B} + \vec{C}$.

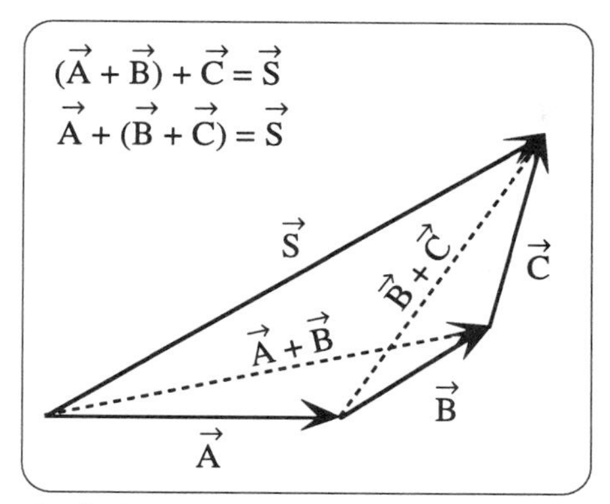

Fig. D-4. Grouping vectors in addition.

Order of addition and numerical multiplication. Does it matter in which order one adds vectors and multiplies them by a number? For example, suppose that one adds two vectors $\vec{A}$ and $\vec{B}$, and then multiplies their sum by the number m. Alternatively, suppose that one first multiplies $\vec{A}$ by m and $\vec{B}$ by m, and then adds the resulting vectors. Is the result the same? As indicated in Fig. D-5, the answer is yes, i.e.,

$$m(\vec{A} + \vec{B}) = m\vec{A} + m\vec{B} . \quad \text{(D-5)}$$

(Thus numerical multiplication of vectors is *distributive.*)

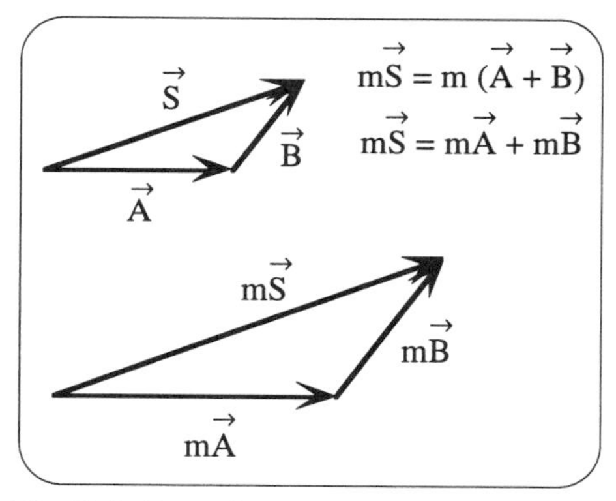

Fig. D-5. Reversing order of addition and multiplication.

Symbolic similarity between vectors and numbers. The sum of vectors has a geometrical interpretation and is thus quite different from a simple sum of numbers. However, the results (D-3), (D-4), and (D-5) indicate that (in operations involving addition and multiplication by numbers) the *symbols* denoting vectors can be manipulated according to the same rules as those denoting ordinary numbers.

➜ ***Go to Sec. 3D of the Workbook.***

E. Difference between vectors

Two numbers can be conveniently compared by finding their *difference*, i.e., the quantity which must be added to one of the numbers to yield the other. Thus one says that the difference 5 − 3 = 2 because 3 + 2 = 5 (i.e., because adding this difference 2 to the subtracted number 3 yields the original number 5).

Definition of vector difference. As illustrated in Fig. E-1, two vectors can be similarly compared by finding the vector which must be added to one to find the other.

Def:

Difference between vectors: The difference $\vec{A} - \vec{B}$ between two vectors is the vector $\vec{D}$ such that $\vec{B} + \vec{D} = \vec{A}$ (i.e., the vector which must be added to the subtracted vector to yield the original vector).	(E-1)

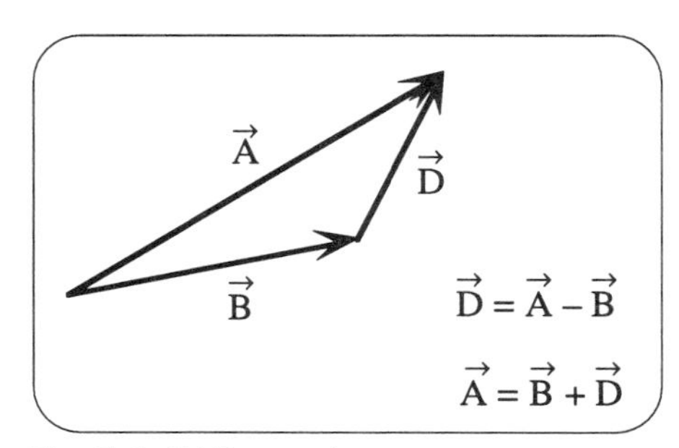

Fig. E-1. Difference between two vectors.

Defining method. The following method specifies *how* one can actually find the difference of two vectors:

Finding the vector difference $\vec{A} - \vec{B}$ (1) Draw the representing arrows from the *same* point (as in Fig. E-1). (2) Draw an arrow from the tip of the subtracted arrow to the tip of the original arrow. (This arrow represents the vector difference $\vec{D} = \vec{A} - \vec{B}$, i.e., one can check that $\vec{A} = \vec{B} + \vec{D}$.)	(E-2)

Relation between vector subtraction and addition. As is apparent from Fig. E-2, the subtraction of a vector is equivalent to the addition of the negative of that vector. Thus one can write

$$\vec{A} - \vec{B} = \vec{A} + (-\vec{B}) \, . \tag{E-3}$$

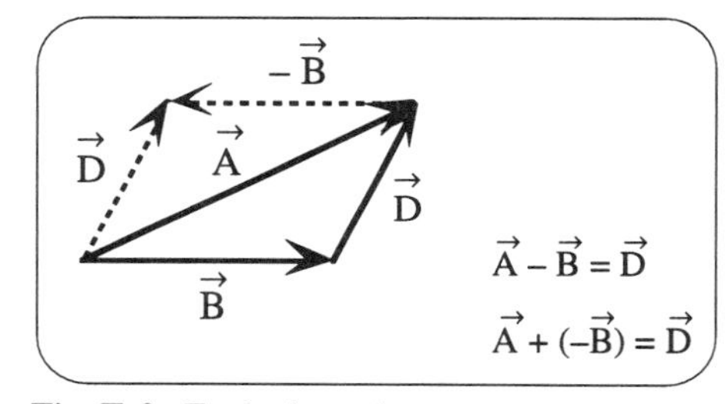

Fig. E-2. Equivalence between vector subtraction and addition of a negative vector.

However, it is ordinarily much more useful to subtract vectors directly by drawing their representing arrows from the *same* point, as in Fig. E-1. (Indeed, the magnitudes and directions of the vectors can then be most easily compared, and the difference of the vectors makes this comparison more quantitative.)

➜ ***Go to Sec. 3E of the Workbook.***

F. Symbolic properties of vectors

Symbolic similarity to numbers. As discussed in Sec. D, when adding vectors or multiplying them by numbers, the symbols denoting vectors can be manipulated according to the same rules as those denoting ordinary numbers. Furthermore, the subtraction of any vector $\vec{V}$ is equivalent to the addition of $-\vec{V}$. Thus one arrives at the following conclusion:

Symbols denoting vectors can be manipulated according to the same rules as ordinary numbers (in all operations involving the addition, subtraction, or numerical multiplication of vectors).	(F-1)

Equations relating vectors can, therefore, often be manipulated and solved in the same way as equations involving ordinary numbers. (However, the symbols denoting vectors have different meanings and represent spatial relationships).

Undefined operations. Note that some operations have not been defined for vectors. For example, we have not defined any way of *multiplying* two vectors (although we shall later define some useful ways of doing this). Furthermore, the notion of division by a vector is completely undefined.

Utility of vectors. The conclusion (F-1) implies that geometrical arguments about spatial relationships can often be carried out by simple familiar manipulations of algebraic symbols. This is why vectors are very useful in scientific work and will be used constantly throughout this book.

➔ ***Go to Sec. 3F of the Workbook.***

G. Displacements and changes of position

Position vector. As mentioned in (A-4), the position of a point or particle can be specified by its displacement from a specified point in a reference frame. This specification can be expressed in terms of vectors as follows:

Def: **Position vector:** The position vector of a point P is its displacement from a specified point O in a reference frame. (G-1)

The position vector is commonly denoted by $\vec{r}$. Since this vector is a displacement, its magnitude r is a length (i.e., the distance from O to P). As usual, the direction of this vector must be specified relative to some reference frame (e. g., relative to the directions $\hat{i}$ and $\hat{j}$ of some coordinate system like that illustrated in Fig. G-1).

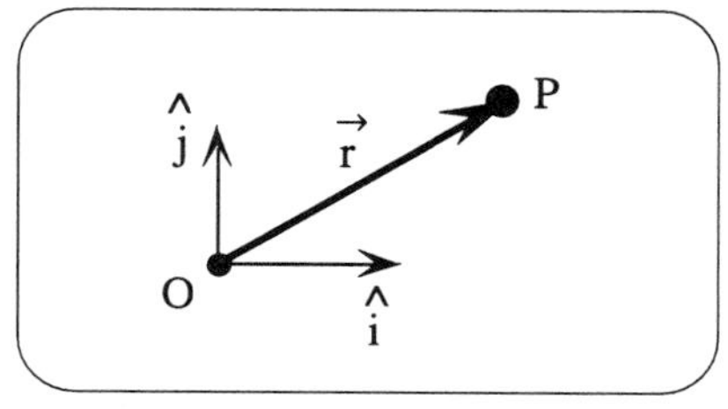

Fig. G-1. Position vector of a point P.

Motion. An object is said to move if its position changes. Since the position of an object must be specified relative to some reference frame, motion is also a *relative* concept. (The idea of motion is thus meaningless without specification of the reference frame relative to which it occurs.)

> ***Example: Motion of a car***
>
> When a car travels along a road, the driver moves relative to the road, but does not move relative to the car. Similarly, a road sign moves relative to the car, but does not move relative to the road.

Displacement resulting from a change of position. If a particle moves from a point P to another point P', its motion can be described by its displacement $\vec{D}$ from P to P'. As is clear from Fig. G-2, this displacement can be expressed in terms of the position vectors of P and P', i.e.,

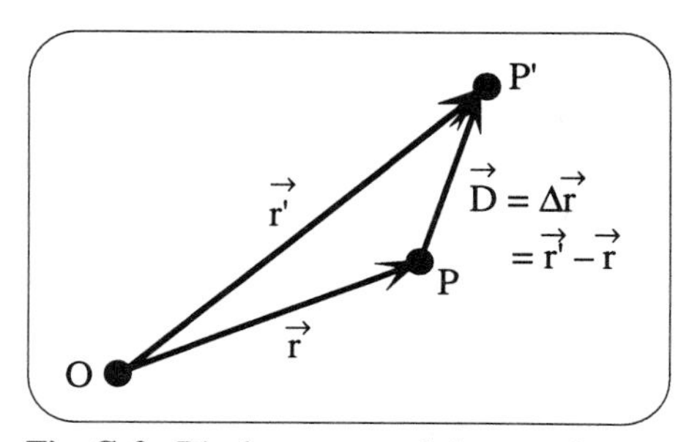

Fig. G-2. Displacement and change of position.

$$\vec{D} = \vec{r}' - \vec{r} = \Delta\vec{r}. \tag{G-2}$$

Here we have used the conventional symbol Δ (the Greek letter "delta") to denote a change defined so that, for any quantity Q,

$$\Delta Q = Q' - Q\,, \tag{G-3}$$

i.e., (change of value) = (new value) – (old value) . (G-4)

Thus (G-2) merely states that *the displacement of a particle is equal to the change of its position vector* (its new position vector minus its original one).

➜ ***Go to Sec. 3G of the Workbook.***

H. Summary

Knowledge about vectors

Vector: A quantity specified jointly by a magnitude and direction.

$m\vec{A}$ Vector having a magnitude $|m|\,|\vec{A}|$. Its direction is along $\vec{A}$ if m is positive, opposite to $\vec{A}$ if m is negative.

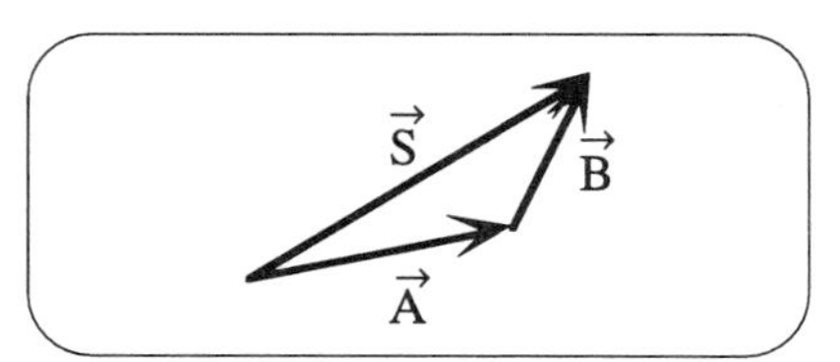

Fig. H-1.

$\vec{A} + \vec{B}$ Vector sum $\vec{S}$ (as indicated in Fig. H-1).

$\vec{A} - \vec{B}$ Vector difference $\vec{D}$ such that $\vec{B} + \vec{D} = \vec{A}$. (See Fig. H-2).

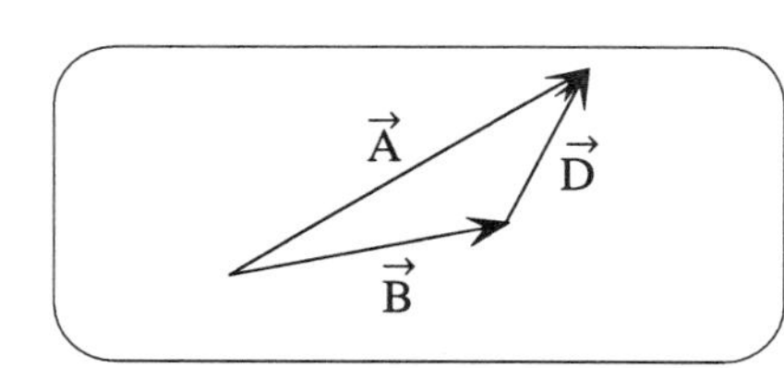

Fig. H-2.

Symbolism: Algebraic vector symbols can be manipulated like numbers (in all operations involving addition, subtraction, and multiplication by numbers).

Position vector: Position vector $\vec{r}$ of a point P relative to some point O is the displacement from O to P. (See Fig. H-3).

Displacement: $\vec{D} = \Delta\vec{r}$ (the displacement from P to P' in Fig. H-3).

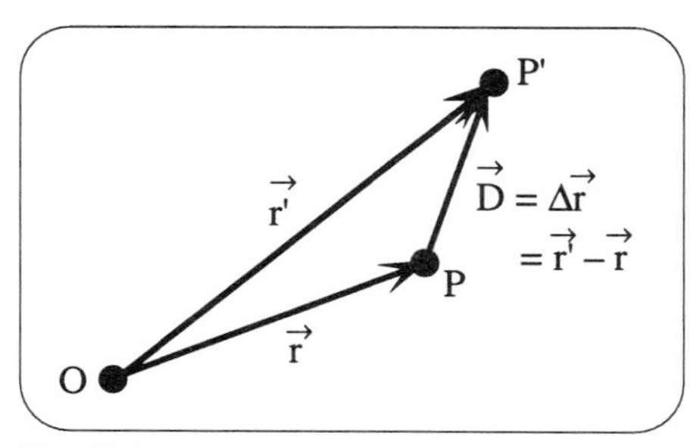

Fig. H-3.

New abilities

You should now be able to do the following:

(1) Add vectors, subtract vectors, and multiply them by numbers.

(2) Translate geometrical information into vector symbolism, and vice versa.

(3) Manipulate vector symbols in algebraic expressions and equations.

➜ ***Go to Sec. 3H of the Workbook.***

4 Velocity and Acceleration

The preceding chapter discussed how motions may be specified by displacements corresponding to changes of position. But how can we describe how the positions of objects change in the course of *time*, or how *fast* objects move? The present chapter deals with these questions. We shall thus be led to introduce the centrally important concepts of "velocity" and "acceleration". These scientific concepts have more precise meanings than in everyday life, allow more detailed descriptions of motion, and are widely useful. (For example, the next four chapters will merely deal with applications of these concepts.) Furthermore, these concepts provide the essential ingredients for a theory capable of predicting the motions of objects.

A. Time variation of position

As discussed previously, one says that a particle *moves* relative to some reference frame if its position relative to this frame changes. Conversely, one says that a particle is *at rest* relative to some reference frame if its position relative to this frame remains unchanged.

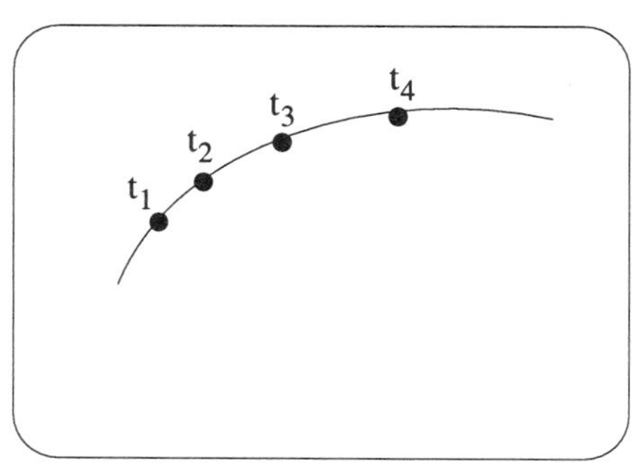

Fig. A-1. Sequence diagram for a moving particle.

To describe in greater detail *how* a particle moves, one needs to specify its position at successive times. For example, one might specify its position at regular time intervals of 0.1 second. This sequence of positions can be displayed in a diagram by points indicating the positions of the particle at these successive times. Such a *sequence diagram* is shown in Fig. A-1. One may also indicate the successive position vectors of the particle relative to some conveniently chosen coordinate system (as illustrated in Fig. A-2).

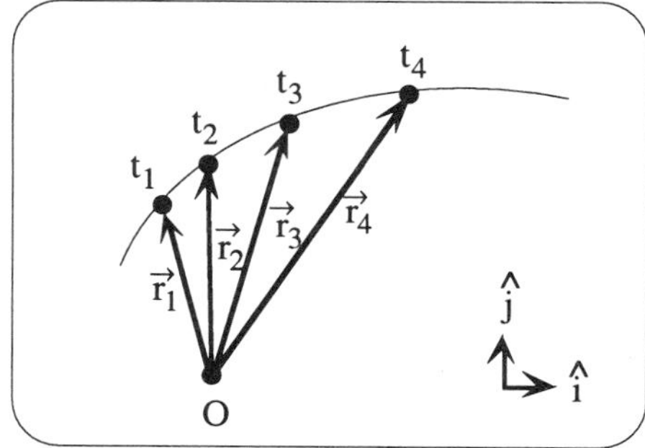

Fig. A-2. Same diagram, with indicated position vectors.

A sequence diagram, like that in Fig. A-1, corresponds closely to the data obtained by observing the motion with a camera focused on the region through which the particle moves. For example, one can keep the camera shutter open while the particle moves through darkness. If one illuminates the scene with a "stroboscope", which emits regularly spaced brief flashes of light, the resulting photograph shows then directly the particle's position at successive times. (Alternatively, one may constantly illuminate the particle, but open the camera shutter very briefly at regularly spaced times.)

> ***Example A-1: Stroboscopic photograph of a falling ball***
>
> Fig. A-3 shows a stroboscopic photograph indicating the successive positions of a falling ball. This photograph shows that, during equal intervals of time, the ball traverses progressively increasing distances. The ball moves, therefore, downward with *increasing* speed.

➜ ***Go to Sec. 4A of the Workbook.***

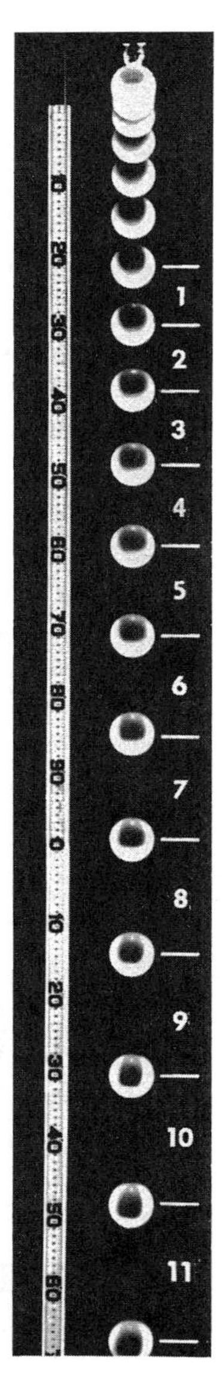

Fig. A-3. Stroboscopic photograph of a falling ball released from rest. [Reproduced with permission from PSSC PHYSICS, 2nd edition, 1965; D.C. Heath and Company with Education Development Center, Inc., Newton, MA.]

Velocity

B. Average velocity

The preceding section described a particle's motion by specifying its successive positions during its *entire* motion. However, a more detailed description is obtained by focusing attention on how the particle moves during any specified *time interval.* In this way one can also describe how fast the particle moves during any such time interval, i.e., how *rapidly* its position changes with time. The following paragraphs discuss how this can be done.

Motion during some time interval. Consider a particle moving along some path. At a particular time t of interest, the particle is located at some position which may be specified by the position vector $\vec{r}$. At a later time t', the particle is located at some other position which may be specified by the position vector $\vec{r}'$. As indicated in Fig. B-1, during the time interval $\Delta t = t' - t$ the displacement of the particle is then $\Delta\vec{r} = \vec{r}' - \vec{r}$.

Definition of average velocity. To describe how rapidly the particle moves, we consider the ratio $\Delta\vec{r}/\Delta t$ obtained by dividing the particle's displacement by the time interval during which it occurs. This ratio properly indicates how fast this displacement occurs. For example, if the particle moves faster (i.e., if the same displacement occurs during a shorter time), the magnitude of this ratio is larger (since the denominator of the ratio is then smaller).

This ratio $\Delta\vec{r}/\Delta t$ is called the *average velocity* of the particle during the time interval between t and t'. If this average velocity is denoted by $\vec{v}_{av}$, one can summarize this definition by writing

Def: ***Average velocity*** *(between specified times):* $$\vec{v}_{av} = \frac{\Delta\vec{r}}{\Delta t}\,. \qquad \text{(B-1)}$$

According to this definition, the average velocity is a *vector* having the units of a length divided by a time (i.e., the SI units of meter/second).

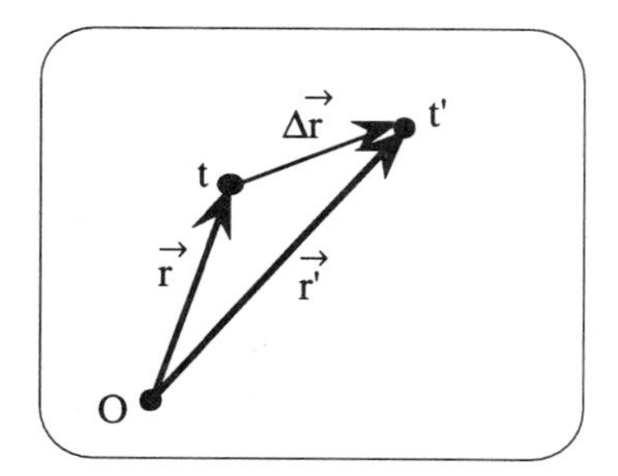

Fig. B-1. Displacement of a particle between two times.

Deficiencies of the average velocity. The average velocity is *not* a very useful concept because it provides only a crude measure of how rapidly a particle moves. For instance, various different motions can occur during a long time interval, yet they may all be described by the same average velocity.

> ***Example***
>
> Suppose that a car starts from rest at some time t, travels around a racetrack, and then comes to rest again at its starting point at some later time t'. The car's displacement during this entire time is then zero. Hence its average velocity during this time is also zero. This would be true regardless of how the car speeds up or slows down during its motion along the track.

➜ ***Go to Sec. 4B of the Workbook.***

C. Velocity at an instant

Consideration of very short time intervals

A more detailed description of motion can be obtained by considering how a particle moves during any *very short time interval.* In this way one can also describe how fast the particle moves at any instant, i.e., the rate of change of its position during any very short time. (For example, such detailed information is provided by a car's speedometer which indicates how fast the car moves at any instant.)

To accomplish this goal, one can compare the particle's position at the time t of interest with its position at a *slightly* later time t'. The time difference $\Delta t = t - t'$ is then small, and the corresponding displacement $\Delta \vec{r}$ of the particle is also small. However, the ratio $\Delta \vec{r}/\Delta t$ (i.e., the average velocity during this very short time interval) is ordinarily *not* small. Indeed, as t' is chosen closer and closer to the original time t (i.e., as the time interval Δt is made progressively smaller), the ratio $\Delta \vec{r}/\Delta t$ approaches some constant value (called a *limiting value)* which becomes independent of the choice of t'.

> ***Limiting value of the average velocity***
>
> For example, Fig. C-1 shows a particle's position at a particular time t of interest. For comparison, it also shows its positions at several later times t_1', t_2', and t_3' chosen successively closer to its original position. The time intervals $\Delta t = t' - t$ for each of these later times are then correspondingly smaller, and the particle's displacements $\Delta \vec{r}$ during these time intervals are also correspondingly smaller. However, the average velocity $\Delta \vec{r}/\Delta t$, during any of these time intervals, is ordinarily *not* small (since the smaller numerator of the ratio is divided by a correspondingly smaller denominator).
>
> As the time interval Δt becomes progressively smaller, the *direction* of the particle's displacement $\Delta \vec{r}$ approaches that of the dashed line in Fig. C-1, i.e., it becomes tangent to the

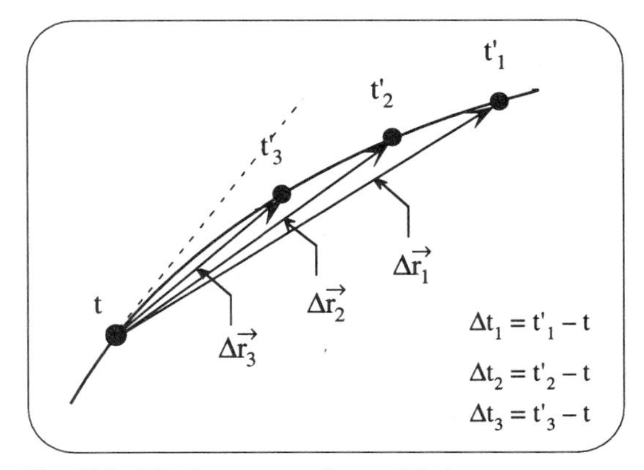

Fig. C-1. Displacements of a particle between some time t and several slightly later times.

particle's path at the time t. Thus the particle's average velocity $\Delta\vec{r}/\Delta t$ becomes also tangent to the path (since the time difference Δt is merely a positive number).

The *magnitude* $|\Delta\vec{r}|$ of the particle's displacement varies smoothly with the time interval Δt, as indicated schematically in the graph of Fig C-2. Since this graph is a smooth curve, it is nearly straight in any sufficiently small region, in particular in the region where Δt is sufficiently small. Correspondingly, the magnitude of the average velocity (i.e., the *ratio* $|\Delta\vec{r}|/\Delta t$) approaches some constant value.

For example, if the time interval Δt has a value Δt_3 half as large as Δt_2, the magnitude $|\Delta\vec{r}_3|$ of the corresponding displacement is half as large as $|\Delta\vec{r}_2|$. The magnitude $|\Delta\vec{r}_3|/\Delta t_3$ of the average velocity during the smaller time interval Δt_3 is then the *same* as the magnitude $|\Delta\vec{r}_2|/\Delta t_2$ of the average velocity during the longer time interval Δt_2. If the time interval Δt is chosen sufficiently small, the magnitude of the average velocity thus approaches a constant limiting value, as schematically illustrated in Fig. C-3.

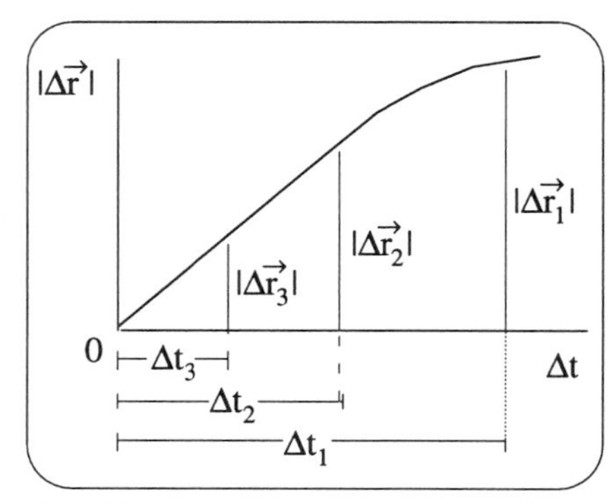

Fig. C-2. Graph showing how the magnitudes of the displacements in Fig. C-1 vary with the corresponding time intervals.

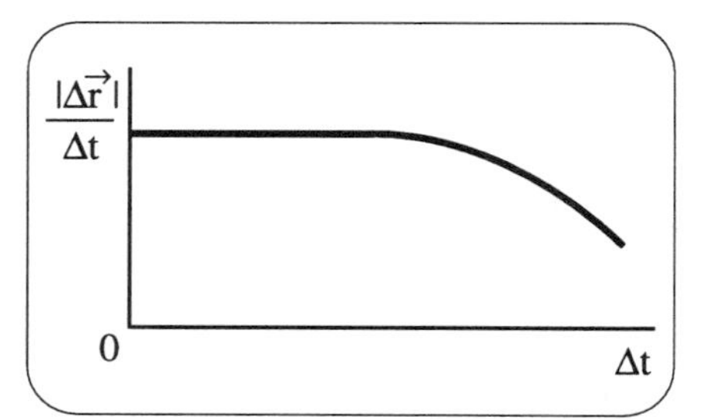

Fig. C-3. Graph indicating how the magnitude of the average velocity varies with the time interval Δt.

Definition of velocity

Infinitesimal time interval. To describe how rapidly the position of a particle changes at any particular instant, it is most useful to compare the particle's position at the particular time t of interest with its position at a very slightly later time t'. If the time t' is chosen sufficiently close to t (so that Δt is sufficiently small), the average velocity $\Delta\vec{r}/\Delta t$ during this small time attains then a limiting value independent of the choice of t'. The differences Δt and $\Delta\vec{r}$ are then so small that they are called *infinitesimal.*

Def: ***Infinitesimal quantity:*** A very small quantity (small enough that some ratio or other specified quantity differs negligibly from its limiting value). (C-1)

Infinitesimal differences are conventionally denoted by the letter *d* to distinguish them from differences of any size (denoted by the letter Δ). For example, the infinitesimal differences of time and position are denoted by dt and $d\vec{r}$, respectively.

Velocity at any instant. The ratio $\Delta\vec{r}/\Delta t$, determined when the time difference $\Delta t = t' - t$ is infinitesimal, is simply called the *velocity at the time t* and is denoted by the symbol $\vec{v}$. Thus we arrive at the following definition

Def: ***Velocity*** *(at some instant):* $\vec{v} = \frac{d\vec{r}}{dt}$ (C-2)

where $d\vec{r}$ is the infinitesimal displacement of the particle during the infinitesimal time interval $dt = t' - t$. The definition (C-2) can be stated in words by saying that *velocity is the rate of change of position with time.*

In more mathematical language, a rate of change is called a *derivative*.

Note. When one talks about the velocity *at* some instant, one implicitly compares the particle's position at this instant with its position at a slightly later time. However, the precise choice of this later time does not matter provided that it is sufficiently close to the original instant.

Velocity versus average velocity. The *average* velocity, between some time t and a somewhat later time t', is ordinarily only approximately equal to the velocity at the particular instant of time t. It is equal to the velocity *at* the time t only if the time t' is chosen very close to t (i.e., sufficiently close so that any closer choice would leave the value of the average velocity unchanged).

Defining method

The statement (C-2) summarizes compactly the definition of velocity. To interpret this definition properly, one must know a method specifying what one must actually *do* to determine a velocity in any particular instance. The main steps of such a method are listed below and illustrated in Fig. C-4.

Finding velocity

(1) ***Original position.*** Identify the position of the particle at the time t of interest.

(2) ***New position.*** Identify the position of the particle at a slightly later time t'.

(3) ***Displacement*** $\Delta\vec{r}$. Find the displacement (i.e., change of position $\Delta\vec{r}$) of the particle during the small time interval $\Delta t = t' - t$.

(4) ***Average velocity*** $\vec{v}_{av} = \Delta\vec{r}/\Delta t$. Find the ratio $\Delta\vec{r}/\Delta t$, the "average velocity" of the particle during the time Δt.

(5) ***Velocity*** $\vec{v} = d\vec{r}/dt$. Determine the limiting value approached by the average velocity if the time t' is chosen very close to t (so that Δt becomes infinitesimally small and can be denoted by dt). The resultant ratio $d\vec{r}/dt$ is called "the velocity of the particle at the time t".

(C-3)

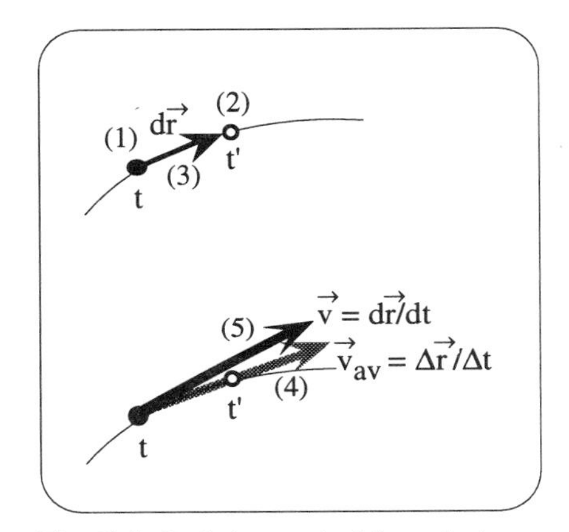

Fig. C-4. Defining method for velocity.

Validity conditions. The preceding definition of velocity is generally valid for any particle moving relative to any reference frame.

Properties of velocity

Velocity is a vector. According to the definition (C-2), the velocity is a *vector* since it is equal to a displacement divided by a time (i.e., a vector divided by a number).

Direction of velocity. The *direction* of the velocity is along the particle's infinitesimal displacement, i.e., tangent to the particle's path along its motion.

Magnitude of velocity (speed). The *magnitude* v of the velocity is the magnitude $|d\vec{r}|$ of the particle's infinitesimal displacement (i.e., the infinitesimal distance ds traversed by the particle) divided by the infinitesimal time dt required to traverse it. Thus

$$v = \frac{|d\vec{r}|}{dt} = \frac{ds}{dt}\,. \qquad \text{(C-4)}$$

The magnitude of the velocity, without attention to its direction, is commonly denoted by the word *speed*, according to the following definition:

Def: | ***Speed:*** Magnitude of the velocity. | (C-5)

Warning. Note that the word "velocity" used in physics has a meaning different from that in everyday life where the word velocity refers roughly to speed, irrespective of direction.

Units. By (C-1), the units of velocity are those of a length divided by a time. Thus the SI units are meter/second.

Complete specification of velocity. To specify a velocity unambiguously, one needs to specify the particle whose motion is being described, the time of interest, and the reference frame relative to which the particle's position is specified. A complete specification thus requires the expression "velocity of *particle* at *time* relative to *reference frame*", where one must specify each italicized entity. (For example, one might talk about the velocity of a boat, at 2 PM, relative to the flowing water of a river.)

➜ ***Go to Sec. 4C of the Workbook.***

D. Applications of velocity

Finding position from velocity

If one knows the velocity $\vec{v}$ of a particle at any instant of time, the definition (C-2) allows one to find the small displacement $d\vec{r}$ of the particle during an infinitesimal time interval dt. Indeed, if one multiplies both sides of (C-2) by dt, one gets

$$d\vec{r} = \vec{v}\,dt\,. \qquad \text{(D-1)}$$

Suppose that the particle is located at the position $\vec{r}$ at the time t. Its position $\vec{r}\,'$ at the slightly later time t' can then readily be found since

$$d\vec{r} = \vec{r}\,' - \vec{r} \qquad \text{(D-2)}$$

as illustrated in Fig. D-1. Hence

$$\vec{r}\,' = \vec{r} + d\vec{r} = \vec{r} + \vec{v}\,dt, \qquad \text{(D-3)}$$

i.e., (new position) = (old position) + (displacement).

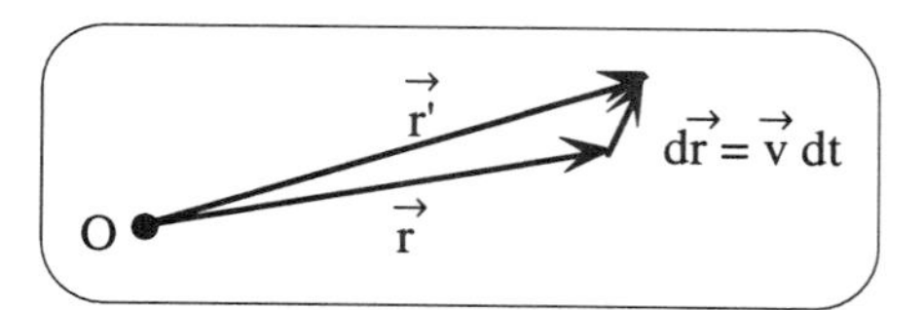

Fig. D-1. Finding position at a slightly later time.

The result (D-3) allows one to find the displacement of the particle during a very short time interval only. However, suppose that a particle's velocity is known at *all* instants during some large time interval. Then one can find the particle's displacement during *every* small time interval, and can then find the total displacement of the particle by simply adding all these successive small displacements. In this way one can exploit available information about infinitesimal time intervals to find the position of the particle at any other time.

Sequence diagrams with velocity

A knowledge of velocity allows one to construct sequence diagrams which include explicit information about the velocity. For example, Fig. D-2 elaborates the previous Fig. A-3 by using arrows to indicate the velocity of the particle at each of the instants shown in the diagram.

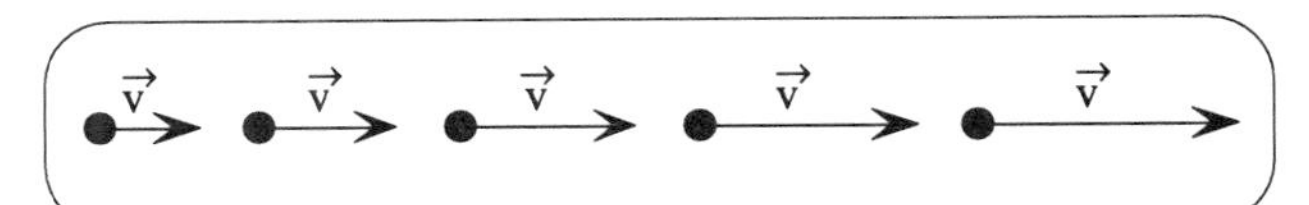

Fig. D-2. Sequence diagram for a moving particle with indicated velocity.

➜ ***Go to Sec. 4D of the Workbook.***

Acceleration

E. Average acceleration

As a particle moves, its velocity ordinarily changes. How can one conveniently describe how *rapidly* its velocity changes?

Consider a particle moving along some path, as illustrated in Fig. E-1. At a particular time t of interest, the particle is located at the indicated position and has a velocity $\vec{v}$. At a later time t' the particle is located at some other position and has a new velocity $\vec{v}'$ ordinarily different from its original velocity $\vec{v}$. This change may be due to a change of the magnitude of the velocity, to a change of its direction, or to a change of both. We should like to describe, in a unified way, how rapidly any such velocity change occurs.

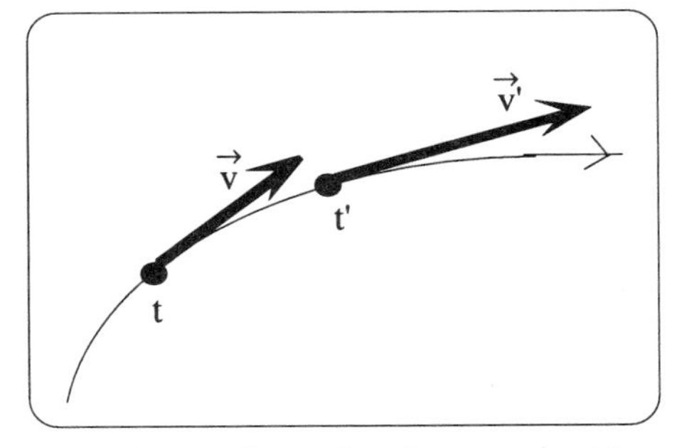

Fig. E-1. Particle moving along a path with changing velocity.

> ***Examples***
> If a car travels along a *straight* road with *increasing* speed, its velocity changes because the *magnitude* of its velocity changes (although the direction of the velocity remains unchanged).
> If a car travels along a *curved* road with *constant* speed, its velocity also changes because the *direction* of its velocity changes (although the magnitude of its velocity remains unchanged).

The actual change of velocity $\Delta\vec{v} = \vec{v}' - \vec{v}$ can be found by subtracting the velocities vectorially, as shown in Fig. E-2. Indeed, this vector diagram indicates properly that

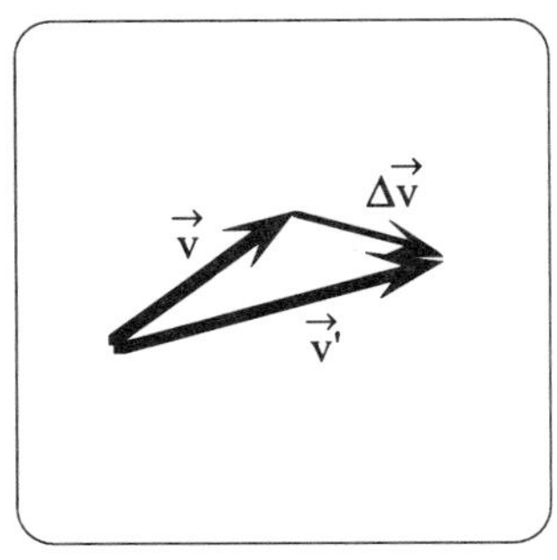

Fig. E-2. Velocity change of the particle in Fig. E-1.

$$(\text{change of velocity } \Delta\vec{v}) = (\text{new velocity } \vec{v}') - (\text{old velocity } \vec{v})$$

or equivalently

$$(\text{new velocity } \vec{v}') = (\text{old velocity } \vec{v}) + (\text{change of velocity } \Delta\vec{v})\,.$$

To describe how *rapidly* the velocity changes, we can proceed in a fashion analogous to that used to define the average velocity in Sec. B. Thus we consider the ratio $\Delta\vec{v}/\Delta t$ obtained by dividing the velocity change by the time interval $\Delta t = t' - t$ during which it occurs. This ratio provides a convenient way of describing how fast the velocity changes. For example, if the same velocity change occurs more rapidly (i.e., if the same velocity change occurs during a shorter time), then the magnitude of this ratio is larger.

This ratio $\Delta\vec{v}/\Delta t$ is called the *average acceleration* of the particle during the time interval between t and t'. If this average acceleration is denoted by $\vec{a}_{av}$, one can summarize this definition by writing

Def: ***Average acceleration*** *(between specified times):* $\vec{a}_{av} = \dfrac{\Delta\vec{v}}{\Delta t}$. (E-1)

According to this definition, the average acceleration is a *vector* having the units of a velocity divided by a time [i.e., the SI units of (m/s)/s = m/s^2].

Deficiencies of the average acceleration. The average acceleration is again *not* a very useful concept because it provides only a crude measure of how rapidly a particle's velocity changes. For instance, various different motions can occur during some long time interval, yet they may all be described by the same average acceleration.

Example

A car, traveling along a straight road with an eastward velocity of 40 km/hour, starts speeding up at some time t_o to pass a truck. After 30 seconds, the car travels again with an eastward velocity of 40 km/hour. Since the initial and final velocities of the car are the same, the car's average acceleration during this 30-second time interval is zero. This is true regardless of *how* the speed of the car was changing during this time interval.

➜ ***Go to Sec. 4E of the Workbook.***

F. Acceleration at an instant

Definition of acceleration

To describe more precisely how rapidly a particle's velocity changes at any instant, one can consider its motion during a very short time interval. Thus one can compare the particle's velocity $\vec{v}$ at the particular time t of interest with its velocity $\vec{v}'$ at a slightly later time t'. If the time t' is chosen so close to t that Δt is sufficiently small, the ratio $\Delta\vec{v}/\Delta t$ (i.e., the average acceleration during this small time) becomes independent of the choice of t'. The differences Δt and $\Delta\vec{v}$ are then infinitesimally small, and can be denoted by dt and $d\vec{v}$ respectively. The ratio of these infinitesimal differences is simply called the *acceleration at the time t*, and is denoted by the symbol $\vec{a}$ according to the following definition:

Def: **Acceleration** *(at some instant):* $\vec{a} = \dfrac{d\vec{v}}{dt}$. (F-1)

Summarized in words, this definition states that the *acceleration is the rate of change of velocity with time.*

Acceleration versus average acceleration. The *average* acceleration, between some time t and a somewhat later time t', is ordinarily only approximately equal to the acceleration at the particular instant of time t. It is equal to the acceleration *at* the time t only if the time t' is chosen very close to t (i.e., sufficiently close so that any closer choice would leave the value of the average acceleration unchanged).

Defining method

The statement (F-1) summarizes compactly the definition of acceleration, but must be properly interpreted in any particular instance. The following method, illustrated in Fig. F-1, specifies what one must actually do to determine the acceleration by comparing the velocity at the time of interest with the velocity at a slightly later time:

Finding acceleration

(1) ***Original velocity*** $\vec{v}$. Identify the velocity of the particle at the time t of interest.

(2) ***New velocity*** $\vec{v}'$. Identify the velocity of the particle at a slightly later time t'.

(3) ***Change of velocity*** $\Delta\vec{v}$. Find the velocity change $\Delta\vec{v} = \vec{v}' - \vec{v}$ of the particle during the small time interval $\Delta t = t' - t$.

(4) ***Average acceleration*** $\vec{a}_{av} = \Delta\vec{v}/\Delta t$. Find the ratio $\Delta\vec{v}/\Delta t$, the "average acceleration" of the particle during the time Δt.

(5) ***Acceleration*** $\vec{a} = d\vec{v}/dt$. Determine the limiting value approached by the average acceleration if the time t' is chosen very close to t (so that Δt becomes infinitesimally small and can be denoted by dt). The resultant ratio $d\vec{v}/dt$ is called "the acceleration of the particle at the time t".

(F-2)

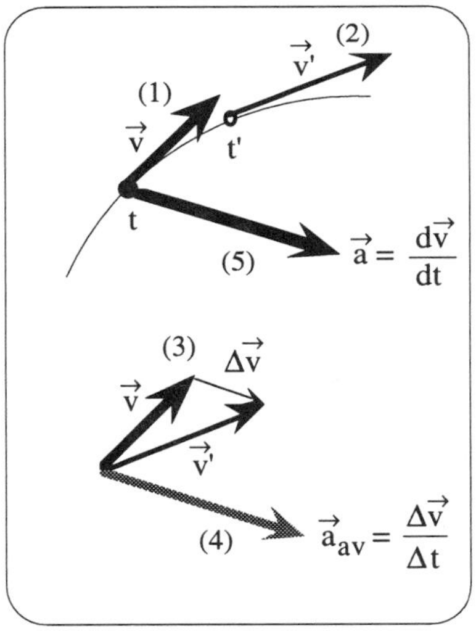

Fig. F-1. Defining method for acceleration.

Validity conditions. The preceding definition of acceleration is generally valid for any particle moving relative to any reference frame.

Properties of acceleration

Vector properties. According to the definition (F-1), acceleration is a *vector* (since it is equal to a vector change of velocity divided by a numerical time interval). The direction of the acceleration is along the particle's infinitesimal change of velocity.

Warning. Note that the word "acceleration" used in physics has a meaning different from that in everyday life (where acceleration describes roughly increases in speed, without attention to any changes in the direction of the velocity).

Units. By (F-1), the units of acceleration are those of a velocity divided by a time. Thus the SI units are (m/s)/s or simply m/s^2.

Complete specification of acceleration. To specify completely an acceleration without ambiguity, one needs to specify the particle whose motion is being described, the time of interest, and the reference frame relative to which the particle's velocity is specified. Complete specification requires thus the expression "acceleration of *particle* at *time* relative to *reference frame*", where one must specify each italicized entity. (For example, one might talk about the acceleration of a falling ball, relative to a moving elevator, two seconds after the elevator started moving up from the ground floor.)

➜ ***Go to Sec. 4F of the Workbook.***

G. Applications of acceleration

Finding velocity from acceleration

If one knows the acceleration $\vec{a}$ of a particle at any one time, the definition (F-1) allows one to find the small velocity change $d\vec{v}$ of the particle during a very short time interval dt. Indeed, (F-1) implies that

$$d\vec{v} = \vec{a}\,dt \ . \qquad \text{(G-1)}$$

If a particle has a velocity $\vec{v}$ at the time t, its velocity $\vec{v}'$ at a slightly later time t' can then readily be found since

$$d\vec{v} = \vec{v}' - \vec{v} \qquad \text{(G-2)}$$

as illustrated in Fig. G-1. Hence

$$\vec{v}' = \vec{v} + d\vec{v} = \vec{v} + \vec{a}\,dt, \qquad \text{(G-3)}$$

i.e., (new velocity) = (old velocity) + (change of velocity). (G-4)

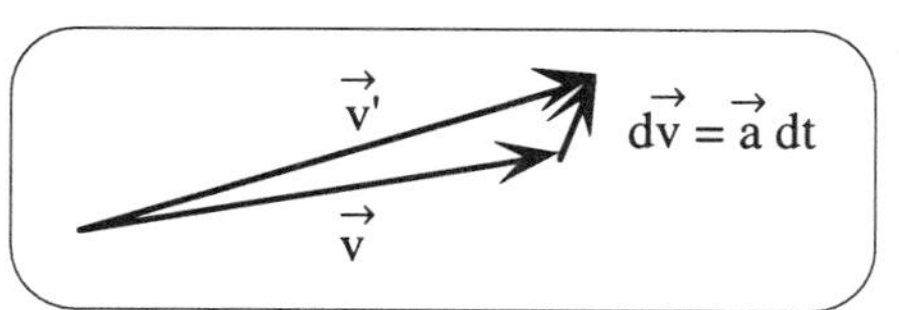

Fig. G-1. Finding velocity at a slightly later time.

The preceding calculation is valid for a very short time interval only. But, if the acceleration is known at all instants during a long time interval, the knowledge of infinitesimal velocity changes can be used repeatedly and thus extended to find the velocity change during the entire long time.

Chapter 9 will discuss Newton's theory which is highly successful in predicting the motion of objects. This theory provides primarily information about the acceleration of a particle at any instant. But, by using arguments such as those in this section and in Sec. D, this information can then be used to predict the velocity and position of the particle at any other time.

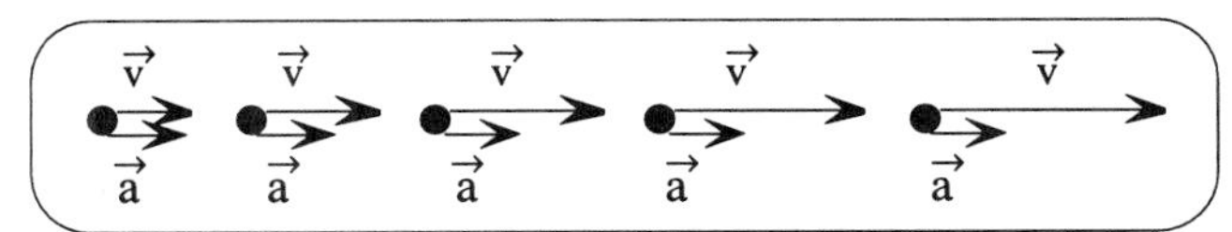

Fig. G-2. Sequence diagram for a moving particle with indicated velocity and acceleration.

Sequence diagrams with acceleration

One can readily construct sequence diagrams which include information about the acceleration as well as about the velocity. To do this, one may use arrows to indicate a particle's acceleration and velocity at each instant. For example, Fig. G-2 illustrates a particle moving with constant acceleration.

> ***Scientific versus mathematical descriptions***
>
> There are distinctive differences in the way mathematics is used in science or in mathematics. The primary goal of a science is to explain and predict observable phenomena; concepts that cannot be related to observations are, therefore, scientifically irrelevant and meaningless. On the other hand, the goal of pure mathematics is to investigate logical relationships among abstract concepts; the relationships of such concepts to any observations are thus mathematically irrelevant.
>
> For example, scientists consider a quantity infinitesimal if it is negligibly small compared to the desired or possible precision of measurements. The mathematical concept of a quantity which is arbitrarily close to zero (and thus immeasurably small) is thus scientifically meaningless, but a useful approximation.

➜ ***Go to Sec. 4G of the Workbook.***

H. Summary

Definitions

Velocity: $\vec{v} = \dfrac{d\vec{r}}{dt}$ (rate of change of position with time)

Acceleration: $\vec{a} = \dfrac{d\vec{v}}{dt}$ (rate of change of velocity with time)

New abilities

You should now be able to do the following:

(1) Interpret properly the preceding definitions by using the defining methods (C-3) or (F-2) to determine a particle's velocity or acceleration in any specific instance.

(2) Use words and/or sequence diagrams to describe the motion of a particle, including information about its velocity and acceleration.

➜ ***Go to Sec. 4H of the Workbook.***

5 Motion Along a Straight Line

These next four chapters discuss how the basic concepts of velocity and acceleration can be applied to various simple but important situations. In particular, the present chapter starts by considering the case of motion along a straight line. We shall be interested in investigating the following questions: What are convenient ways of describing a particle's motion along a straight line? How can information about the motion be used to determine quantitatively the velocity or acceleration at any instant? Conversely, how can mere information about the velocity or acceleration be used to predict the motion during any period of time?

A. Component description of motion

Useful coordinate system. When a particle P moves along a straight line, the position of the particle is most simply described by choosing a coordinate system whose origin O lies on this line and whose reference direction $\hat{i}$ is parallel to this line, as indicated in Fig. A-1.

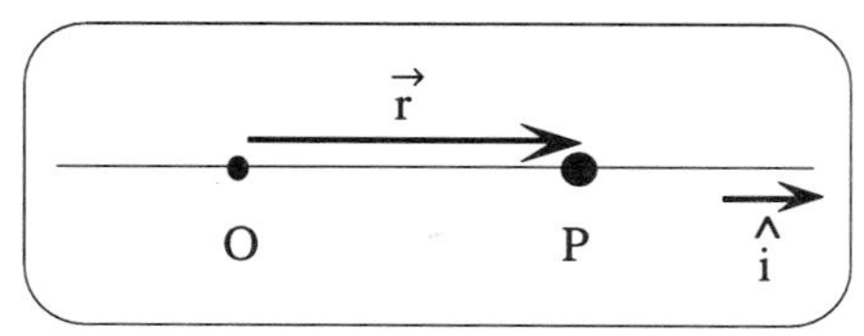

Fig. A-1. Coordinate system for straight-line motion.

As the particle moves along this line, its position vector $\vec{r}$, measured from O, is always parallel to this line (i.e., directed either along $\hat{i}$ or opposite to it). All displacements of the particle are obviously parallel to this line. Hence the velocity and acceleration of the particle also have directions parallel to this line. Since all vectors describing the particle's motion are parallel to this line, the situation is then especially simple.

Component of a vector

Parallel vector specified by a number. The simplicity becomes apparent by noting that any vector $\vec{A}$ parallel to $\hat{i}$ can be expressed as a multiple of $\hat{i}$. Thus one can write

$$\vec{A} = A_i \, \hat{i} \qquad \text{(A-1)}$$

where A_i is some number. The magnitude of this number is simply equal to the magnitude of the vector $\vec{A}$ (since the unit vector $\hat{i}$ has magnitude 1). The sign of this number is positive if $\vec{A}$ is directed along $\hat{i}$, and is negative if $\vec{A}$ is directed opposite to $\hat{i}$. Hence this *number* A_i can be used to specify completely the magnitude and direction of the vector $\vec{A}$.

Definition of component. The number A_i is called the *component* of the vector $\vec{A}$ along the direction $\hat{i}$.

Def:

> ***Component of a vector.*** If a vector is parallel to a given direction, the *component* of the vector along this direction is a *number* whose magnitude is equal to the magnitude of the vector. The sign of this component is positive if the vector is along the specified direction, and negative if it is opposite to this direction. (A-2)

Components of motion quantities. The component of the position vector along the $\hat{i}$ direction is conventionally denoted by x and is also called the *position coordinate*. Thus one can write

$$\vec{r} = x\,\hat{i}\,. \tag{A-3}$$

Correspondingly, one conventionally uses v_x and a_x to denote the components of the velocity and acceleration along the $\hat{i}$ direction. Thus one can write

$$\vec{v} = v_x\,\hat{i}\,, \tag{A-4}$$

$$\vec{a} = a_x\,\hat{i}\,. \tag{A-5}$$

Example

Relative to the coordinate system indicated in Fig. A-2, the position vector of particle A is (2 m)$\hat{i}$ and its position coordinate x is 2 m. The position vector of the particle B is (−3 m)$\hat{i}$ and its position coordinate x is −3 m.

If a particle moves along the $\hat{i}$ direction with a speed of 7 m/s, its velocity component v_x is 7 m/s. But, if it moves with the same speed in the opposite direction, its velocity component v_x is −7 m/s.

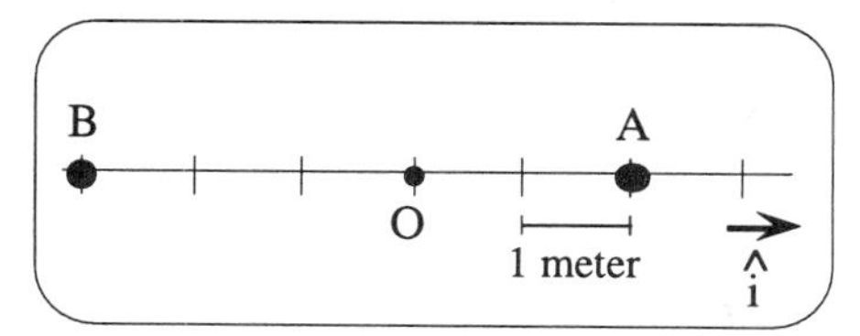

Fig. A-2. Position coordinates.

Component forms of basic definitions

The definition of velocity can be compactly summarized by the relation (4C-2), i.e.,

$$\vec{v} = \frac{d\vec{r}}{dt}\,. \tag{A-6}$$

This relation between vectors implies a corresponding numerical relation between their components. Indeed, the small displacement $d\vec{r}$ can be easily expressed in terms of its components

$$d\vec{r} = \vec{r}\,' - \vec{r} = x'\,\hat{i} - x\,\hat{i} = (x' - x)\,\hat{i} = dx\,\hat{i}\,.$$

By using (A-4), the definition (A-6) can then be written as

$$v_x \hat{i} = \frac{dx}{dt} \hat{i} .$$

But, if two vectors are equal, their components along any direction must also be equal. Hence the last relation implies that

$$\boxed{v_x = \frac{dx}{dt}} \qquad \text{(A-7)}$$

which is analogous to (A-6).

Similarly, the definition (4F-1) of acceleration

$$\vec{a} = \frac{d\vec{v}}{dt} \qquad \text{(A-8)}$$

implies the following corresponding relation between the components of these vectors

$$\boxed{a_x = \frac{dv_x}{dt} .} \qquad \text{(A-9)}$$

➜ ***Go to Sec. 5A of the Workbook.***

B. Alternate descriptions of linear motion

The use of vector components allows one to describe motion along a straight line entirely in terms of *numbers* rather than vectors, with considerable simplification. It then becomes easy to describe the motion quantitatively, as well as qualitatively by words or diagrams.

Table description

For example, one may describe the motion of a particle by using a table to specify the position coordinate x of the particle at various times t, as schematically illustrated in Fig. B-1. (Such a table may also include numerical information about the components of the particle's velocity and acceleration.)

t (second)	x (meter)
0	0
1	5
2	20
3	45

Fig. B-1. Tabular description of motion.

Graph description

An alternative way of describing the motion is by plotting the numerical values in a graph of x versus t, like the one in Fig. B-2. The advantage is that such a graph displays at a glance information about the entire motion of the particle. (However, a graph cannot specify individual values as precisely as actual numbers can.)

Graph of position coordinate versus time. A graph of position coordinate versus time also makes apparent the particle's velocity at any time t. Indeed, consider a slightly later time t', and the correspondingly slightly different value x' of the position coordinate at that time. During the short time dt = t'– t the position coordinate then changes by dx = x' – x. The ratio dx/dt (which is equal to the velocity component v_x) indicates in Fig. B-2 the *slope* of the curve at the time t. If the slope has a larger magnitude (i.e., if the curve is steeper), the magnitude of the velocity component v_x is larger. Furthermore, if the slope of the curve is positive (i.e., if x increases with time so that dx is positive), v_x is positive. Conversely, if the slope is negative (i.e., if x decreases with time so that dx is negative), v_x is negative.

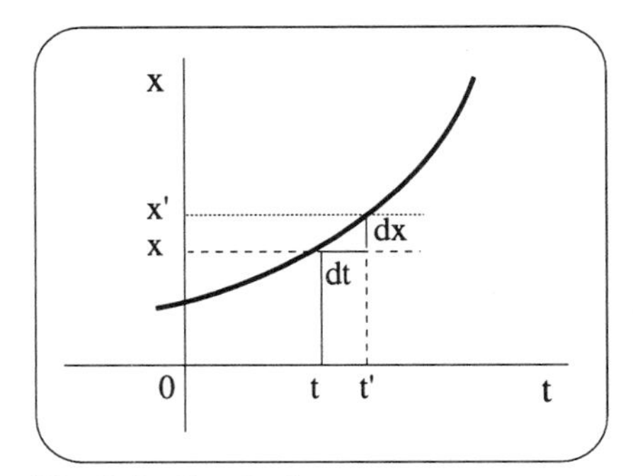

Fig. B-2. Graph of x versus t.

Graph of velocity component versus time. One can similarly display information about the velocity of a particle by means of a graph showing how the velocity component v_x depends on the time t. The slope of this graph is then equal to dv_x/dt, i.e., it is equal to the acceleration component a_x of the particle.

Equation description

Information about a particle's motion can often conveniently be summarized by means of an equation. If all vectors are expressed in terms of their components, such an equation simply relates numerical quantities. For instance, the motion of a particle along a straight line may be completely described by an equation which specifies how its position coordinate x depends on the time t. Thus the equation

$$x = (5\ \text{m/s}^2)\ t^2 \tag{B-1}$$

describes compactly the same motion as that previously described by the table in Fig. B-1.

➜ ***Go to Sec. 5B of the Workbook.***

Local information from global

The preceding sections discussed how one can describe a particle's motion *globally*, i.e., during an entire period of time. However, one is often more interested in describing the motion *locally*, i.e., in the immediate vicinity of any instant of time. Indeed, this interest motivates one to specify the *rate of change* at any instant of time, e.g., the velocity or acceleration. Hence we shall now examine, in the simple case of motion along a straight line, how global information about a particle's motion can be used to determined quantitatively its velocity or acceleration.

C. Changes and rates

Consider some quantity Q (e.g., a particle's position coordinate x or velocity component v_x). How can one calculate the change of this quantity, or its *rate* of change? For example, if one has an equation specifying how the *position* of a particle varies with time, how can one find its velocity? Similarly, if one has an equation specifying how the *velocity* of a particle varies with time, how can one find its acceleration?

Basic definitions

Finite and infinitesimal changes. A change of a quantity Q is denoted by ΔQ and is defined as

$$\Delta Q = Q' - Q \tag{C-1}$$

where Q' denotes the new value of the quantity and Q its original value. In the special case where the change is infinitesimally small, such a change is denoted by dQ (rather than ΔQ).

Rate of change. Suppose that a quantity Q varies with the time t. The rate of change of the quantity is then defined as

In more mathematical language, the rate of change of a quantity is called the *derivative* of this quantity.

$$\text{rate of change of Q} = \frac{dQ}{dt} \, . \tag{C-2}$$

In other words, one considers the value Q of the quantity at any particular time t of interest and compares it with its value Q' at a slightly later time t'. The rate of change (C-2) is then the ratio of the infinitesimal difference $dQ = Q' - Q$ compared to the infinitesimal time difference $dt = t' - t$. As previously discussed, these quantities are considered infinitesimal if a choice of t' closer to t (resulting in smaller values of dt and dQ) leaves the value of the ratio dQ/dt unchanged. This ratio is then called the *rate of change* of Q with time, at the particular time of interest.

General relations between changes or rates

When quantities are simply related, their changes (or rates of change) are related in correspondingly simple ways. The following are generally useful relations.

Change of a multiple of a quantity. If c is some constant, how does the quantity cQ change when Q itself changes?

The change of cQ is equal to

$$\Delta(cQ) = cQ' - cQ = c\,(Q' - Q)$$

or

$$\boxed{\Delta(cQ) = c\,\Delta Q \, .} \tag{C-3}$$

In other words, if a quantity is multiplied by a constant, the change of this quantity is multiplied by the same constant.

In particular, the result (C-3) is also true for infinitesimal changes. Suppose that dQ is the infinitesimal change of the quantity Q caused by an infinitesimal change dt. Division of (C-3) by dt then yields the following simple relation between corresponding rates

$$\boxed{\frac{d(cQ)}{dt} = c\frac{dQ}{dt}\,.} \tag{C-4}$$

Change of a sum of quantities. How is the change of the sum of two quantities Q_1 and Q_2 related to the individual changes of these quantities?

By the definition of a change,

$$\Delta(Q_1 + Q_2) = (Q_1' + Q_2') - (Q_1 + Q_2) = (Q_1' - Q_1) + (Q_2' - Q_2)$$

or

$$\boxed{\Delta(Q_1 + Q_2) = \Delta Q_1 + \Delta Q_2\,.} \tag{C-5}$$

In other words, the change of a sum is simply equal to the sum of the changes.

If the changes are infinitesimal and due to an infinitesimal change of t, division of (C-5) by dt then yields the following corresponding relation between rates

$$\boxed{\frac{d(Q_1+Q_2)}{dt} = \frac{dQ_1}{dt} + \frac{dQ_2}{dt}\,.} \tag{C-6}$$

Simple infinitesimal changes and rates

Let us now examine infinitesimal changes, and corresponding rates, of some simple (but frequently occurring) time-dependent quantities.

Case where Q = c. The simplest situation is that where $Q = c$ (i.e., where a quantity Q is merely equal to some constant c, as indicated by the graph in Fig. C-1). Such a constant quantity obviously does *not* change. Hence

$$\boxed{\Delta c = 0 \quad \text{and} \quad \frac{dc}{dt} = 0\,.} \tag{C-7}$$

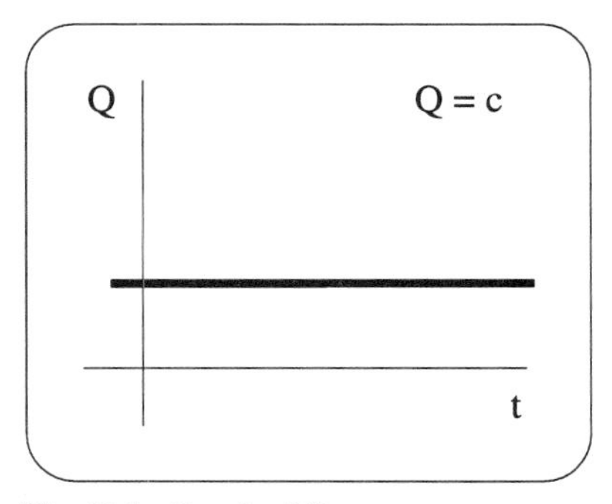

Fig. C-1. Graph of Q = c.

Case where Q = t. If $Q = t$, the rate of change of this quantity is simply

$$\boxed{\frac{dt}{dt} = 1\,.} \tag{C-8}$$

This result is immediately applicable to the slightly more complex case where $Q = ct$ and c is any constant. (The corresponding graph is a straight line like that shown in Fig. C-2.) Then the rate dQ/dt is just

$$\frac{d(ct)}{dt} = \frac{c\,dt}{dt} = c\,.$$

The rate of change is thus the same for any value of t (corresponding to the fact that the slope of the graph in Fig. C-2 is everywhere the same).

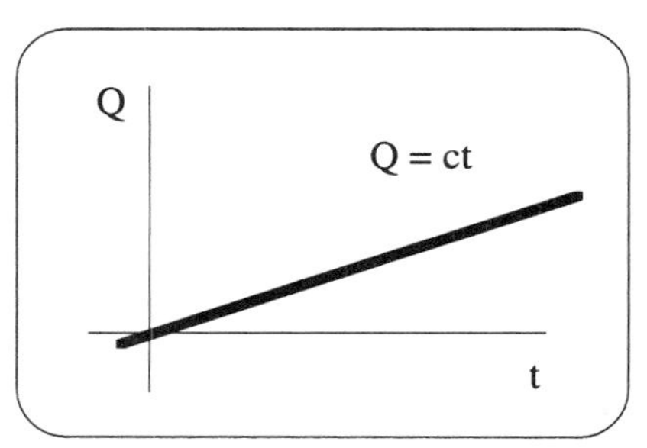

Fig. C-2. Graph of Q = ct.

Case where $Q = t^2$. Lastly, consider the slightly more complex case where $Q = t^2$. An infinitesimal change of this quantity, corresponding to an infinitesimal change $dt = t' - t$, is then equal to

$$d(t^2) = t'^2 - t^2 = (t + dt)^2 - t^2$$

since $t' = t + dt$. Thus one gets

$$d(t^2) = [t^2 + 2\,t\,dt + (dt)^2] - t^2 = 2\,t\,dt + (dt)^2 = (2t + dt)\,dt$$

Since dt is infinitesimally small, it is utterly negligible compared to 2t. Hence

$$\boxed{d(t^2) = 2t\,dt \quad \text{or} \quad \frac{d(t^2)}{dt} = 2t\,.} \qquad \text{(C-9)}$$

This result is immediately applicable to the slightly more complex case where $Q = ct^2$ and c is any constant. (The corresponding graph is a curved line like that shown in Fig. C-3.) Then the rate dQ/dt is just

$$\frac{d(ct^2)}{dt} = c\,\frac{d(t^2)}{dt} = 2ct\,.$$

In this case the rate of change of the quantity is *different* for different values of t. Correspondingly the slope of the curve in Fig. C-3 is *not* constant.

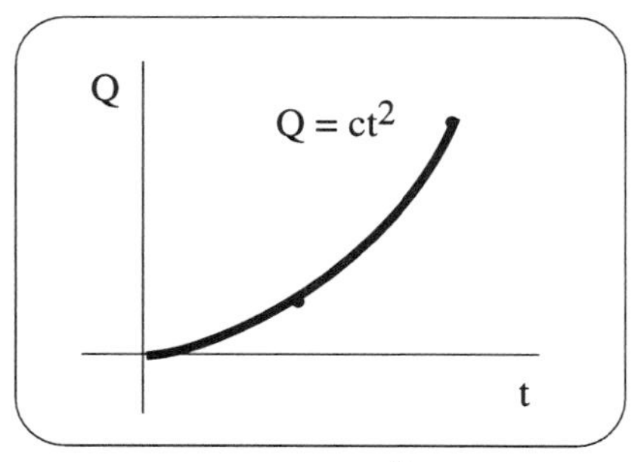

Fig. C-3. Graph of $Q = ct^2$.

General relation for rate of change

Reasoning, similar to that leading to (C-9), shows more generally that, for any number n,

$$d(t^n) = n\,t^{n-1}\,dt \quad \text{or} \quad \frac{d(t^n)}{dt} = n\,t^{n-1}\,. \qquad \text{(C-10)}$$

The results (C-7), (C-8), and (C-9) are just special cases of this general relation. Note that *the exponent of t in the rate is always one less than the exponent of t in the original quantity.*

Example C-1: Finding velocity from position information

The following equation specifies the position coordinate x of a car at any time t after it starts from rest traveling east along a straight road:

$$x = (1.5\ \text{m/s}^2)\,t^2\,. \qquad \text{(C-11)}$$

What then is the velocity of the car 5.0 seconds after starting from rest?

By (C-9) we obtain

$$v_x = dx/dt = (1.5\ \text{m/s}^2)\,(2t) = (3.0\ \text{m/s}^2)\,t$$

which specifies the car's velocity component at any time. Hence, at the particular time t = 5 second,

$$v_x = (3.0\ \text{m/s}^2)\,(5.0\ \text{s}) = 15\ \text{m/s}. \qquad \text{(C-12)}$$

The velocity of the car at this time is, therefore, 15 m/s east.

➜ ***Go to Sec. 5C of the Workbook.***

D. Calculating velocity or acceleration

The results of the last section allow one to find the rates of change of many quantities whose time-dependence is specified by equations. Thus one can use information about the position of a particle to find its velocity, or use information about the velocity of a particle to find its acceleration. The following example provides an illustration.

> ***Example: Velocity and acceleration of a falling object***
>
> A stone is dropped from a tall building. Its position coordinate x (measured from the ground along the upward $\hat{i}$ direction indicated in Fig. D-1) varies with the time t after its release according to the equation
>
> $$x = b - ct^2 \qquad \text{(D-1)}$$
>
> where $b = 60$ m, $c = 5.0$ ms^{-2}.
>
> (a) How high above the ground is the stone 3.0 s after its release?
>
> (b) What are the velocity and acceleration of the stone at that instant?

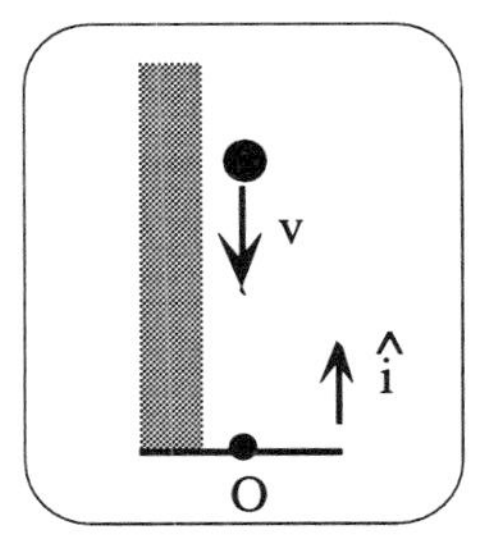

Fig. D-1. Falling stone

By (D-1), the position of the stone is then

$$x = (60\text{ m}) - (5.0\text{ ms}^{-2})\,(3.0\text{ s})^2 = 15\text{ m}\,,$$

i.e., the stone is then 15 meter above the ground.

The stone's velocity at any time can be found from the definition of velocity, expressed in the component form (A-7). Using (D-1), and our knowledge of rates from Sec. C, we get

$$v_x = \frac{dx}{dt} = \frac{db}{dt} - c\,\frac{d(t^2)}{dt} = 0 - 2ct\,. \qquad \text{(D-2)}$$

Thus, for t = 3.0 s,

$$v_x = -2\,(5.0\text{ m/s}^2)\,(3.0\text{ s}) = -30\text{ m/s}\,.$$

As indicated by the minus sign, the stone's velocity is then 30 m/s *downward*.

The acceleration of the stone at any time can similarly be found from the component form (A-9) of its definition and the rates found in Sec. C. Thus the information (D-2) about the velocity implies that

$$a_x = \frac{dv_x}{dt} = -2c\,. \qquad \text{(D-3)}$$

As indicated by the minus sign, the acceleration is then directed downward and has the same constant value at *any* time. Thus its numerical value is

$$a_x = -2\,(5.0\text{ m/s}^2) = -10\text{ m/s}^2$$

at any instant while the stone is falling.

→ ***Go to Sec. 5D of the Workbook.***

Global information from local

The preceding sections discussed how global information about motion can be used to find local information about its rate of change, e.g., about velocity or acceleration. However, the task of making predictions usually involves the *opposite* problem, i.e., how can one use local information about rates of change to predict the entire motion during a long period of time?

This question is at the heart of an extremely useful approach to scientific problems. It is difficult or impossible to predict events that will occur after a long time (e.g., to predict the positions of all the planets a hundred years from now). But, knowing a given situation, it is often easy to predict what will happen after a very short time (e.g., to predict the positions of the planets a millisecond from now). Such local information can, however, be used to make predictions about much longer times.

Indeed, suppose that knowledge about a situation allows one to predict the situation one millisecond later. Knowledge of the *present* situation allows one then to predict the situation one millisecond from now. This knowledge allows one then, in turn, to predict the situation one millisecond later, i.e., two milliseconds from now. This knowledge allow one then, in turn, to predict the situation one millisecond later, i.e., three milliseconds from now. Continuing repeatedly in this way, one can then predict the situation at any later time, even a century from now.

The preceding method, used to generate global information from local information, dates back to Newton and is extremely useful in all the sciences. [It is especially effective with the aid of some mathematical tools (like calculus) and of modern computers.] In the next couple of sections we shall begin to apply this method to particles moving along a straight line with constant velocity or constant acceleration. Thus we shall address the following question: How can we use local information about a particle's velocity or acceleration to predict its entire motion?

E. Motion with constant velocity

Description of the problem. Consider the situation where a particle moves along a straight line with *constant velocity*. The particle's velocity component v_x along this line has then the same value at every instant, i.e.,

$$v_x = \text{constant} \tag{E-1}$$

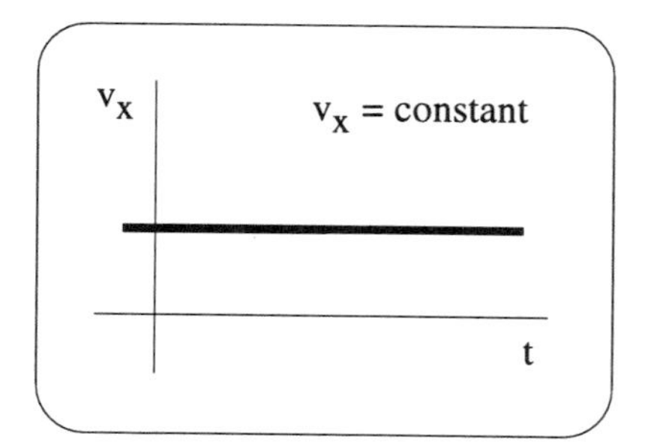

Fig. E-1. Graph of velocity versus time.

as illustrated by the graph in Fig. E-1. We also suppose that the position coordinate x_o of the particle is known at some instant. This instant can be called the *initial time* t = 0 if one starts measuring the time from this instant (e.g., by using a stopwatch). Then one knows that

$$\text{when } t = 0, \qquad x = x_o\,. \tag{E-2}$$

Any other time t denotes then the time elapsed since that initial instant.

How can one use the preceding known information to find the position coordinate x of the particle at any other time t?

We shall illustrate two alternative methods for solving this problem.

Guess-and-check method

This method is widely useful when one wants to find information about quantities from information about their infinitesimal changes (i.e., when one wants to solve *differential equations*).

Quite a few problems can be solved by the following method: (a) Systematically *guessing* the answer on the basis of available knowledge (*not* just randomly guessing by relying on sheer luck). (b) Carefully *checking* that the guessed answer is correct and complete.

This method is easily applied to the present problem to find the particle's position at any time.

Infinitesimal change of position. The definition of velocity provides a relation between the particle's position and the time. In its component form (A-7), this definition states that $v_x = dx/dt$. Thus

$$\frac{dx}{dt} = v_x \tag{E-3}$$

where the right side is the known velocity component. This equation relates corresponding *infinitesimal* changes of position and time. What then is the *total* change $x - x_o$ of position during the entire elapsed time t, i.e., what is the total displacement component

$$D_x = \Delta x = x - x_o\ ? \tag{E-4}$$

Position at any time. The present situation is particularly simple since the velocity is constant. Thus the ratio dx/dt has the *same* value v_x for *every* infinitesimal time interval. It is then a sensible guess that this ratio should also have the same value when calculated for the *entire* elapsed time t, i.e., that

$$\frac{x - x_o}{t} = \frac{dx}{dt} = v_x\ .$$

Thus we obtain the following relation between the displacement component $D_x = x - x_o$ and the elapsed time t:

$$\text{rel(disp, t):} \qquad \boxed{D_x = x - x_o = v_x t} \tag{E-5}$$

[Here rel(disp, t) is a convenient abbreviation indicating that this is the desired *relation between the displacement and elapsed time*.]

Check of the result. One can readily check the result (E-5) by finding the rates of change of both sides of this relation. Since x_o is merely a constant, one thus gets

$$\frac{dx}{dt} - 0 = v_x$$

which agrees properly with (E-3). Furthermore, (E-5) implies that, if $t = 0$, $x = x_o$, in proper agreement with (E-2).

Implications. The relation (E-5) merely indicates that the particle's *displacement is proportional to the elapsed time.* Correspondingly, its position coordinate changes with time at a constant rate specified by the constant velocity (as illustrated by the constant slope of the graph in Fig. E-2).

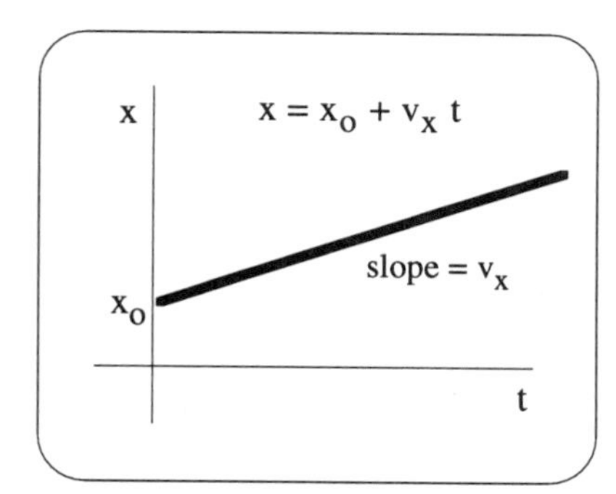

Fig. E-2. Predicted graph of position versus time.

> ***Example***
>
> Suppose that a car travels north with a *constant* speed of 30 m/s. Its velocity component v_x along the northern direction is then 30 m/s. After 20 seconds, the component D_x of the car's displacement along this direction would then be simply (30 m/s) (20 s) = 600 m.

Addition method

Alternatively, the particle's total displacement can be found by simply adding all its successive infinitesimal displacements.

Indeed, the definition (E-3) implies that the particle's *infinitesimal* displacement dx during any infinitesimal time interval dt is given by

$$dx = v_x \, dt \,. \qquad \text{(E-6)}$$

The particle's total displacement, during the entire elapsed time t, can then be found by adding all such successive infinitesimal displacements, as illustrated in Fig. E-3. Thus

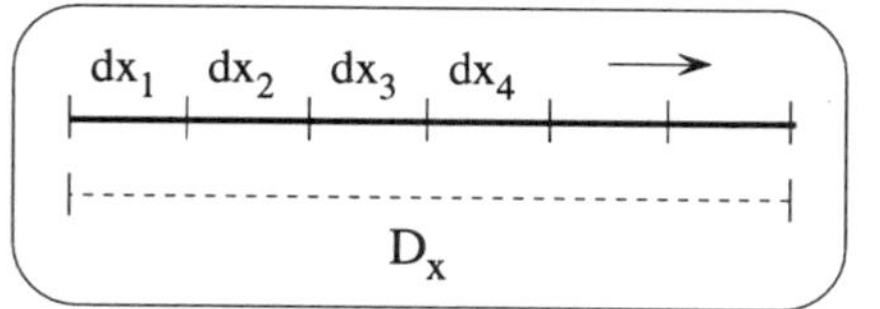

Fig. E-3. Successive infinitesimal displacements of a particle moving with constant velocity.

$$D_x = (dx)_1 + (dx)_2 + (dx)_3 + \ldots$$

$$= v_x \, (dt)_1 + v_x \, (dt)_2 + v_x \, (dt)_3 + \ldots \,.$$

This sum is easily calculated because v_x is merely a constant. Thus one gets

$$D_x = v_x \, [(dt)_1 + (dt)_2 + (dt)_3 + \ldots]$$

or

$$D_x = v_x \, t$$

since the sum of all the successive infinitesimal time intervals is merely the total elapsed time t. Thus we obtain the same result (E-5) as that obtained previously.

→ ***Go to Sec. 6E of the Workbook.***

F. Motion with constant acceleration

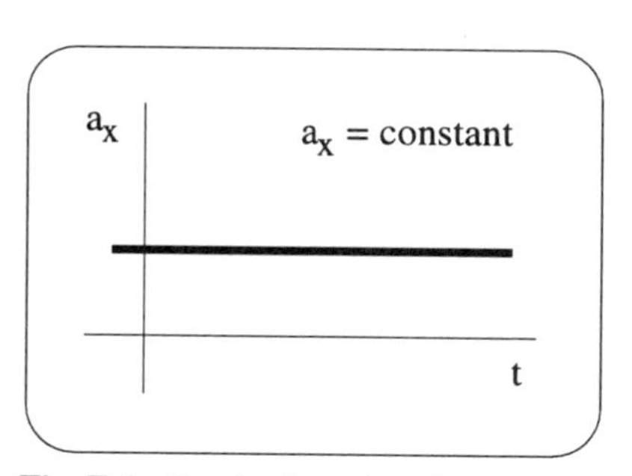

Fig. F-1. Graph of acceleration versus time.

Consider next the somewhat more complex situation where a particle moves along a straight line with a *constant acceleration.* The particle's acceleration component a_x along this line has then the same value at every instant, i.e.,

$$a_x = \text{constant} \qquad \text{(F-1)}$$

as illustrated by the graph in Fig. F-1. We also suppose that the particle's velocity component v_{xo} and position coordinate x_o are known at some initial time (which may be called $t = 0$ if one starts measuring elapsed time from that instant). Thus we know that,

$$\text{when } t = 0, \qquad v_x = v_{xo} \text{ and } x = x_o \,. \qquad \text{(F-2)}$$

How can one use the preceding known information to find the velocity component v_x and position coordinate x of the particle at any other time t?

Finding velocity

Description of the problem. In the present situation, the desired velocity component v_x of the particle is related to its known acceleration component a_x by the definition (A-9) of acceleration. Thus we know that $a_x = dv_x/dt$ or that

$$\frac{dv_x}{dt} = a_x \qquad \text{(F-3)}$$

where the right side is simply a constant. This equation relates corresponding infinitesimal changes of velocity and time. What then is the total change of velocity during the entire elapsed time t?

Velocity at any time. The problem here is analogous to the previous problem where we used the relation (E-3) about constant velocity to find the corresponding relation (E-4) about change of position. Since the acceleration is constant, the ratio dv_x/dt has the same value a_x for any infinitesimal time interval dt. It is then an obvious guess that this ratio should also have the same value when calculated for the *entire* elapsed time t, i.e., that

$$\frac{v_x - v_{xo}}{t} = \frac{dv_x}{dt} = a_x \,.$$

Thus we obtain the result

$$\text{rel(vel, t):} \qquad \boxed{v_x - v_{xo} = a_x t \,.} \qquad \text{(F-4)}$$

[Here we have used the abbreviation rel(vel, t) to indicate that this is a *relation between the velocity and the elapsed time.*] This relation indicates that the velocity changes with time at a constant rate specified by the constant acceleration (as illustrated by the constant slope of the graph in Fig. F-2).

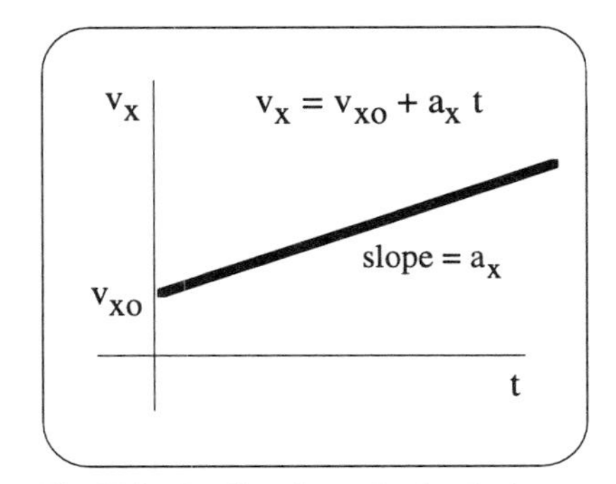

Fig. F-2. Predicted graph of velocity versus time.

Check. One can readily check that the relation (F-4) is correct. Indeed, its rate of change is

$$\frac{dv_x}{dt} - 0 = a_x$$

which agrees with (F-3). Furthermore, (F-4) implies that $v_x = v_{xo}$ when $t = 0$, in agreement with the known information (F-2).

> ***Example F-1: Accelerating car***
>
> A car travels north along a straight road with a speed of 18 m/s. To pass a truck, it starts accelerating with a *constant* acceleration of 1.5 m/s^2. What is the velocity of the car 6.0 s after it starts accelerating?

The relations (F-4) and (F-7) can also be derived by the addition method.

Choose the $\hat{\imath}$ direction pointing north. The velocity component v_x of the car is then such that the velocity change

$$v_x - 18\text{ m/s} = (1.5\text{ m/s}^2)\,(6.0\text{ s}) = 9.0\text{ m/s}.$$

Hence $v_x = 27$ m/s, i.e., the velocity of the car is then 27 m/s north.

Finding displacement or position

Description of the problem. What is the position of the particle at any time if its velocity varies with the time as specified by (F-4)? To relate the position to the velocity, we start from the definition of velocity $v_x = dx/dt$ (expressed in terms of components). By (F-4), this implies that, at any time t,

$$\frac{dx}{dt} = v_x = v_{xo} + a_x t\,. \tag{F-5}$$

This equation involves corresponding *infinitesimal* changes of position and time. What then is the total change $x - x_o$ of position during some elapsed time t?

Position at any time. The answer is easily guessed. If the acceleration were zero, the particle would constantly move with its initial velocity v_{xo}. According to (E-5), the change of its position would then be simply equal to $v_{xo}t$. But the velocity (F-5) involves also the extra term $a_x t$. From our knowledge (C-9) of rates we know that such a rate of change proportional to t comes from a quantity proportional to t^2. Hence it is a good guess that the change of position should also involve an extra term proportional to t^2, i.e., that

$$x - x_o = v_{xo}t + ct^2 \tag{F-6}$$

where the extra term involves some constant c. The actual value of this constant can be determined by making sure that the rate of change of (F-6) agrees with the known velocity (F-5). This rate of change of (F-6) is

$$\frac{dx}{dt} - 0 = v_{xo} + 2ct$$

or

$$v_x = v_{xo} + 2ct\,.$$

This agrees with (F-5) provided that $c = \frac{1}{2}a_x$. Furthermore (F-6) properly implies, in agreement with (F-2), that $x = x_o$ when $t = 0$.

Hence (F-6) satisfies all requirements and specifies correctly the particle's change of position (i.e., its displacement component $D_x = x - x_o$) during any elapsed time t. Thus

rel(disp, t):

$$\boxed{D_x = x - x_o = v_{xo}t + \frac{1}{2}a_x t^2} \tag{F-7}$$

where rel(disp, t) is a convenient abbreviation indicating that this is the desired *relation between the displacement and the elapsed time.*

Implications. The relation (F-7), and corresponding curved graph in Fig. F-3, properly indicate that the particle's position coordinate x changes with time at a *non-constant* rate. This must be so since the particle's velocity changes, i.e., because its acceleration is non-zero.

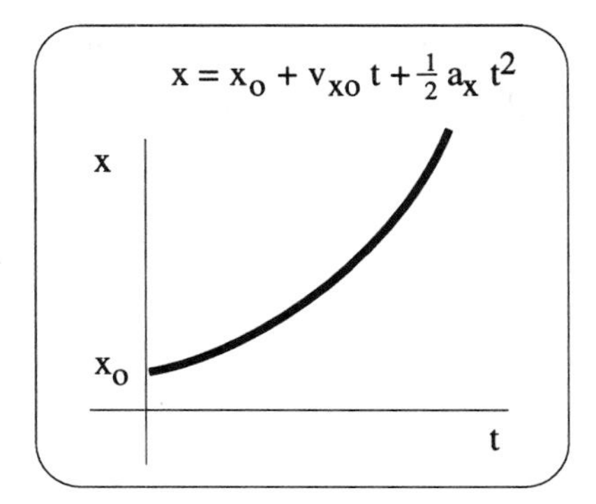

Fig. F-3. Predicted graph of position versus time.

Stated in words, the relation (F-7) says that the ⟨displacement D_x of the particle⟩ = ⟨its displacement $v_{xo}t$ *if* it moved with a constant velocity v_{xo} without acceleration⟩ + ⟨an additional displacement $\frac{1}{2}a_x t^2$ due to its acceleration⟩.

The relation (F-7) includes also the special case where the particle moves with constant velocity. Indeed, in this case $a_x = 0$ and the relation (F-7) becomes identical to the relation (E-5) previously derived for motion with constant velocity.

Example F-2: Accelerating car

Consider again the Example F-1 where a car travels north with a speed of 18 m/s and then starts accelerating north with a *constant* acceleration of 1.5 m/s². What then is the car's displacement during the next 6.0 seconds?

Choose the $\hat{i}$ direction again along the northern direction. Then (F-7) implies that

$$D_x = (18 \text{ m/s})\,(6.0 \text{ s}) + \frac{1}{2}(1.5 \text{ m/s}^2)\,(6.0 \text{ s})^2 = 135 \text{ m},$$

i.e., the car's displacement is 135 m north.

Example F-3: Particle starting from rest at origin

Suppose that a particle moves with constant acceleration after starting from rest at the origin. Then $v_{xo} = 0$ and $x_o = 0$. The relations (F-4) and (F-7) then imply that

$$v_x = a_x t \quad \text{and} \quad x = \frac{1}{2} a_x t^2 .$$

In this case, the velocity is thus simply proportional to the elapsed time, and the distance traveled is proportional to the *square* of the elapsed time. For example, if the elapsed time is 3 times as large, the attained velocity is 3 times as large, but the distance traveled is 9 times as large.

Relation between velocity and displacement

If a particle moves with constant acceleration, the relations (F-4) and (F-7) are direct implications of the definitions of acceleration and velocity. These relations specify the velocity and displacement of the particle at any time. Hence they describe completely the particle's motion and can also readily be used to obtain other pertinent information.

For example, one may sometimes not be interested in how long a *time* a particle has traveled, but in knowing how its velocity is related to its displacement. (For instance, one might want to know what velocity an accelerating car attains after traveling a certain distance, but may *not* be interested in knowing how long a time the car has traveled.)

It is easy to find a direct relation between the velocity and displacement (without mention of the time). To do this, one only needs to eliminate the time between the relations (F-4) and (F-7). Thus (F-4) can be used to find the time t elapsed before the particle attains a particular velocity v_x. This yields

$$t = \frac{v_x - v_{xo}}{a_x} . \qquad \text{(F-8)}$$

By substituting this value of t into (F-7), one can then find the corresponding displacement D_x. Some straightforward algebra leads then to the following relation between the particle's velocity and displacement:

rel(vel, disp): $$\boxed{v_x^2 - v_{xo}^2 = 2a_x D_x} \qquad \text{(F-9)}$$

Algebra leading to the preceding result

Substitution of (F-8) into (F-7) yields

$$D_x = v_{xo}t + \frac{1}{2}a_x t^2$$

$$= [v_{xo} + \frac{1}{2}a_x t\,]\,t$$

$$= \left[v_{xo} + \frac{1}{2}a_x\left(\frac{v_x - v_{xo}}{a_x}\right)\right]\left(\frac{v_x - v_{xo}}{a_x}\right)$$

$$= [v_{xo} + \frac{1}{2}(v_x - v_{xo})]\left(\frac{v_x - v_{xo}}{a_x}\right)$$

$$= \frac{1}{2}[v_x + v_{xo}]\left(\frac{v_x - v_{xo}}{a_x}\right)$$

$$D_x = \frac{(v_x^2 - v_{xo}^2)}{2a_x} .$$

This last result can be rewritten in the convenient form (F-9).

Summary of motion with constant acceleration

The preceding discussion can be briefly summarized as follows: The definitions of acceleration and velocity imply immediately the corresponding relations (F-4) and (F-7) showing how the velocity and displacement depend on the elapsed time. (As we have seen, these relations are probably more easily derived than specially remembered.) By using simple algebra to eliminate the time between these relations, it is then also easy to obtain the corresponding relation (F-9) between velocity and displacement. (Indeed, it is often easier to do such algebra than to remember this relation.) The relations thus obtained are the following:

rel(vel, t): $\boxed{v_x - v_{xo} = a_x t}$, (F-10)

rel(disp, t): $\boxed{D_x = v_{xo}t + \frac{1}{2}a_x t^2}$, (F-11)

rel(vel, disp): $\boxed{v_x^2 - v_{xo}^2 = 2a_x D_x}$. (F-12)

The elapsed time t in these relations may also be negative (if it indicates the time which did elapse before the stopwatch was started).

(The preceding relations are also valid in the special where the velocity is constant so that $a_x = 0$.)

→ ***Go to Sec. 6F of the Workbook.***

G. Gravity near the surface of the earth

Gravitational interaction. It is a commonplace observation that any object, released near the surface of the earth, falls vertically downward toward the ground (i.e., in a direction toward the center of the earth). This behavior would not occur if the object were somewhere in outer space very far from the earth. Thus the object is influenced by the presence of the neighboring earth, i.e., the observed motion is due to the object's *interaction* with the neighboring earth. This interaction is called *gravitational* interaction.

Downward acceleration of falling objects. Closer observations (e.g., by means of stroboscopic photographs like those mentioned in Sec. 4A) reveal that any particle falls downward with increasing speed, i.e., with a downward acceleration. To observe the motion under conditions where the particle interacts *solely* with the earth, one can make sure that the particle does not touch any other objects and does not interact with the surrounding air. (For example, one can observe the particle moving in a vacuum.)

For example, if a particle is released *from rest*, it moves with a downward acceleration so that it falls downward with constantly increasing speed. On the other hand, suppose that a particle is initially *thrown vertically upward*. Then it also moves with a downward acceleration. However, as a result the particle now first moves up with decreasing speed, then momentarily attains zero speed, and then moves down with increasing speed.

Gravitational acceleration independent of particle properties. Detailed observations and some highly precise experiments reveal the following remarkable fact illustrated in Fig. G-1:

> Any particle, interacting solely with the earth, moves at any point with a downward acceleration *independent of the nature of the particle*. (G-1)

In other words, the acceleration of any particle is the *same* — irrespective of its size, shape, or material! For example, in a vacuum, a steel ball, a gold coin,

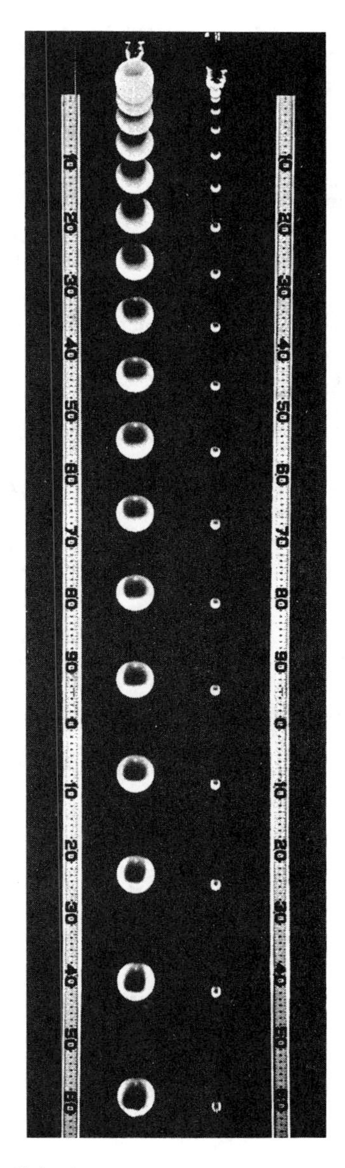

Fig. G-1. Stroboscopic photograph of two different falling balls released from rest at the same time. [Reproduced with permission from PSSC PHYSICS, 2nd edition, 1965; D.C. Heath and Company with Education Development Center, Inc., Newton, MA.]

or a feather all fall with exactly the same acceleration! Thus (G-1) describes a very remarkable property of our universe. (One could certainly conceive of a world where this would not be true.) This property has far-reaching implications and is also a cornerstone of Einstein's general theory of relativity.

The immediate implication of (G-1) is that one can simply talk about the gravitational acceleration *at a point,* i.e., about the gravitational acceleration of *any* particle located at this point. This gravitational acceleration is commonly denoted by the symbol $\vec{g}$.

Magnitude of the gravitational acceleration. The magnitude of $\vec{g}$ is somewhat different at different points. For example, if a particle is located farther from the earth, its interaction with the earth is somewhat smaller. Correspondingly, the magnitude g of the gravitational acceleration at the top of a high mountain is slightly smaller than at sea level. However, g is nearly constant close to the surface of the earth (i.e., at points whose height above the surface is much smaller than the radius of the earth). Measurements show that the approximate value of the gravitational acceleration is

near sea level, $\vec{g} \approx 9.80\ \text{m/s}^2$ downward. (G-2)

Any object, interacting solely with the earth near its surface, moves thus with this known constant acceleration $\vec{g}$. Hence the entire discussion of the preceding Section F is immediately applicable to many practical problems involving the motion of such objects.

➜ ***Go to Sec. 6G of the Workbook.***

H. Summary

Rates of change (velocity & acceleration)

Components of motion quantities along a fixed direction:

Components of position vector, of velocity, and of acceleration:

$x,\quad v_x = dx/dt,\quad a_x = dv_x/dt.\quad (\text{Displacement } D_x = x - x_0)$

Relations between changes or between rates of change:

$\Delta(cQ) = c\,\Delta Q\,,\qquad \dfrac{d(cQ)}{dt} = c\,\dfrac{dQ}{dt}$

$\Delta(Q_1 + Q_2) = \Delta Q_1 + \Delta Q_2\,,\qquad \dfrac{d(Q_1+Q_2)}{dt} = \dfrac{dQ_1}{dt} + \dfrac{dQ_2}{dt}$

Rates of change of some simple quantities:

$$\frac{dc}{dt} = 0 \ \{\text{if c is a constant}\}, \quad \frac{dt}{dt} = 1, \quad \frac{d(t^2)}{dt} = 2t$$

Motion with constant velocity or acceleration

All the following relations are simple implications of the definitions of velocity ($v_x = dx/dt$) and of acceleration ($a_x = dv_x/dt$).

Motion with constant velocity (v_x = constant)

rel(disp, t): $D_x = v_x t$ (relates displacement & time)

Motion with constant acceleration (a_x = constant)

rel(vel, t): $v_x - v_{xo} = a_x t$ (relates velocity & time)

rel(disp, t): $D_x = v_{xo}t + \frac{1}{2} a_x t^2$ (relates displacement & time)

rel(vel, disp): $v_x^2 - v_{xo}^2 = 2a_x D_x$ (relates velocity & displacement)

Gravity near the surface of the earth

Acceleration = $\vec{g}$, same for *every* particle at a given position.

$g \approx 9.80 \text{ m/s}^2$ near the surface of the earth.

New abilities

You should now be able to do the following:

(1) Describe the motion of a particle along a straight line qualitatively in terms of words or sequence diagram — or more quantitatively in terms of tables, graphs, or equations.

(2) If a simple equation specifies how a particle's position varies with time, find its velocity at any instant. If a simple equation specifies how its velocity varies with time, find its acceleration at any instant.

(3) If a particle moves along a straight line with constant velocity, find its position at any time. If it moves with constant acceleration (e.g., because of gravity near the earth), find its velocity or position at any time.

➜ ***Go to Sec. 5H of the Workbook.***

6 Problem Solving

The knowledge gained in the preceding chapter can, in principle, be used to solve many diverse problems dealing with objects moving along straight lines. However, many physics problems (as well as many problems in other sciences or engineering) are often difficult or impossible to solve without a systematic approach. This chapter examines, therefore, why problem solving can be difficult and then introduces a systematic problem-solving method that can help to overcome these difficulties. This method can be used to deal with a wide range of problems, in physics and beyond. In the present chapter the method will be applied to relatively simple problems involving motion along straight lines. With minor modifications, it will then repeatedly be used to deal with the more complex problems encountered later in the book.

A. Introduction to problem solving

What is a problem?

A problem is a task which requires one to devise a sequence of actions leading from some initial situation to some specified goal. A well-specified sequence of such legitimate actions constitutes a *solution* of the problem.

A problem may be schematically represented by a diagram like that in Fig. A-1. Here the point A represents the initial situation, the point G represents the goal, and a line linking two points represents a legitimate action leading from one situation to another. The highlighted sequence of links leading from A to G represents a solution of this problem.

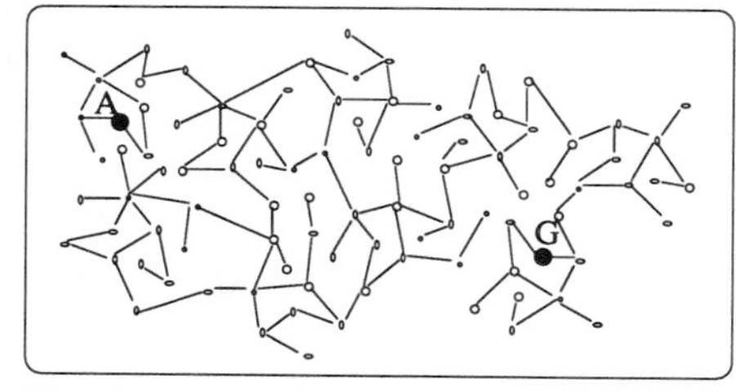

Fig. A-1. Schematic representation of a problem.

Problems arise constantly in science (as well as in everyday life) because we are so often interested in pursuing various goals. In particular, problem solving is essential in science to attain the central scientific goals of predicting or explaining observable phenomena.

What is a solution?

In order to solve problems, one must clearly understand what is meant by a solution.

Def: **_Solution of a problem:_** A sequence of well-specified legitimate actions leading from the initial situation to the desired goal. (A-1)

This definition has several important implications.

Solution versus answer. The *solution* of a problem specifies the entire sequence of actions leading to the goal (e.g., all the reasoning steps in an argument). By contrast, the *answer* to the problem is merely the attained goal (i.e., a specification of the result without a specification of the process leading to it). For example, sheer luck may occasionally lead to the correct *answer* of a problem, even though the *solution* is wrong. The following is a simple illustration.

Example: Right answer, but wrong solution

A young student, asked to divide 64 by 16, simply "canceled" the 6 in the numerator and in the denominator, as indicated in Fig. A-2. Thus the student obtained the correct answer 4. However, the student's solution is completely wrong.

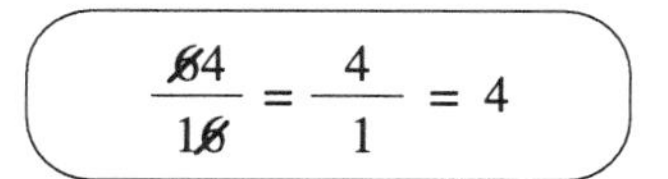

Fig. A-2. Wrong solution leading to correct answer.

Legitimate actions. Only certain kinds of actions are permissible in any domain. For example, the rules of chess permit only certain kinds of moves, and the laws of our society permit certain kinds of behaviors but not others.

The purpose of science is to predict or explain a large range of phenomena on the basis of very few basic premises. Accordingly, the legitimate actions used for problem solving in science must all involve logical steps based on a few well-specified basic scientific relations (definitions or principles).

Intuitive guesses may sometimes be useful in *suggesting* a solution (but can also be misleading or wrong). However, they need to be checked and, unless put on a firmer basis, do *not* constitute legitimate steps in the solution of a scientific problem.

For example, the basic scientific relations available to us up to now include mainly the definitions of velocity and acceleration, the relations (5F-10) through (5F-12) implied by them, and the properties of the gravitational acceleration near the earth. Hence the solution of all motion problems discussed in this chapter can involve only arguments based on these basic relations.

Well-specified actions. It is essential that the actions specified by a solution are sufficiently well-specified that they can be readily understood and checked by other people. (Otherwise it is not possible to verify that a solution is correct, to use it reliably, or to modify it appropriately if necessary.)

Illegible scribbles, scattered over a page, clearly do *not* constitute a solution. Even a neatly arranged sequence of correct equations do *not* constitute a solution. Indeed, such equations merely show the implementations of certain actions, but do not indicate *why* these actions were performed. A mere string of equations is largely incomprehensible to an outside reader (and sometimes even to the author looking at them after a few days).

A solution thus needs to include not only equations showing the implementations of various actions, but also *explanatory comments* indicating the decisions leading to these actions. It is only then that the *design* of the solution becomes clear so that the solution can be properly understood, verified, and used.

Unexplained problem "solutions" might be tolerated in childish school settings, but are not accepted in any actual scientific or work environment. (For example, a computer program consisting merely of lines of code, without explanatory comments, is almost useless in real life.)

Central difficulties of problem solving

Problem solving can be intellectually demanding because it requires one to cope with the following difficulties.

Decision making. As is apparent from Fig. A-1, trying to solve a problem is somewhat like trying to find one's way through a maze. One is originally in some situation A and wants to get to the goal G. Starting from A, there are many permissible actions one can take (as indicated by the lines in Fig. A-2), but most cause one to get lost or stuck without ever getting to the goal. Hence one faces the following difficulty: How can one make judicious decisions so as to choose, out of the very many possible action sequences which lead to nowhere, the one (or very few) which lead to the desired goal?

The actual difficulty presented by a problem depends not only on the problem itself, but also on the knowledge possessed by the person trying to solve it. For example, an *unfamiliar* problem may be quite difficult for someone. But the same problem may be very easy if the person has already encountered its solution previously.

Initial problem analysis. Before one can decide about alternative actions, one needs to address a prior difficulty: How can one initially describe and analyze a problem so as to identify the useful possible actions among which one can choose?

Indeed, a good initial analysis of a problem can greatly facilitate the task of finding its solution. Conversely, if the initial analysis of a problem is deficient, it may be impossible to find its solution despite all subsequent efforts. (Fig. A-3 illustrates schematically a poorly analyzed problem which does not even indicate the available pathways through the maze).

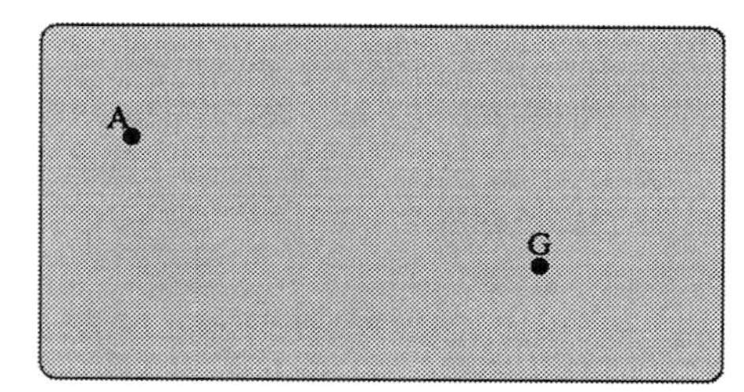

Fig. A-3. Schematic representation of a poorly analyzed problem.

Need for a systematic problem-solving method

Most problems in physics can *not* be solved by reaching haphazardly for various equations (just as it is highly improbable that random choices will allow one to find one's way through a complex maze).

Mere acquaintance with particular examples of problem solutions is also of limited utility. Any such solution can certainly be examined and assessed for its correctness. But it reveals little about the *process* needed to construct the solution — about how to make just the right decisions and how to recover from impasses when one gets stuck. Yet, it is precisely such knowledge about the solution *process* which is required if one is to deal with unfamiliar problems.

To help one cope with the difficulties of problem solving, one thus needs a systematic general method for dealing with diverse and unfamiliar problems.

As indicated in Fig. A-4, such a problem-solving method must address the following essential needs: (a) Analyzing the problem so as to facilitate its subsequent solution. (b) Constructing the solution by choosing judiciously what basic knowledge to apply and then implementing these choices. (c) Checking the solution — and then revising it to correct its deficiencies and ensure that it is satisfactory.

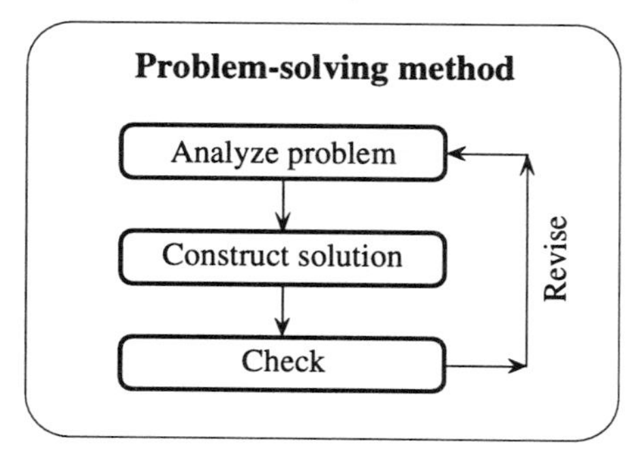

Fig. A-4. Main steps of the problem-solving method.

Note that the preceding needs are *not* met if one jumps into a problem by immediately trying to construct its solution. Instead, one must carefully look before leaping in, and carefully look back afterwards. The time spent on such preparation and review is usually amply repaid by avoiding mistakes and fruitless floundering.

The following sections examine more fully the major steps of the problem-solving method outlined in Fig. A-4. Elaborated in more detail, this method is widely applicable (even outside of physics) and will be used throughout the rest of the book.

➜ ***Go to Sec. 6A of the Workbook.***

B. Initial analysis of a problem

The purpose of the initial analysis of a problem is to bring the problem into a form facilitating its subsequent solution.

Basic description. The initial analysis of a problem must start by clearly specifying the problem. As indicated in Fig. B-1, such a basic problem description can be achieved by doing the following:

* ***Situation.*** Describe the situation by summarizing the known information. Do this by drawing diagram(s) accompanied by some words, and by introducing useful symbols. (Diagrams are extremely important to help one visualize a situation.)
* ***Goals.*** Specify compactly the goals of the problem.

Refined description. The basic description of a problem can usefully be refined by analyzing the problem further in the following ways:

* ***Time sequence & intervals.*** Specify the *time-sequence* of events (e.g., by visualizing the motion of objects as they might be observed in successive movie frames). Also identify *time intervals* where the description of the situation is distinctly different (e.g., where the acceleration of an object is different).
* ***Physics description.*** Describe the situation in terms of important physics concepts (e.g., by specifying information about velocity and acceleration).

Importance of adequate analysis. The initial analysis of a problem should be good enough that there should be no further need to refer to the original problem statement. This initial analysis is very important since it can greatly facilitate the solution of the problem. Conversely, if the initial analysis is deficient or wrong (e.g., if it omits important information or makes unwarranted assumptions), no amount of subsequent work can lead to a successful solution.

Analysis of a problem

* **Basic description**
    * ***Situation***
        Known information
        Diagram & useful symbols
    * ***Goals***
* **Refined description**
    * ***Time sequence & intervals***
    * ***Physics description***
        (e.g., velocity, acceleration)

Fig. B-1. Initial analysis of a problem.

Example: Analysis of a motion problem

The following illustrates the initial analysis of a typical motion problem.

Statement of the problem (ball thrown from a balcony). A girl, standing on the balcony of a building, throws a ball vertically upward. She observes that the ball reaches its maximum height 1.20 s after leaving her hand, and some time later hits the ground 13.0 m below her hand. (Air resistance can be assumed negligible.)

(a) With what speed did the ball leave the girl's hand?
(b) How high above the balcony did the ball rise?
(c) With what speed did the ball hit the ground?

Analysis of the problem. The analysis of this problem is shown below in a compact form where most of the relevant information is indicated in the diagram. (The right margin includes some explanatory remarks.)

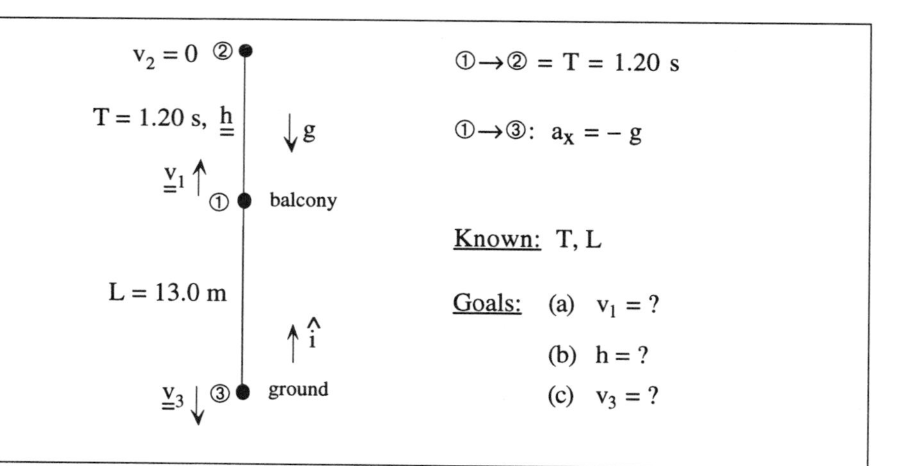

Situation diagram. Time sequence is indicated by the successive important instants ①, ②, ③, ...[①→② indicates the time elapsed between instants ① and ②.]

①: Ball leaves hand with speed v_1.
②: Ball at highest point, height h.
③: Ball at ground, distance L below hand.

There is only one distinct time interval since the ball's acceleration is equal to $\vec{g}$ during the entire time ①→③.

Velocity at highest point is zero since height there momentarily neither increases nor decreases, i.e., does not change.

Wanted unknown quantities are doubly underlined in the diagram.

➜ ***Go to Sec. 6B of the Workbook.***

C. Construction of a solution

Decomposition into subproblems

A good strategy for constructing the solution of a problem is to "divide and conquer". This can be done by successively solving simpler *subproblems* (i.e., subsidiary problems which facilitate the solution of the original problem), and doing this repeatedly until the original problem has been solved.

Decomposition process. As schematically indicated in Fig. C-1, the solution can be achieved by repeated application of the following two steps: (1) Choosing a useful subproblem, and (2) implementing the solution of this subproblem to obtain useful information.

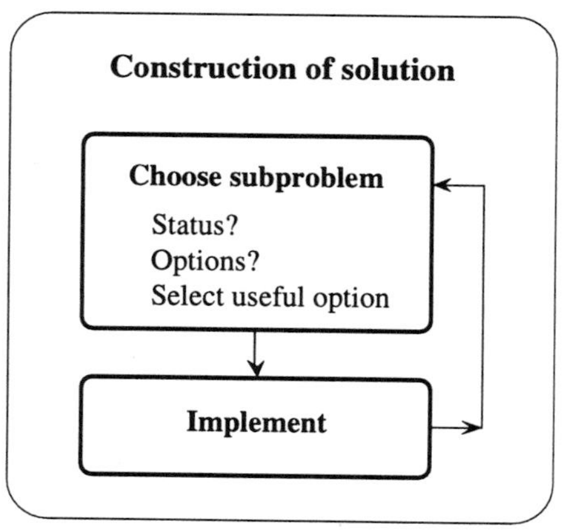

Fig. C-1. Constructing a solution by repeated decomposition into subproblems.

Choice of a useful subproblem. How can one make wise decisions to choose a useful subproblem? Fig. C-1 indicates how this may be done:

(1) Examine the status of the problem at any stage by identifying the available known and unknown information, and the obstacles hindering a solution. (2) Identify available options for subproblems that can help overcome the obstacles. (3) Select a useful subproblem among these options.

Main kinds of subproblems

The preceding choice process is fairly easy if there are only few alternative options among which one needs to decide. Indeed, this is usually the case if one's available knowledge is well organized.

Finding useful relations. The major obstacle hindering the solution of any problem is ordinarily a lack of useful information (particularly at the beginning of the problem when little such information is yet available). To overcome this obstacle, one addresses the subproblem aiming to find some useful relation. The method for doing this is the following (also summarized in Fig. C-2):

> Apply a *basic relation* (from general physics knowledge)
> to some *object*
> at some *time* (or between some *times*)
> along some *direction*.

Here each of the underlined entities represents a specific choice that needs to be made. (Which specific basic relation should be applied? To which specific object in the problem should it be applied? At which specific time in the problem should it be applied? Along which direction?)

In the case of problems involving a particle moving with constant acceleration along a straight line, the basic physics relations are very few: They are just the definitions of velocity and acceleration, or equivalently the three motion relations (5F-10), (5F-11), and (5F-12) implied by them. Furthermore, we know under what conditions they are useful (i.e., that one is useful to relate velocity and time, another to relate displacement and time, etc.). Choosing between these few available basic relations is thus rather simple.

For example, a typical choice made to find a specific useful relation in a motion problem might be the following: Apply rel(vel, t) [i.e., the relation between velocity and time], to the ball, 2.0 s after it is thrown, in the upward direction.

Eliminating unwanted quantities. The other main obstacle in problem solving occurs when available useful relations contain unwanted unknown quantities. One must then address the subproblem of eliminating such an unwanted quantity. This is readily done by combining two relations containing this quantity, using familiar methods of algebra.

If only one such relation is available, one needs to find more information about the quantity. One is then again faced with the task of finding a useful relation. (After this has been done, one has enough relations which can be combined to eliminate the unwanted quantity.)

Fig. C-2 summarizes the preceding two main kinds of subproblems that need to be addressed in problem solving.

Implications for explanatory comments. As pointed out in Sec. A, a problem solution must include enough explanatory comments so that it can be understood and checked by others. These explanatory comments need to indicate the main decisions made in constructing the solution (i.e., what particular subproblems were addressed and implemented in the solution).

Subproblem options

{lack useful info} → ***Find useful relation*** Apply *basic relation* to *object* at *time* (or between *times*) along *direction*

{unwanted quantity} → ***Eliminate quantity*** Combine *relations* → {lack info about unknown}

Fig. C-2. Main problem-solving obstacles and available options for subproblems to overcome these obstacles.

The preceding paragraphs indicate that the needed comments are typically of the following two kinds: (a) Comments about what particular basic relation is being applied to what object (at what time, along what direction) to find a useful relation. (b) Comments about what particular relations are being combined to eliminate some specified quantity.

Example: Solution of the previous motion problem

The process of constructing a solution is illustrated below by considering the problem previously analyzed in Sec. B and summarized by Fig. C-3. The explanatory comments specify the choices implemented by the equations. The remarks in the right margin indicate also some of the considerations leading to such a choice [e.g., what obstacle is being addressed (indicated inside curly brackets) and why the particular choice has been made among available options].

$v_2 = 0$ ② ; $T = 1.20$ s, $\underline{\underline{h}}$; g ; $\underline{\underline{v_1}}$ ① balcony ; $L = 13.0$ m ; $\hat{i}$; $\underline{\underline{v_3}}$ ③ ground

Goals: $v_1 = ?$ $h = ?$ $v_3 = ?$

Fig. C-3. Previous analysis of the problem in Sec. B.

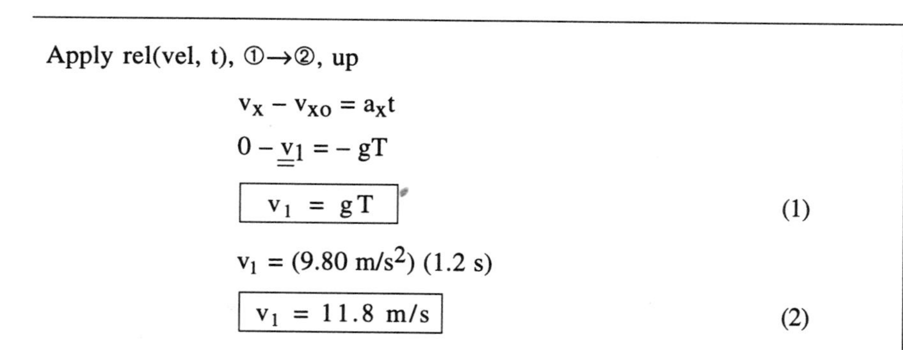

Apply rel(vel, t), ①→②, up

{Lack info about v_1.} Need to relate velocities and known time T. Hence apply rel(vel,t).

$$v_x - v_{xo} = a_x t$$

$$0 - \underline{\underline{v_1}} = -gT$$

Motion is described relative to the upward $\hat{i}$ direction.

$$\boxed{v_1 = gT} \qquad (1)$$

Result in terms of symbols.

$$v_1 = (9.80\ \text{m/s}^2)\ (1.2\ \text{s})$$

$$\boxed{v_1 = 11.8\ \text{m/s}} \qquad (2)$$

Numerical result.

Apply rel(disp, t), ①→②, up

{Lack info about h.} Need to relate upward displacement h to the known elapsed time T. Hence apply rel(disp, t).

$$D_x = v_{xo}t + \frac{1}{2}a_x t^2$$

$$\underline{\underline{h}} = v_1 T - \frac{1}{2}gT^2 = (gT)T - \frac{1}{2}gT^2$$

By (1):

$$\boxed{h = \frac{1}{2}gT^2} \qquad (3)$$

Result in terms of symbols.

$$h = \frac{1}{2}(9.80\ \text{m/s}^2)\ (1.2\text{s})^2$$

$$\boxed{h = 7.1\ \text{m}} \qquad (4)$$

Numerical result.

Apply rel(vel, disp), ①→③, down

$$v_x{}^2 - v_{xo}{}^2 = 2a_xD_x$$

$$v_3{}^2 - v_1{}^2 = 2\,g\,L$$

$$\boxed{v_3 = \sqrt{v_1{}^2 + 2gL}} \qquad (5)$$

$$v_3 = \sqrt{(11.8\text{ m/s})^2 + 2\,(9.80\text{ m/s}^2)\,(13.0\text{ m})}$$

$$\boxed{v_3 = 19.9\text{ m/s}} \qquad (6)$$

{Lack info about v_3.} Need to relate velocities to displacement, without being interested in required time. Hence apply rel(vel, disp).

Since we decided to measure quantities downward, $a_x = +g$ and $D_x = +L$.

➜ ***Go to Sec. 6C of the Workbook.***

D. Checking a solution

General remarks

The initial solution of a problem is rarely free of errors or other deficiencies. Any solution should, therefore, be regarded as provisional until it has been checked and appropriately revised.

The following questions, summarized in Fig. D-1, are generally useful to check any solution.

(1) **Goals attained?**
- * Has *all* wanted information been found?

(2) **Well-specified?**
- * Are answers expressed in terms of known quantities?
- * Are units specified?
- * Are both magnitudes and directions of vectors specified?

(3) **Self-consistent?**
- * Are units in equations consistent?
- * Are signs (or directions) on both sides of an equation consistent?

(4) **Other-consistent?** (i.e., consistent with other known information)
- * Are values sensible (e.g., consistent with known magnitudes)?
- * Are answers consistent with special cases (e.g., with extreme or specially simple cases)?
- * Are answers consistent with known dependence (e.g., with knowledge of how quantities increase or decrease)?
- * Are answers same as those obtained by other solution methods?

(5) **Optimal?**
- * Are answers and solution as clear and simple as possible?
- * Is answer a general algebraic expression rather than a mere number?

Checks

Goals attained?
All wanted info found?
Well-specified?
Answers in terms of knowns?
Units?
Vector magnitudes and directions?
Self-consistent?
Units?
Signs or directions?
Other-consistent?
Sensible values?
Special cases?
Known dependence?
Other solutions?
Optimal?
Clear and simple?
General?

Fig. D-1. Checks for a solution.

Example: Checking the solution of the previous problem

Fig. D-2 summarizes two of the answers obtained by solving the problem in Sec. C. The following illustrate some useful ways of checking these results.

$$v_1 = gt \quad (1)$$

$$h = \frac{1}{2} gt^2 \quad (3)$$

v_1 = initial speed

h = maximum height

Fig. D-2. Two of the answers obtained by solving the problem in Sec. C.

Self-consistency of units. The units on both sides of each of the relations in Fig. D-2 are properly consistent. For example, in the result (3),

Units of left side = m.

Units of right side = $(m/s^2)\, s^2 = m$.

Consistency with special cases. Suppose that the time T required for the ball to reach its maximum height is very large. In this special extreme case we expect that the ball would have been thrown with a very large initial speed v_1 and would reach a very large height h. The results (1) and (3) in Fig. D-2 are consistent with this expectation. Indeed, they imply that, if T is very large, both v_1 and h are very large.

Note that consistency with special cases can only be checked if the answers to a problem are not merely special numbers, but algebraic expressions applicable for any values of the relevant quantities. Indeed, ease of checking is one reason why it is useful to obtain problem answers in general algebraic form.

➜ ***Go to Sec. 6D of the Workbook.***

E. Problem-solving practice

Problem solving can only be learned by means of adequate practice. Thus you should try to solve as many of the problems in the Workbook as you can. However, practice is only effective it helps you acquire good work habits. Thus you would be well-advised to solve all problems by always using the systematic problem-solving method discussed in the preceding sections.

The following are some practical suggestions to help you solve problems more easily and correctly.

Good form. Mere attention to good form can help one to keep track of important information and to avoid mistakes. Hence it is useful to heed the following guidelines: (a) Write legibly and arrange your work on the page in a well-organized way. (b) Number all important equations so that you can easily refer to them. (c) Highlight important results and final answers by surrounding them with boxes.

Use of symbols. It is ordinarily best to introduce useful symbols, to work with these symbols to obtain an algebraic result, and to substitute numbers only at the end. Doing this has many advantages: (a) The solution is then generally useful, applicable to any special case without need for recalculation. (b) The general result often reveals important relationships and qualitative insights not apparent from specific numbers. (c) The result can be readily checked by assessing its consistency with known special cases. (d) Actual calculations are often greatly facilitated because one minimizes the need to deal with cumbersome numbers and units, reduces the chances for arithmetic mistakes, and can exploit cancellations and simplifications.

Identifying unknown quantities. It is useful to identify clearly the unknown quantities in a problem since these require particular attention. To do this, it is sometimes helpful to underline the symbols denoting unknown quantities (so as to distinguish them from other symbols denoting known quantities). In particular, it is helpful to underline *twice* wanted unknowns which ultimately need to be found, and to underline *once* unwanted unknowns which ultimately need to be eliminated.

➜ ***Go to Sec. 6E of the Workbook.***

F. Summary

Problem-solving method

The method involves the following main steps:

* Analyzing the problem. (See Fig. F-1.)
* Constructing the solution. (See Fig. F-2.)
  Main subproblem options (needed choices underlined):
    * Find useful info by applying *basic relation*, to *object*, at *time* (or between *times*), along *direction*.
    * Eliminate unwanted quantity by combining specified *relations*.
* Checking the solution. (See Fig. F-3. More details in Fig. D-1.)
* Revising the solution to improve it.

Analysis of a problem

* **Basic description**
    * ***Situation***
      Known information
      Diagram & useful symbols
    * ***Goals***
* **Refined description**
    * ***Time sequence & intervals***
    * ***Physics description***
      (e.g., velocity, acceleration)

Fig. F-1. Initial analysis of a problem.

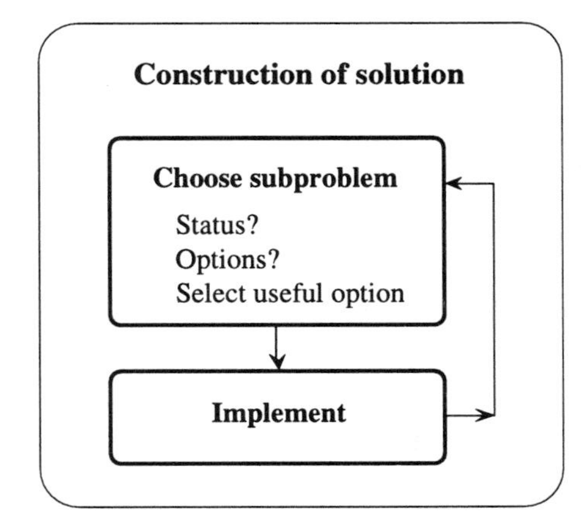

Fig. F-2. Construction of solution.

Checks

Goals attained?
Well-specified?
Self-consistent?
Other-consistent?
Optimal?

Fig. F-3. Checking a solution.

New abilities

You should be able to apply the preceding problem-solving method to solve diverse problems involving particles moving along a straight line with constant velocity or constant acceleration.

➜ ***Go to Sec. 6F of the Workbook.***

7 Curved Motion with Constant Acceleration

The preceding two chapters applied the concepts of velocity and acceleration to deal with motion along a straight line. Building on this preparation, how can we apply these concepts to more complex and commonplace situations where objects move along *curved* paths? In particular, how can we deal with such situations in the simple case where an object moves with *constant* acceleration (e.g., if it is a ball thrown along a curved path near the surface of the earth)? The present chapter addresses these questions.

A. Constant acceleration and projectile motion

There are many cases where a particle moves through space with some constant acceleration $\vec{a}$. For example, the particle might be a baseball or projectile moving near the surface of the earth after it has been launched from the ground. In this case the acceleration of the particle is simply the gravitational acceleration $\vec{g}$ (if air resistance is negligible so that the particle interacts solely with the earth).

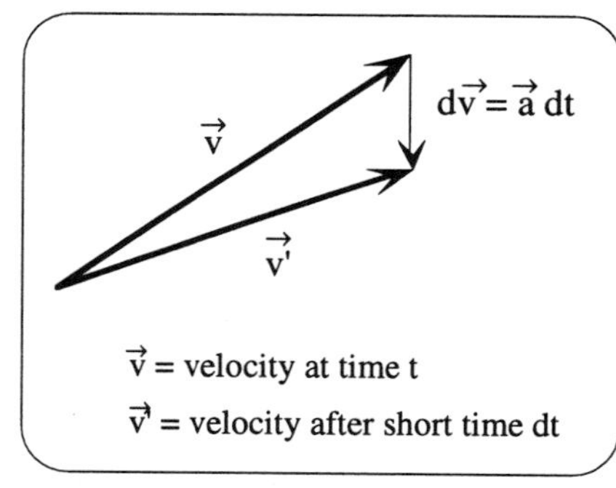

Fig. A-1. Small velocity change produced by a downward acceleration.

If a particle moves with a downward acceleration $\vec{a}$, its change of velocity $d\vec{v}$ during any infinitesimal time interval (e.g., during a millisecond) is directed downward, as indicated in Fig. A-1. If the acceleration is constant, this velocity change is also the same during any millisecond. As is apparent from Fig. A-1, the downward acceleration constantly changes the direction of the velocity so as to turn it toward the downward direction. The net result is that the path of the particle is bent into the shape of a curve, like that in Fig. A-2. Thus the result is a trajectory familiar to anyone who has watched baseballs or other thrown objects.

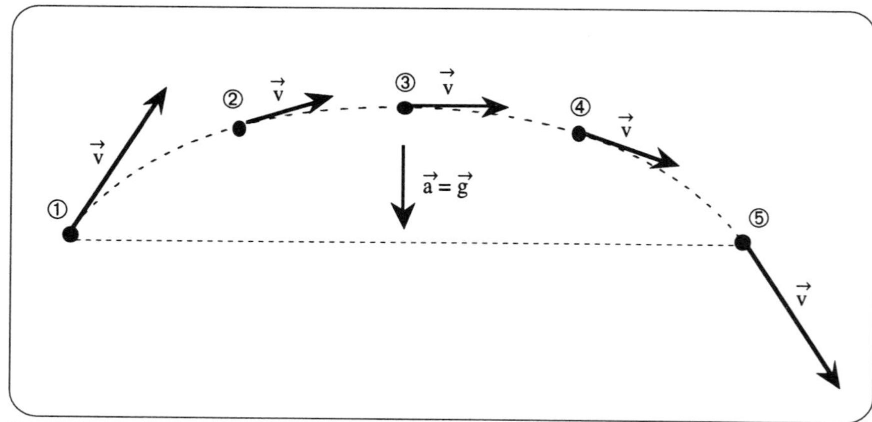

Fig. A-2. Particle path resulting from a constant downward acceleration.

If one knows the particle's velocity and position at some instant, one can add vectors (as in Fig. A-1) to find its velocity and position at a slightly later time. One can then successively repeat this process to trace out, point-by-point, the entire subsequent trajectory of the particle. The preceding process is, however, cumbersome and makes quantitative calculations difficult. The next sections discuss how the motion can be analyzed much more easily.

➜ ***Go to Sec. 7A of the Workbook.***

B. Component vectors

Motion along a curved path is complicated because the velocity and acceleration vectors have differing directions (as illustrated in Fig. A-2). The situation can, however, be analyzed very simply if the acceleration is constant so that its direction remains always the same.

Definition of component vectors. Suppose that there is a direction of special interest, e.g., the direction of the constant acceleration or the direction specified in Fig. B-1. Any vector $\vec{A}$ (lying in the same plane as this direction) can then be expressed as the sum of two other vectors parallel and perpendicular to this direction. For example, in Fig. B-1 these vectors are denoted by $\vec{A}_{\parallel}$ and $\vec{A}_{\perp}$, respectively. Thus we arrive at the following definition:

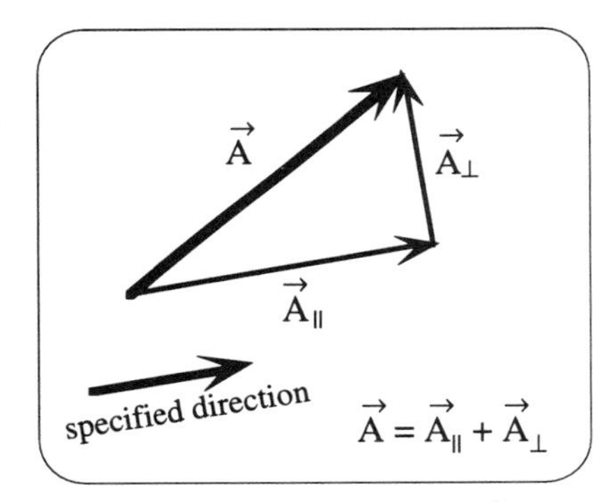

Fig. B-1. Component vectors of $\vec{A}$ parallel and perpendicular to some specified direction.

Def: ***Component vectors:*** Any vector $\vec{A}$ can be expressed as the sum of two vectors parallel and perpendicular to any specified direction in the same plane. These two vectors are called the "component vectors of $\vec{A}$ parallel and perpendicular to the specified direction". (B-1)

For example, in Fig. B-1 the vectors $\vec{A}_{\parallel}$ and $\vec{A}_{\perp}$ are, respectively, the component vectors of $\vec{A}$ parallel and perpendicular to the specified direction. (It may sometimes be convenient to represent all these vectors by arrows drawn from the same point, as indicated in Fig. B-2.)

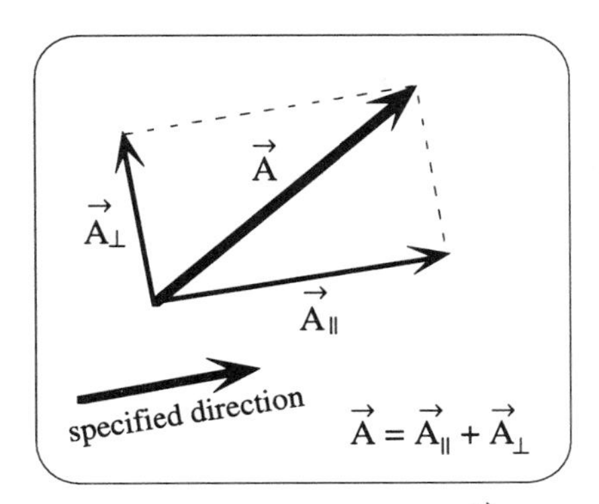

Fig. B-2. Component vectors of $\vec{A}$, represented by arrows drawn from the same point.

By expressing a vector in terms of its component vectors one can deal with these separately and thus often simplify many problems.

Defining method. The definition (B-1) leads to the following method (illustrated in Fig. B-3) specifying what one must actually *do* to determine the component vectors of a vector:

Finding component vectors

(1) ***Parallel guideline.*** Through one end of the arrow representing the vector $\vec{A}$, draw a guideline parallel to the specified direction.

(2) ***Perpendicular guideline.*** Through the other end of the arrow representing the vector $\vec{A}$, draw a guideline perpendicular to the specified direction. [The intersection of these lines determines the perpendicular sides of a right triangle.] (B-2)

(3) ***Component vectors.*** Along these sides, draw the two successive arrows representing the vectors whose sum is the original vector $\vec{A}$. [These are the component vectors of $\vec{A}$ parallel and perpendicular to the specified direction.]

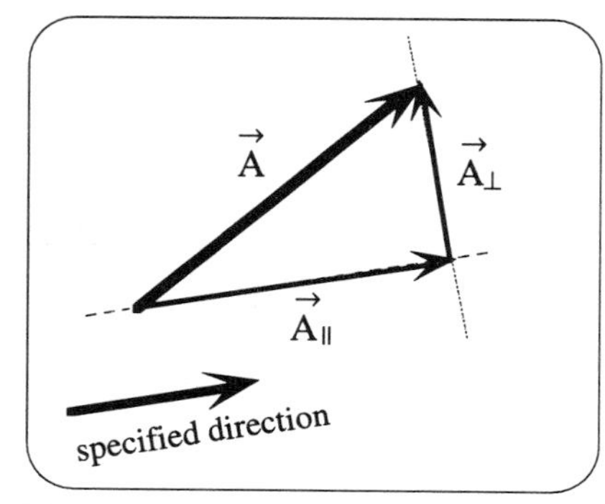

Fig. B-3. Determining component vectors.

➜ ***Go to Sec. 7B of the Workbook.***

C. Numerical components of a vector

Numerical specification of a component vector

A component vector parallel to a given direction can be specified very simply in terms of a *number*. (We already did this in Chapter 5 when discussing the motion of a particle along a straight line.) For example, suppose that the direction of interest in Fig. C-1 is specified by the unit vector $\hat{i}$. Then the component vector of $\vec{A}$ parallel to this direction can be written as

$$\vec{A}_\parallel = A_i\,\hat{i} \qquad \text{(C-1)}$$

where A_i is some positive or negative number called the *numerical component of the vector $\vec{A}$ along the direction $\hat{i}$*. This numerical component (often simply called *component*) has the properties specified by the following definition already introduced in Sec. 5A:

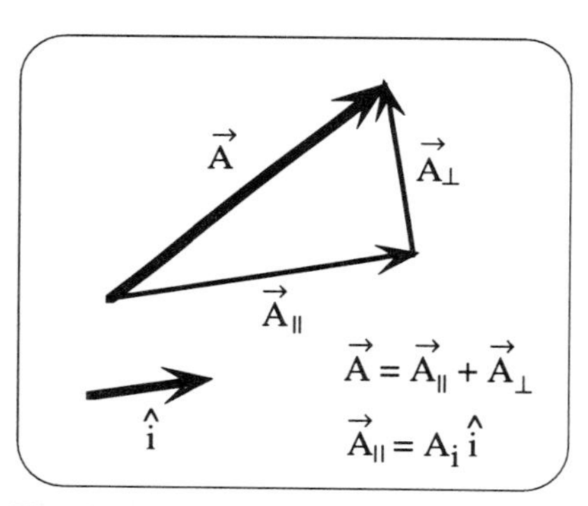

Fig. C-1. Component vectors parallel and perpendicular to a unit vector.

Def: ***Component of a vector along a specified direction:*** A *number* whose magnitude is equal to the magnitude of the component vector along this direction. The sign of this number is positive if the component vector is along the specified direction, and negative if it is opposite to this direction. (C-2)

Note that the *component vector* of a vector is a *vector*, but that the *component* of this vector is a *number* used to specify the component vector.

Example C-1: Component vectors and components

In Fig. C-2, the velocity $\vec{v}$ has a magnitude of 10 m/s. Its *component vector* parallel to the eastern direction is 6 m/s east (or 6 m/s$\hat{i}$ if $\hat{i}$ is a unit vector along the eastern direction). Its *component* along the eastern direction is simply 6 m/s. (Its component along the *western* direction is −6 m/s.)

The component vector of the velocity $\vec{v}$ *perpendicular* to the eastern direction is 8 m/s north. Its component along the northern direction is simply 8 m/s.

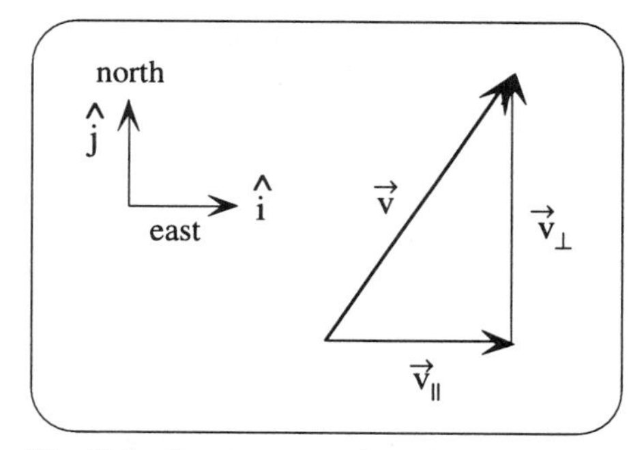

Fig. C-2. Components of a velocity.

Relations between corresponding components

If two vectors are equal, their component vectors along *any* direction must also be equal. Consequently, their numerical components along any direction must also be equal. In short,

$$\text{for any direction } \hat{i}, \quad \boxed{\text{if } \vec{A} = \vec{B}, \qquad A_i = B_i\,.} \tag{C-3}$$

This result is very useful because it implies that a vector equation is equivalent to several simpler numerical equations.

It is also apparent from Fig. C-3 that, along any direction, the component of the sum of two vectors is equal to the sum of the components of these vectors. Thus,

$$\text{for any direction } \hat{i}, \quad \boxed{\text{if } \vec{S} = \vec{A} + \vec{B}, \qquad S_i = A_i + B_i\,.} \tag{C-4}$$

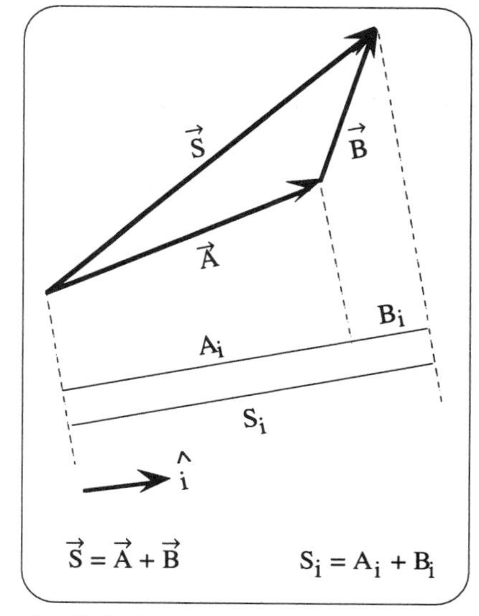

Fig. C-3. Relation between the sum of vectors and the sum of their components.

Components relative to coordinate directions

Any vector can be completely specified by its components relative to the coordinate directions of some conveniently chosen coordinate system. Indeed, suppose that the coordinate directions are specified by the mutually perpendicular unit vectors $\hat{i}$ and $\hat{j}$ indicated in Fig. C-4. Then the position vector $\vec{r}$ of any point P can be expressed as the sum of its component vectors along these directions, and hence also in terms of its numerical components x and y along these directions. (These components are also called the *position coordinates* of the point.) Thus one can write

$$\vec{r} = x\,\hat{i} + y\,\hat{j}\,. \tag{C-5}$$

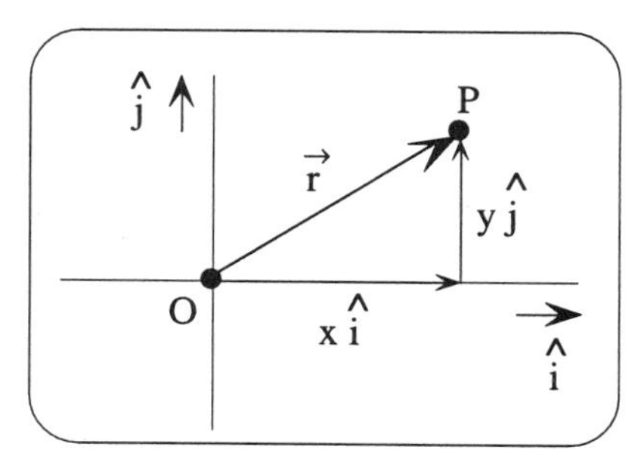

Fig. C-4. Components of a position vector relative to coordinate directions.

Any other vector can be similarly expressed in terms of its components along these coordinate directions. For example, as indicated in Fig. C-5, the velocity $\vec{v}$ of a particle can be described by its numerical components (conventionally denoted by v_x and v_y). Thus one can write

$$\vec{v} = v_x\,\hat{i} + v_y\,\hat{j}\,. \tag{C-6}$$

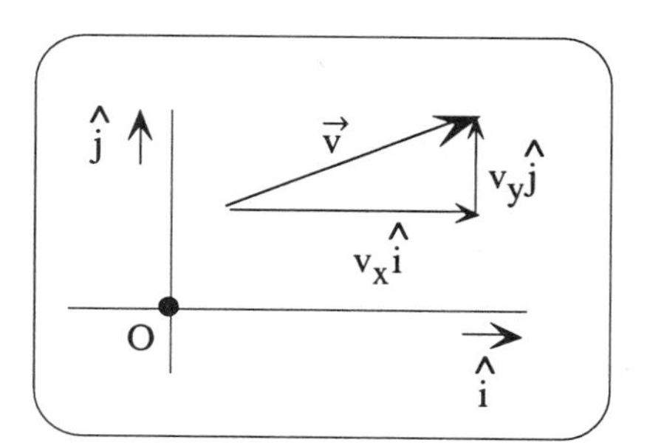

Fig. C-5. Components of velocity.

Example C-2: Components of position vectors

In Fig. C-6, the position vector $\vec{r}$ of the point P is $-(3\text{ m})\hat{i} + (2\text{ m})\hat{j}$.

The *component vectors* of this position vector, along the $\hat{i}$ and $\hat{j}$ directions, are then respectively $(-3\text{ m})\hat{i}$ and $(2\text{ m})\hat{j}$.

The *components* of this position vector, along these directions, are respectively -3 m and 2 m.

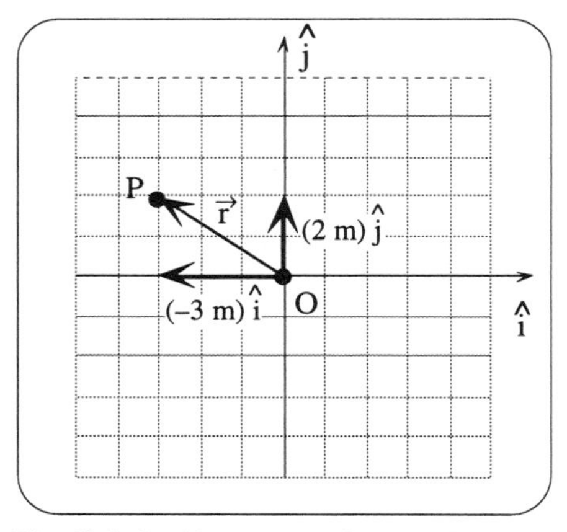

Fig. C-6. Position vector of a point P.

Alternative descriptions of a vector

A vector may be equally well described either by its magnitude and direction, or by its components relative to some conveniently chosen coordinate directions.

For example, the vector $\vec{A}$ indicated in Fig. C-7 may be specified by its magnitude A and by the angle θ specifying its direction relative to the $\hat{i}$ direction. Alternatively, it may be described by its components A_x and A_y relative to the coordinate directions $\hat{i}$ and $\hat{j}$. Thus one can write

$$\vec{A} = A_x\,\hat{i} + A_y\,\hat{j}\,. \qquad \text{(C-7)}$$

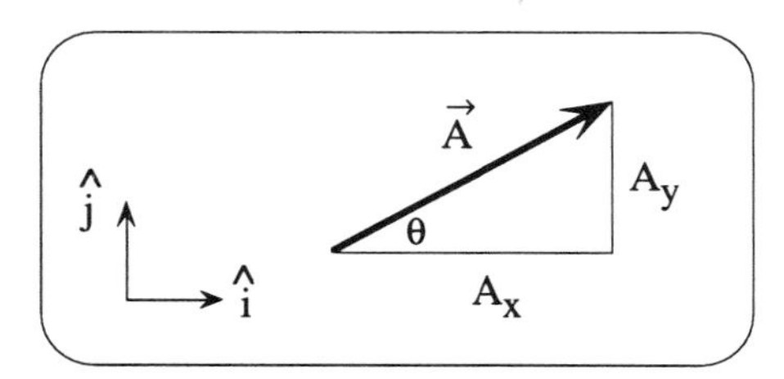

Fig. C-7. Components of a vector.

Components from magnitude and direction. If one knows the magnitude and direction of a vector, one can readily find its components. Thus the components of the vector $\vec{A}$ in Fig. C-7 are

$$A_x = A\cos\theta\,, \qquad A_y = A\sin\theta\,. \qquad \text{(C-8)}$$

Magnitude and direction from components. Conversely, if one knows the components of a vector, one can easily find its magnitude and direction. For example, the magnitude A of the vector in Fig. C-7 can be found by using the Pythagorean theorem for the right triangle. Thus

$$A^2 = A_x{}^2 + A_y{}^2 \qquad \text{(C-9)}$$

Furthermore, the angle θ in Fig. C-7 is such that

$$\tan\theta = \frac{A_y}{A_x}\,. \qquad \text{(C-10)}$$

➜ ***Go to Sec. 7C of the Workbook.***

D. Component description of projectile motion

Motion with constant acceleration is particularly simple because both the direction and magnitude of the acceleration remain unchanged. A very common example is the motion of a projectile near the surface of the earth. Indeed, the projectile's acceleration (neglecting air resistance) is then always equal to the constant downward gravitational acceleration $\vec{g}$. Thus

$$\vec{a} = \vec{g}\,. \qquad \text{(D-1)}$$

Component vectors describing the motion

Since the direction of the acceleration is constant, all vectors can be described in terms of their components parallel and perpendicular to this special direction.

For example, in the case of projectile motion, one can describe all vectors in terms of their vertical component vectors parallel to the gravitational acceleration, and their horizontal component vectors perpendicular to this acceleration. The relation (D-1) then reveals a basic simplicity. (a) Since the acceleration is vertically downward, the horizontal component vector of the acceleration is zero. Hence the *horizontal* component vector of the velocity remains *unchanged* throughout the motion. (b) Since the non-zero vertical component vector of the acceleration is equal to $\vec{g}$, the *vertical* component vector of the velocity *does* change. Indeed, it changes in exactly the same way as if the particle were moving along a vertical line.

These simple features of the motion are illustrated in the top part of Fig. D-1. This top part is identical to the previous Fig. A-2, but shows the component vectors of the velocity parallel and perpendicular to the downward direction. The bottom part of Fig. D-1 shows these same component vectors separately and illustrates that the horizontal component vector of the velocity remains unchanged.

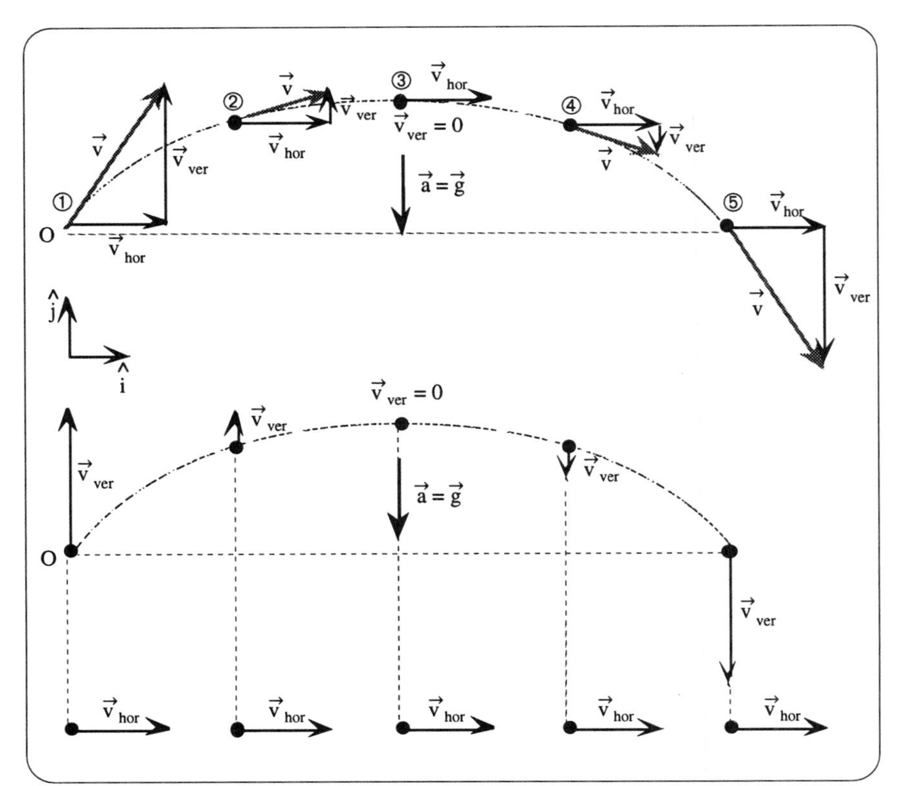

Fig. D-1. Horizontal and vertical component vectors of a projectile's velocity.

Numerical components describing the motion

The simplicity of the component description can be fully exploited by expressing all the vectors in terms of their numerical components along the vertical and horizontal directions. To do this, one needs only to introduce a coordinate system whose coordinate directions are vertical and horizontal. For example, in Fig. D-1 we might choose a coordinate system with an $\hat{i}$ direction pointing horizontally to the right and a $\hat{j}$ direction pointing vertically upward.

Separation into horizontal and vertical motions. The equality of the vectors in (D-1) implies, according to (C-3), that their components along any direction are also equal. Thus the components of $\vec{a}$ and $\vec{g}$ along the horizontal $\hat{i}$ direction must be equal, and their components along the vertical $\hat{j}$ direction must also be equal. In other words, the vector equation (D-1) implies the following two corresponding numerical equations for the components of the vectors:

$$\boxed{a_x = 0, \quad a_y = -g} \; . \tag{D-2}$$

Here the minus sign is due to the fact that the gravitational acceleration $\vec{g}$ is directed downward, opposite to the chosen $\hat{j}$ direction.

The relations (D-2) show that the description of the motion in terms of components is very simple since one can discuss separately the horizontal motion along the $\hat{i}$ direction and the vertical motion along the $\hat{j}$ direction. Furthermore, each of these motions is the same as one of the straight-line motions previously discussed in Chapter 5.

Horizontal motion. For the motion along the horizontal $\hat{i}$ direction, (D-2) states that the acceleration is zero. Hence the horizontal velocity and position components at any time t can be immediately inferred from their definitions since $dv_x/dt = a_x$ and $dx/dt = v_x$. Correspondingly, the relations of Sec. 5F imply that

$$\left.\begin{aligned} &a_x = 0, \\ &v_x = v_{xo} = \text{constant}, \\ &D_x = v_x t \,. \end{aligned}\right] \tag{D-3}$$

These indicate that the horizontal velocity simply remains constant, equal to its initial value. Hence the particle's horizontal displacement increases with time at a constant rate.

Vertical motion. For the motion along the vertical $\hat{j}$ direction, (D-2) states that the acceleration has a constant magnitude equal to that of the gravitational acceleration. Hence the vertical velocity and position components at any time t can be immediately inferred from their definitions since $dv_y/dt = a_y$ and $dy/dt = v_y$. Correspondingly, the relations of Sec. 5F imply that

$$a_y = -g\,,$$
$$v_y - v_{yo} = -gt\,, \qquad \text{(D-4)}$$
$$D_y = v_{yo}t - \frac{1}{2}gt^2\,.$$

These indicate that the vertical motion is exactly the same as that of a particle moving vertically under the influence of gravity.

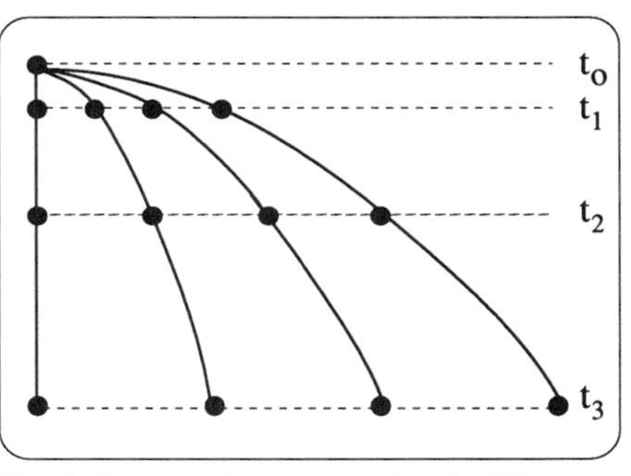

Fig. D-2. Particles projected with different horizontal velocities.

Relation between horizontal and vertical motions. The particle's separate horizontal and vertical motions are connected through the common time. For example, while the particle traverses a horizontal displacement during some time, it traverses a corresponding vertical displacement during the *same* time.

Particles projected with different horizontal velocities

To illustrate the independence of the horizontal and vertical motions, consider a situation where several particles are projected from the same point with different *horizontal* velocities. Then each such particle starts at the same height with zero vertical velocity, and moves with the same gravitational acceleration downward. Hence the vertical motion of each particle is exactly the same. Indeed, (D-4) implies that, at any instant, the vertical velocity of each particle should be the same, and the vertical height of each particle should also be the same.

The motions of these particles thus differ only because they move with different horizontal velocities. While each particle falls through the *same* vertical distance, it traverses a different horizontal distance (as illustrated in Fig. D-2 and in the stroboscopic photograph of Fig. D-3). For instance, a particle starting from rest with zero horizontal velocity falls straight down, while particles with larger horizontal velocities traverse larger horizontal distances.

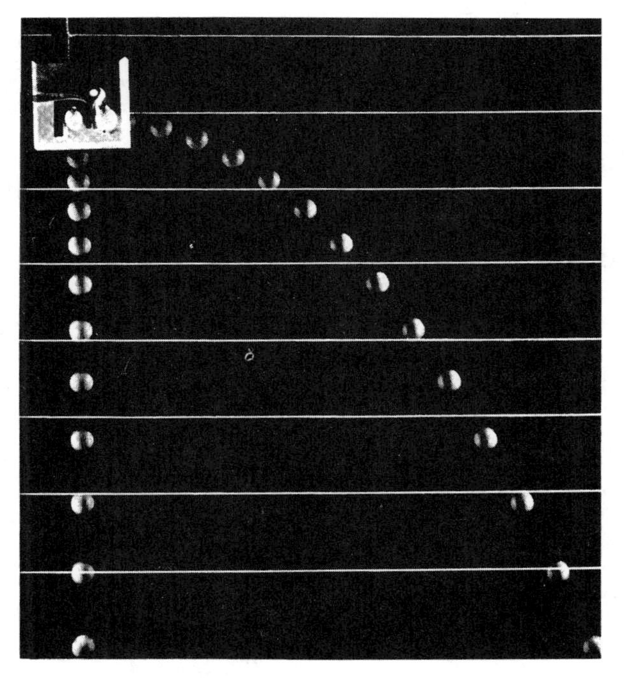

Fig. D-3. A stroboscopic photograph showing two balls released at the same instant, one with zero horizontal velocity and the other with an appreciable horizontal velocity. [Reproduced with permission from PSSC PHYSICS, 2nd edition, 1965; D.C. Heath and Company with Education Development Center, Inc., Newton, MA.]

➜ ***Go to Sec. 3A of the Workbook.***

E. Projectile problems

The preceding discussion indicates that any problem of curved motion with constant acceleration can be reduced to separate simpler problems of motions along straight lines. The following problem is a typical example.

Problem statement. A projectile is launched from the horizontal ground with a speed v_o at an angle θ with respect to the horizontal. What is the range of this projectile if air resistance is negligible?

Solution. The following is the solution of the preceding problem.

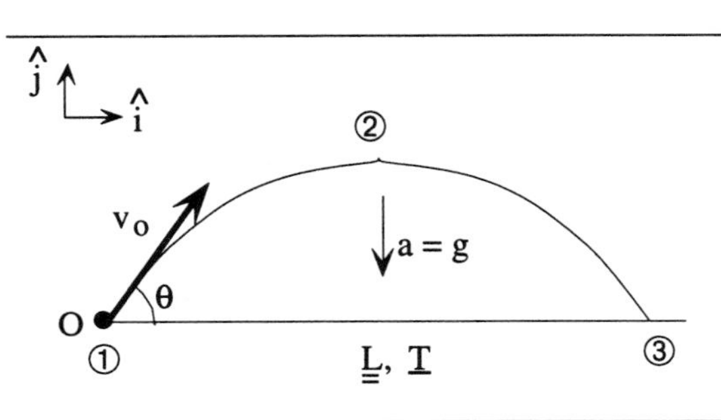

L = range

T = ①→③ = flight time

Known: v_o, θ.

Goals: L = ?

Analysis of problem

Wanted unknowns are underlined twice, unwanted ones underlined once.

Along $\hat{i}$, ①→③

$$a_x = 0$$

Apply rel(vel, t): $v_x = v_o \cos\theta = \text{constant}$

Apply rel (disp, t): $D_x = v_x t$

$$\underline{\underline{L}} = v_o \cos\theta\, \underline{T} \qquad (1)$$

Construction of solution

{Lack information about horizontal motion.}

Along $\hat{j}$, ①→③

$$a_y = -g$$

Apply rel(disp, t): $D_y = v_{yo}t + \frac{1}{2} a_y t^2$

$$0 = v_o \sin\theta\, T + \frac{1}{2}(-g)\, T^2$$

$$0 = v_o \sin\theta - \frac{1}{2} g\underline{T}$$

$$\underline{T} = \frac{2v_o \sin\theta}{g} \qquad (2)$$

{Lack information about T.} Find information about vertical motion.

T is the time needed by projectile to move up and return to ground. The range is the horizontal distance traveled during this time.

The projectile's vertical displacement is zero when it hits the ground.

Eliminate T by (1) & (2):

$$\boxed{L = \frac{2\, v_o^2 \cos\theta \sin\theta}{g}} \qquad (3)$$

{T is unwanted.}

Checks:

Units of (3): Left side: m

Right side: $\frac{(m/s)^2}{m/s^2} = m$ OK

If v_o large: Expect large L. Eq. (3) OK.

If g large: Expect short flight time, small L. Eq. (3) OK.

If $\theta \approx 0$: Expect $T \approx 0$, $L \approx 0$. Eq. (3) OK.

If $\theta \approx 90°$: Expect vertical motion, $L \approx 0$. Since $\cos\theta \approx 0$, Eq. (3) OK

Checks (done in writing or mentally)

Check of unit consistency.

Special cases.

Hypothetical case where gravity would be larger than in actuality.

The projectile immediately hits the ground.

Maximum range of a projectile

As indicated by the last two checks, the range of the projectile is zero if the launch angle θ is 0° or 90°. Hence the range of the projectile is maximum for some value of θ between these two extremes. Closer examination of the result (3) shows that this maximum occurs if $\theta = 45°$.

➜ ***Go to Sec. 7E of the Workbook.***

F. Summary

Component vectors of a vector

The component vectors of a vector $\vec{A}$, parallel and perpendicular to some specified direction, are the vectors $\vec{A}_{\parallel}$ and $\vec{A}_{\perp}$ in Fig. F-1.

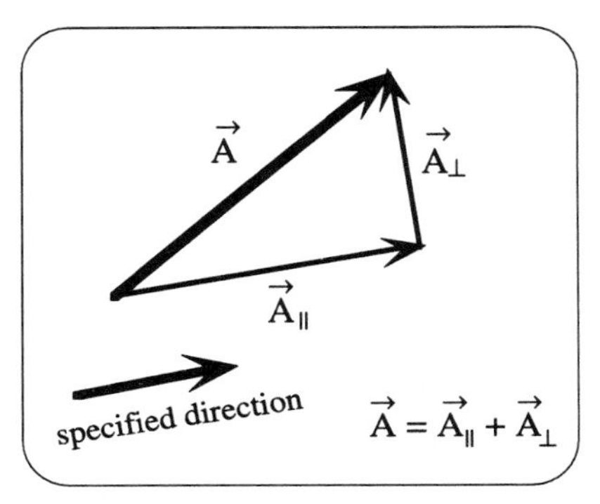

Fig. F-1. Component vectors of a vector.

Component of a vector

The component (i.e., *numerical* component) of a vector $\vec{A}$ along a direction is the *number* specifying the component vector of $\vec{A}$ along this direction.

Magnitude of component = magnitude of component vector.

Sign of component: + if component vector points along direction,
− if it points opposite to this direction.

Vector components relative to a coordinate system

Any vector $\vec{A}$ can be described by its components A_x and A_y relative to some coordinate system (as indicated in Fig. F-2).

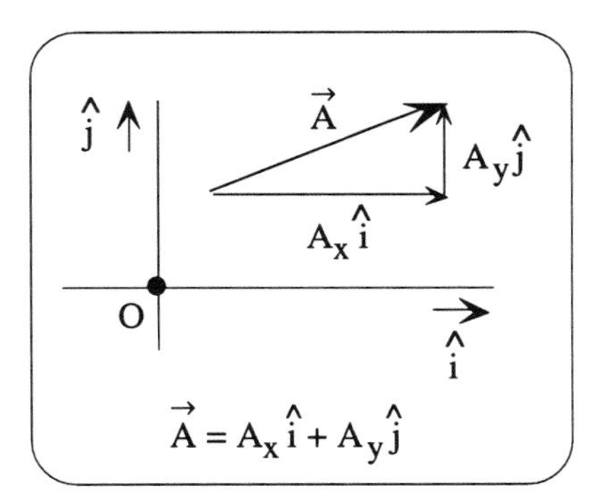

Fig. F-2. Vector components relative to a coordinate system.

Motion with constant acceleration

If a particle moves with a constant acceleration $\vec{a}$, the components of the acceleration along any direction are also constant. The motion along each direction can then be discussed separately like straight-line motion. If the $\hat{i}$ and $\hat{j}$ directions are chosen perpendicular and parallel to $\vec{a}$ (e.g., horizontal and vertical in the case of gravity), then $a_x = 0$ and a_y = constant. The motion is then with constant *velocity* along $\hat{i}$ and with constant *acceleration* along $\hat{j}$.

New abilities

You should now be able to do the following:

(1) Find the component vectors, or numerical components, of any vector along any specified direction.

(2) Specify a vector alternatively in terms of its magnitude and direction, or in terms of its components along the directions of some coordinate system.

(3) If a particle moves with constant acceleration, predict its motion by describing it separately in terms of components parallel and perpendicular to this acceleration.

➜ ***Go to Sec. 7F of the Workbook.***

8 Circular and Relative Motions

The definitions of velocity and acceleration can also easily be applied to some curved motions where the acceleration is *not* constant. An important such situation occurs when a particle moves around a circular path. As discussed in the next few sections, the particle's acceleration can then readily be calculated to deal with various quantitative or qualitative problems. Finally, this chapter will conclude by addressing the following general questions: How can the same motion be described from the point of view of different observers (e.g., from the point of view of a person on the ground or one in a moving vehicle)? How are these different descriptions related and how can they be applied to deal with some commonly occurring kinds of motions?

A. Circular motion with constant speed

Determination of acceleration. Consider first the simple situation where a particle moves with *constant* speed v along a circular arc (or complete circle) of radius r. The *magnitude* of the particle's velocity remains then unchanged. Hence the particle's acceleration is due solely to the change of *direction* of its velocity. To determine the acceleration quantitatively, we need only go back to the definition of acceleration

$$\vec{a} = \frac{d\vec{v}}{dt} \qquad \text{(A-1)}$$

and apply it systematically according to the defining method (F-2) of Chapter 4.

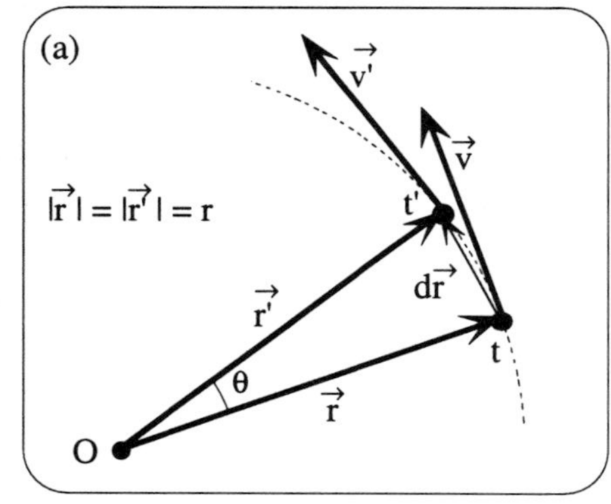

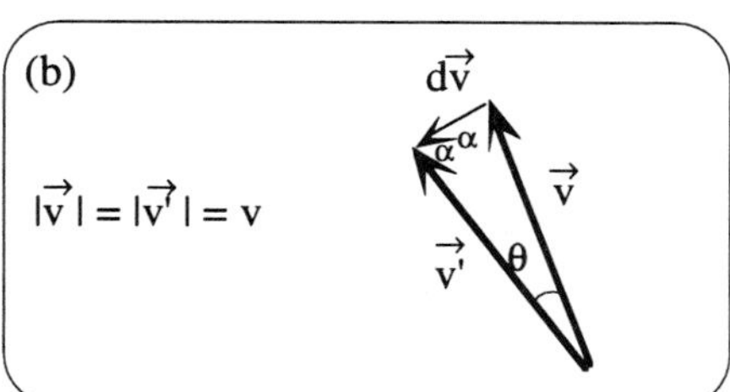

Fig. A-1. Particle moving along a circular arc with constant speed.

Fig. A-1a shows, at some time t, the particle's position vector $\vec{r}$ (relative to the center O of the circular arc) and the particle's velocity $\vec{v}$. The diagram also shows, at a slightly later time t', the particle's position vector $\vec{r}'$ and its velocity $\vec{v}'$. These position vectors have the *same* magnitude r equal to the radius of the circular arc. Furthermore, these velocities are both tangent to the circular path, and they have the *same* magnitude v since the particle moves with constant speed.

Direction of the acceleration. Fig. A-1b shows the vector triangle used to compare these velocities so as to find the particle's change of velocity $d\vec{v} = \vec{v}' - \vec{v}$ during the small time interval dt = t' - t. In this triangle, the angles the same since the velocities $\vec{v}$ and $\vec{v}'$ have the same magnitude. The time interval dt, and thus also the angle θ, is supposed to be infinitesimally small. Since the sum $2\alpha + \theta$ of the angles in the triangle is $180°$, the angle α is thus essentially $90°$. Hence the small velocity change $d\vec{v}$ is *perpendicular* to the velocity $\vec{v}$. The acceleration $\vec{a}$, which has the same direction as this velocity change, is thus also perpendicular to the velocity. (This is actually true for motion with constant speed along *any* curved path, as previously discussed in Chapter 4.)

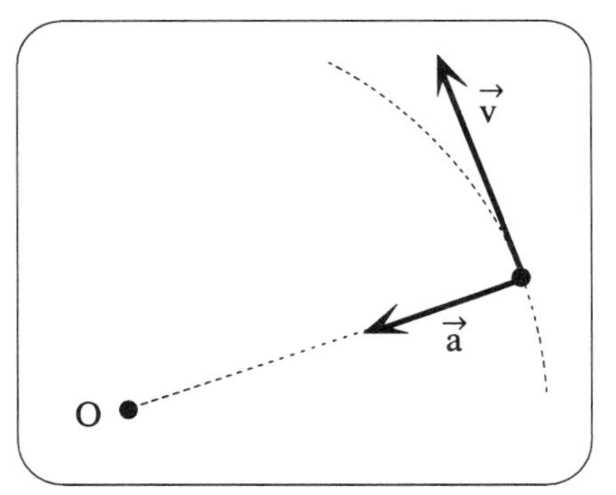

Fig. A-2. Acceleration of a particle moving along a circular arc with constant speed.

The particle's acceleration can be represented by an arrow drawn from the particle's position. This arrow must then be perpendicular to the path and have a direction *inward toward the center* of the circular arc, as illustrated in Fig. A-2.

$$\boxed{\text{Direction of } \vec{a}\text{: toward the center.}} \qquad \text{(A-2)}$$

The term *centripetal acceleration* has historically been used to denote an acceleration directed toward the center. This term is, however, unnecessary and misleading since it suggests something more fancy than the plain English phrase *toward the center.* (It would be just as foolish to talk about a *sinistral* acceleration if one merely means an acceleration directed to the left.)

Magnitude of the acceleration. By (A-1), the *magnitude* $a = |\vec{a}|$ of the particle's acceleration is

$$a = \frac{|d\vec{v}|}{dt} \qquad \text{(A-3)}$$

and can thus be found by determining the magnitude of the small velocity change $|d\vec{v}|$. To do this, one needs only note that the triangle formed by the position vectors in Fig. A-1a is similar to the triangle formed by the velocities in Fig. A-1b. (Indeed, each triangle has equal sides. Furthermore, the angle θ between these sides is the same because the velocities are perpendicular to the corresponding position vectors.) Hence the ratio of the side opposite to the angle θ, divided by the side adjacent to this angle, must be the same for both triangles. Thus

$$\frac{|d\vec{v}|}{v} = \frac{|d\vec{r}|}{r}$$

or

$$|d\vec{v}| = \frac{v}{r}\,|d\vec{r}|\,.$$

Hence (A-3) becomes

$$a = \frac{|d\vec{v}|}{dt} = \frac{v}{r}\,\frac{|d\vec{r}|}{dt} = \frac{v}{r}\,v$$

since $|d\vec{r}|/dt$ is simply the magnitude v of the particle's velocity. Thus we arrive at the conclusion that the magnitude *a* of the particle's acceleration is simply

$$\boxed{a = \frac{v^2}{r}\,.} \qquad \text{(A-4)}$$

➜ ***Go to Sec. 8A of the Workbook.***

B. Period and frequency

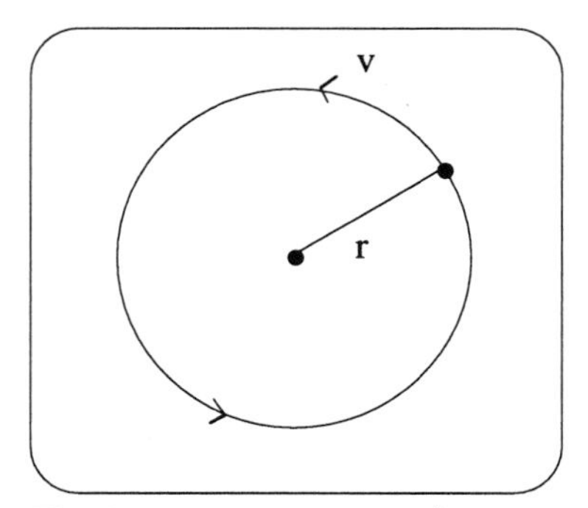

Fig. B-1. Motion around a circle with constant speed.

Definitions. Consider the motion of a particle moving with constant speed around a complete circle, like that in Fig. B-1. The motion can then be described not only by the particle's speed v, but also by specifying its *period* T or rotation *frequency* f. These quantities are defined as follows:

Def: **Period:** Time required for one revolution. (B-1)

Def: **Frequency:** Number of revolutions per unit time. (B-2)

The phrase *per unit time* means *divided by the corresponding time.* Thus

$$\text{frequency f} = \frac{\text{number of revolutions}}{\text{time}} \;. \qquad \text{(B-3)}$$

For example, if a particle requires 0.1 second to go once around its circular path (i.e., if its period T = 0.1 s), then its rotation frequency is 10 revolutions per second (i.e., $f = 0.1\ s^{-1}$).

Relation between speed and period. All these quantities are related to the speed v of the particle. By its definition, the speed (i.e., magnitude of the velocity) is

$$v = \frac{d|\vec{r}|}{dt} \qquad \text{(B-4)}$$

where $d|\vec{r}|$ is the magnitude of the small displacement (i.e., of the small arc length) traversed by the particle during an infinitesimal time dt. If the speed v is constant, the ratio in (B-4) is the *same* for every small time interval. Hence it is also the same as the ratio of the total distance (i.e., circumference) $2\pi r$ around the circle, divided by the total time T needed to traverse it. Thus

$$\boxed{v = \frac{2\pi r}{T}} \qquad \text{(B-5)}$$

which shows how the speed v is related to the period T.

Relation between frequency and period. According to its definition (B-2) or (B-3), the frequency is closely related to the period. Thus the number of revolutions during any time interval Δt, much longer than the period T, is simply $\Delta t/T$ (i.e., the total time interval divided by the time required for one revolution). The definition (B-3) implies then that

$$f = \frac{\Delta t/T}{\Delta t}$$

or

$$\boxed{f = \frac{1}{T}\,.} \qquad \text{(B-6)}$$

Relation between speed and frequency. Conversely, (B-6) implies that T = (1/f). Hence the relation (B-5) can also be written as

$$\boxed{v = 2\pi r f.} \quad \text{(B-7)}$$

This relation between the speed v and frequency f merely states that ⟨the distance traveled per second⟩ is equal to ⟨the distance $2\pi r$ traveled per revolution⟩ multiplied by ⟨the number of revolutions per second⟩.

➜ ***Go to Sec. 8B of the Workbook.***

C. General circular motion

The previous sections discussed the simple special case of circular motion with constant speed. However, the speed of a particle moving along a circular path is ordinarily *not* constant, but changes with time. The velocity $\vec{v}$ of the particle at some time t is then *different* from its velocity $\vec{v}'$ at a slightly later time t'. (For example, if the speed of the particle increases, the magnitude of its later velocity $\vec{v}'$ is slightly larger than that of its earlier velocity $\vec{v}$, as indicated in Fig. C-1a.) The corresponding small velocity change $d\vec{v}$ of the particle during the infinitesimal time dt is indicated in Fig. C-1b. (This vector diagram is shown enlarged so that the velocity change can be examined in greater detail.)

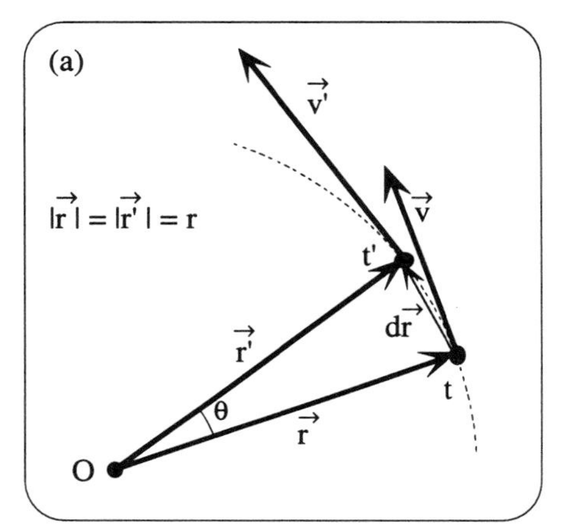

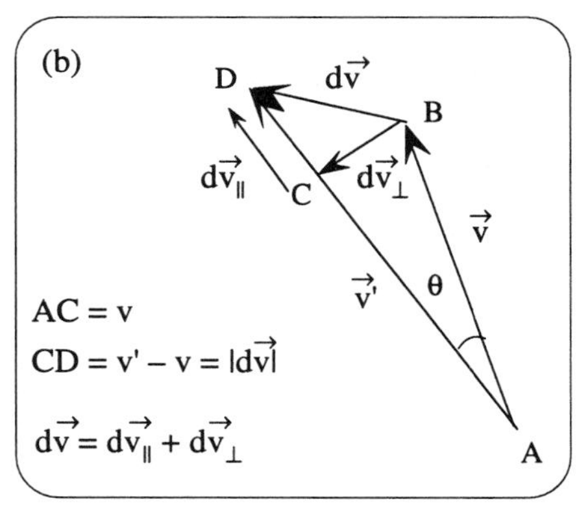

Fig. C-1. (a) Particle moving along a circular arc with changing speed. (b) Enlarged diagram comparing the velocities.

Separable parts of the velocity change. The velocity between the time t and the slightly later time t' has changed for two reasons: (a) The *magnitude* of the velocity has changed by a small amount dv, from its initial magnitude v to its later magnitude v'. (b) The *direction* of the velocity has changed through a small angle θ. (All these changes are very small since the time interval dt is supposed to be infinitesimal. But they have been exaggerated in the diagram to make them more clearly visible.)

To indicate these two kinds of changes separately in Fig. C-1b, the arrow representing the later velocity $\vec{v}'$ has been subdivided into two parts by choosing the point C so that AC has the same magnitude v as that of the original velocity. Then CD represents just the change $dv = v' - v$ of the *magnitude* of the velocity. The velocity change $d\vec{v}$ can then be expressed as the sum of two vectors

$$\begin{aligned} d\vec{v} &= \overrightarrow{BC} + \overrightarrow{CD} \\ &= d\vec{v}_{\perp} + d\vec{v}_{\parallel}. \end{aligned} \quad \text{(C-1)}$$

Here $\overrightarrow{BC}$ denotes the vector drawn from B to C. It represents the velocity change which would occur if the speed remained constant and only the direction of the velocity changed (i.e., it is just the velocity change previously discussed in Sec. A). Since θ is infinitesimal, this vector is perpendicular to $\vec{v}$ and has accordingly been denoted by $d\vec{v}_{\perp}$. The other vector $\overrightarrow{CD}$, drawn from C to D, is essentially parallel to $\vec{v}$ since the angle θ is infinitesimal; hence this vector has been denoted by $d\vec{v}_{\parallel}$.

Component vectors of acceleration. To find the particle's acceleration $d\vec{v}/dt$, one needs merely to divide both sides of (C-1) by the infinitesimal time interval dt during which the velocity change occurs. Thus one can write

$$\vec{a} = \vec{a}_{\perp} + \vec{a}_{\parallel} \tag{C-2}$$

where $\vec{a}_{\perp}$ and $\vec{a}_{\parallel}$ are the component vectors of the acceleration perpendicular and parallel to the velocity.

This result can be most conveniently expressed in terms of the *numerical* components of the acceleration.

Perpendicular component of the acceleration. The acceleration component *perpendicular* to the velocity (i.e., toward the center of the circular path) can be denoted by a_c. It describes how rapidly the *direction* of the particle's velocity changes with time. This component is just equal to v^2/r, i.e., to the magnitude of the acceleration previously calculated in Sec. A for the case of circular motion with constant speed.

Parallel component of the acceleration. The acceleration component *along* the velocity can be denoted by a_v. It describes how rapidly the *magnitude* of the particle's velocity changes with time. From Fig. C-1b, the magnitude of this velocity change is equal to

$$|d\vec{v}_{\parallel}| = CD = |dv| \tag{C-3}$$

where $dv = v' - v$ is the change of the *magnitude* of the velocity (i.e., of the speed) from its original value v to its later value v'. If the later speed is larger than the original speed v (as is true in Fig. C-1), the parallel velocity change $d\vec{v}_{\parallel}$ is directed along $\vec{v}$. If the later speed v' is smaller than the original speed, it is directed opposite to $\vec{v}$. Hence the component a_v of the acceleration, *along* the velocity, is simply equal to dv/dt, the rate of change of *speed* with time. (This component is positive if the speed increases, negative if it decreases, and zero if it is constant.)

Summary of acceleration components. The preceding results provide the following useful description of the acceleration in terms of its components:

Along velocity:	$a_v = \dfrac{dv}{dt}$	(C-4)
Perpendicular to velocity (toward center):	$a_c = \dfrac{v^2}{r}$	

Applicability to any path. Any smooth curve can be approximated, near any of its points, by a circle tangent to it (as illustrated in Fig. C-2). The radius of this circle is called the *radius of curvature* of the curve at this point. Hence (C-4) can be applied to find a particle's acceleration components at any point of *any* curved path (provided that r is understood to be the path's radius of curvature at this point).

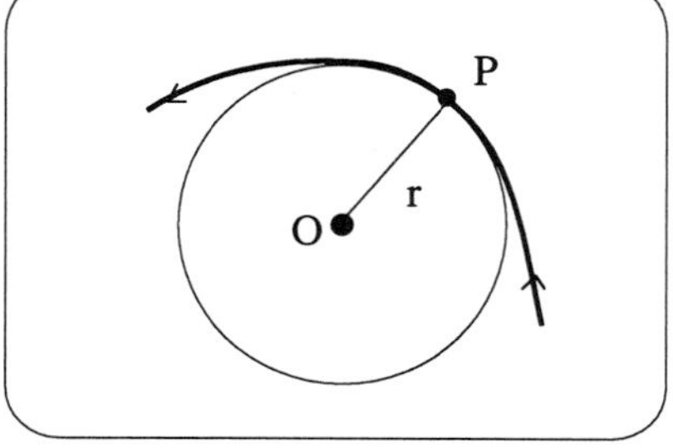

Fig. C-2. Circle approximating a curve near a particular point.

→ ***Go to Sec. 8C of the Workbook.***

D. Projectiles and satellites

Different motions with gravitational acceleration. Imagine that a projectile is launched horizontally from a high tower at the surface of the earth. If it is launched with zero horizontal velocity, it falls straight down toward the center of the earth (as illustrated by the path *a* in Fig. D-1). If it is launched with a non-zero horizontal velocity, it traverses a curved trajectory before hitting the surface of the earth (path *b*). If it is launched with a larger horizontal velocity, it also traverses a curved trajectory, but travels a larger horizontal distance before it hits the surface of the earth (path *c*). Finally, one can imagine that the projectile is launched with such a large horizontal velocity that it travels around the earth in a circular orbit without ever hitting its surface (path *d*). The projectile would then be an artificial satellite orbiting around the earth. (If the initial velocity is larger, the satellite could also travel around the earth in an elliptical orbit.)

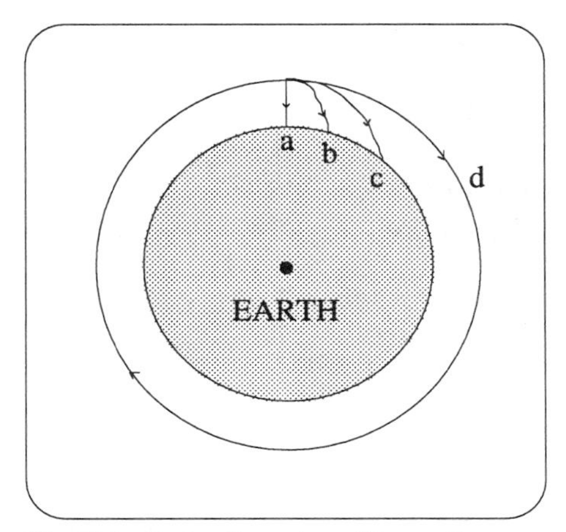

Fig. D-1. Projectile and satellite motions.

All the preceding motions have one thing in common: In all cases the projectile (neglecting air resistance) is influenced only by gravitational interaction with the earth and thus travels with an acceleration of magnitude g directed toward the center of the earth. Thus an artificial satellite is merely a special case of a projectile (a fact well recognized by Newton). However, it is only in recent years that technology has advanced to the point where artificial satellites can be launched with the requisite high speed.

Speed and period of an earth-orbiting satellite. The preceding insights can be used to predict the speed v with which an artificial satellite must travel if it is to circle the earth close to its surface. For example, the satellite might travel at a height of 300 km above the earth's surface so as to be unaffected by the earth's atmosphere. Such a height is much smaller than the earth's radius R (about 6,370 km). Hence the magnitude of the gravitational acceleration, produced by the earth at the position of the satellite, does not differ appreciably from the magnitude g of this acceleration at the earth's surface. As the satellite moves in its circular orbit with constant speed, its acceleration $\vec{a}$ is then directed toward the center of the earth and has a magnitude approximately equal to g. The relation (A-4) then implies that

$$\frac{v^2}{R} = g \qquad \text{(D-1)}$$

since the radius of the satellite's orbit is approximately the same as the radius R of the earth. Thus the speed of such a satellite should be equal to

$$v = \sqrt{gR}\,. \qquad \text{(D-2)}$$

The time T, required by the satellite to travel once around the earth, can then be found by (B-5). Thus

$$T = \frac{2\pi R}{v} = 2\pi\sqrt{\frac{R}{g}}\,. \qquad \text{(D-3)}$$

The meter was originally defined (during the French revolution) to be such that the distance along the earth's surface, from the north pole to the equator, would be exactly 10,000 km. Even after later redefinitions, the circumference $2\pi R$ of the earth is thus approximately 40,000 km. This is an easily remembered number from which the earth's radius can readily be found.

Using the known values of $g = 9.8\ \text{m/s}^2$ and $R = 6.37 \times 10^6$ m, the result (D-2) yields the following numerical value for the required speed of the satellite,

$$v = 7900 \text{ m/s} = 7.9 \text{ km/s} . \tag{D-4}$$

By (D-3) the period of the satellite is then

$$T = 5070 \text{ s} = 84 \text{ min} \approx 1.4 \text{ h} . \tag{D-5}$$

This last value agrees well with the observed periods of satellites or space shuttles orbiting close to the earth.

→ ***Go to Sec. 8D of the Workbook.***

E. Motion relative to different frames

The position of a particle (and hence also its velocity and acceleration) must always be described relative to some specified reference frame. *Any* such reference frame may be chosen. However, descriptions relative to some frames may be more convenient than relative to some others. (For example, when a flight attendant walks down along the aisle of a flying plane, it is easier to describe her motion relative to the plane than relative to the ground.) Hence one must be able to relate motion descriptions relative to different frames.

Consider some particle P whose position may be described relative to two different reference frames with associated coordinate systems whose origins are at the points A and B. These frames may move relative to each other. (However, we assume that they don't *rotate* relative to each other.) For example, one reference frame might be the ground and the other reference frame might be a train moving relative to the ground.

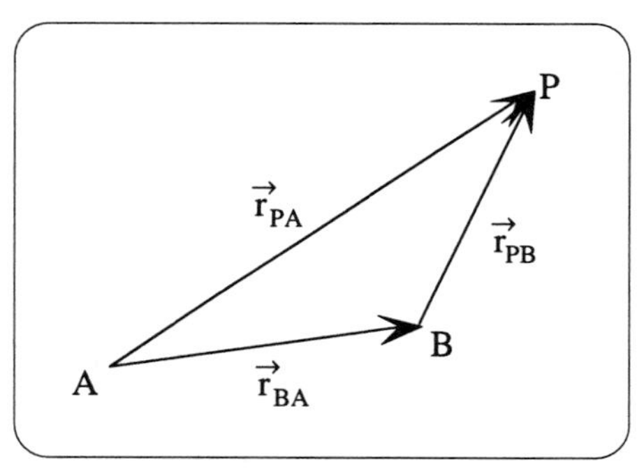

Fig. E-1. Position described relative to different reference frames.

Relation between position vectors. The position of the particle P may be described relative to either one of these reference frames. Thus the position of P relative to A may be described by the position vector $\vec{r}_{PA}$ indicated in Fig. E-1. Similarly, the position of P relative to B may be described by the position vector $\vec{r}_{PB}$. Furthermore, the position of one reference frame relative to the other may be described by the position vector $\vec{r}_{BA}$ of B relative to A.

As is apparent from Fig. E-1, these position vectors are simply related since $\vec{r}_{PA} = \vec{r}_{BA} + \vec{r}_{PB}$. In slightly more convenient order this relation can be written

$$\boxed{\vec{r}_{PA} = \vec{r}_{PB} + \vec{r}_{BA} .} \tag{E-1}$$

Expressed in words, this relation states that the ⟨position vector of particle P relative to A⟩ is equal to the ⟨position vector of P relative to B⟩ plus the ⟨position vector of B relative to A⟩. The preceding sequence of letters is easily remembered. (Indeed, as indicated in Fig. E-2, insertion of the intermediate letter B in the original subscript PA yields the two successive subscripts PB and BA.)

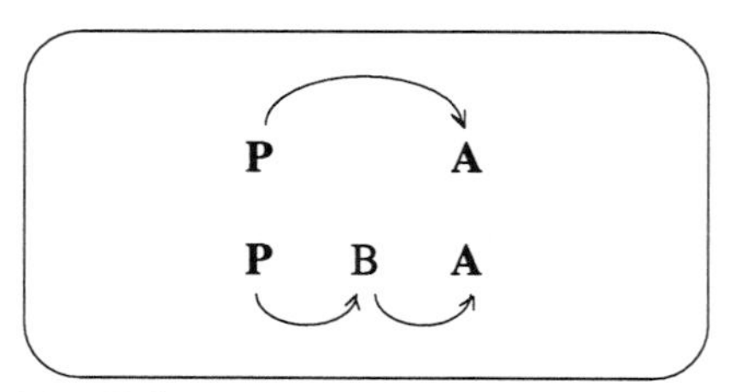

Fig. E-2. Sequence of subscripts in the relation (E-1).

Relation between velocities. The relation (E-1) between the position vectors is valid at any instant of time t. Thus it is also valid at a slightly later time t' so that

$$\vec{r}'_{PA} = \vec{r}'_{PB} + \vec{r}'_{BA} . \tag{E-2}$$

By subtracting (E-1) from (E-2), one then finds a corresponding relation between the small changes of the position vectors, namely

$$d\vec{r}_{PA} = d\vec{r}_{PB} + d\vec{r}_{BA} . \tag{E-3}$$

Dividing both sides of (E-3) by the infinitesimal time interval $dt = t' - t$ then yields a corresponding relation between the velocities. Thus we get

$$\boxed{\vec{v}_{PA} = \vec{v}_{PB} + \vec{v}_{BA} .} \tag{E-4}$$

In words, this merely asserts that the ⟨velocity of P relative to A⟩ is equal to the ⟨velocity of P relative to B⟩ plus the ⟨velocity of B relative to A⟩. For example, if a person walks on a train, the velocity of the person relative to the ground is equal to the velocity of the person relative to a train, plus the velocity of the train relative to the ground.

Relation between accelerations. The relation (E-4) is again valid at any instant of time. By focusing on the corresponding changes in these velocities one then obtains [in a manner completely similar to that leading from (E-1) to (E-4)] the corresponding relation between the accelerations

$$\boxed{\vec{a}_{PA} = \vec{a}_{PB} + \vec{a}_{BA} .} \tag{E-5}$$

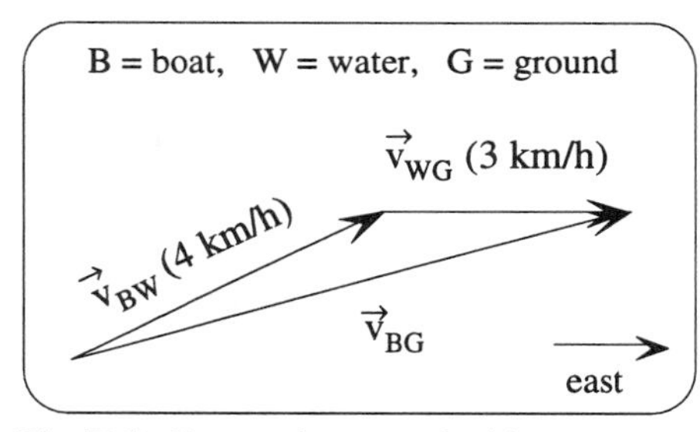

Fig. E-3. Boat and water velocities.

Example: Boat velocities relative to water or ground

The water in a river flows east relative to the ground with a speed of 3 km/h. A boat travels relative to the water of the river with a speed of 4 km/h. What then is the velocity of the boat relative to the ground under various conditions?

The information is summarized in Fig. E-3. By (E-4), the velocities of interest are related so that

$$\vec{v}_{BG} = \vec{v}_{BW} + \vec{v}_{WG} ,$$

a relationship schematically indicated by the vector diagram in Fig. E-3.

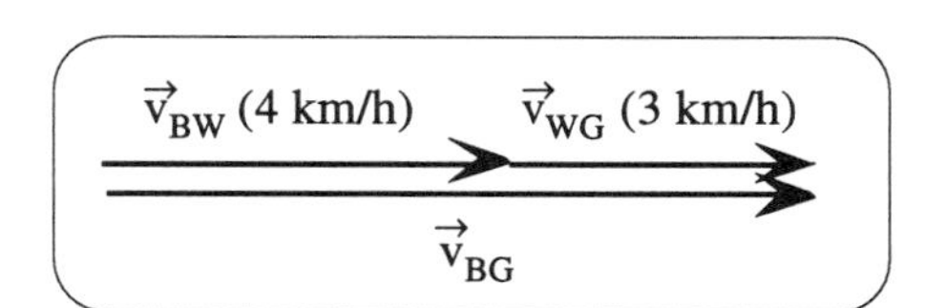

Fig. E-4. Boat velocity along water flow.

Boat velocity along water flow. Suppose that the boat travels east, i.e., in the same direction as the water flow. Then the vector diagram of Fig. E-3 becomes that illustrated in Fig. E-4. Thus the velocity of the boat relative to the ground is 7 km/h east.

Boat velocity opposite to water flow. Suppose that the boat travels west, i.e., opposite to the water flow. Then the vector diagram of Fig. E-3 becomes that illustrated in Fig. E-5. Thus the velocity of the boat relative to the ground is 1 km/h west.

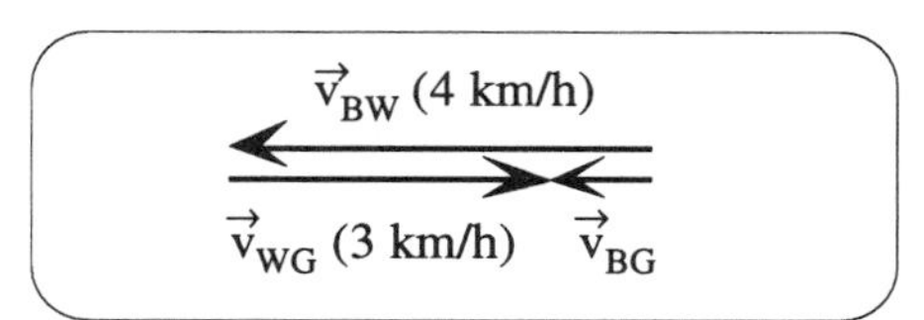

Fig. E-5. Boat velocity opposite to water flow.

Boat velocity perpendicular to water flow. Suppose that the boat travels north, i.e., perpendicular to the water flow. Then the vector diagram of Fig. E-3 becomes that illustrated in Fig. E-6. Thus the velocity of the boat relative to the ground is 5 km/h in a direction 37° east of north.

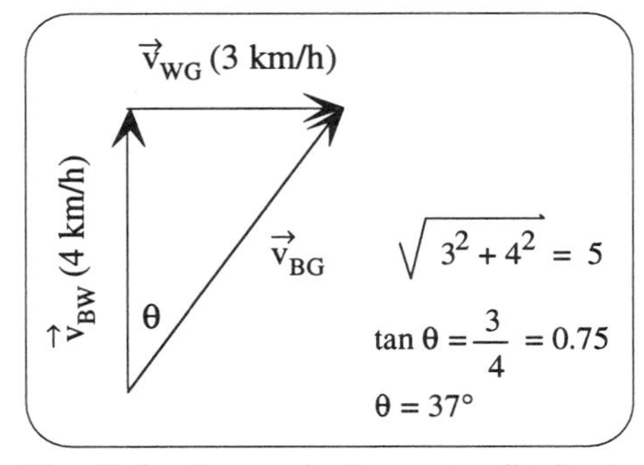

Fig. E-6. Boat velocity perpendicular to water flow.

➜ ***Go to Sec. 8E of the Workbook.***

F. Summary

Circular motion

Components of acceleration (speed v, radius r)

Along velocity $\vec{v}$: $a_v = \frac{dv}{dt}$ (due to change of magnitude of $\vec{v}$)

$\perp$ to velocity (toward center): $a_c = \frac{v^2}{r}$ (due to change of direction of $\vec{v}$)

Period and frequency

For motion around a circle with constant speed v:

$$v = \frac{2\pi r}{T} = 2\pi r\, f \qquad (T = \text{period},\ f = \text{frequency})$$

Motion relative to different frames (See Fig. F-1.)

$$\vec{r}_{PA} = \vec{r}_{PB} + \vec{r}_{BA}$$

$$\vec{v}_{PA} = \vec{v}_{PB} + \vec{v}_{BA}$$

$$\vec{a}_{PA} = \vec{a}_{PB} + \vec{a}_{BA}$$

Example: The velocity of a person P relative to the Atlantic Ocean A is equal to the velocity of the person P relative to a boat B, plus the velocity of the boat B relative to the Atlantic Ocean A.

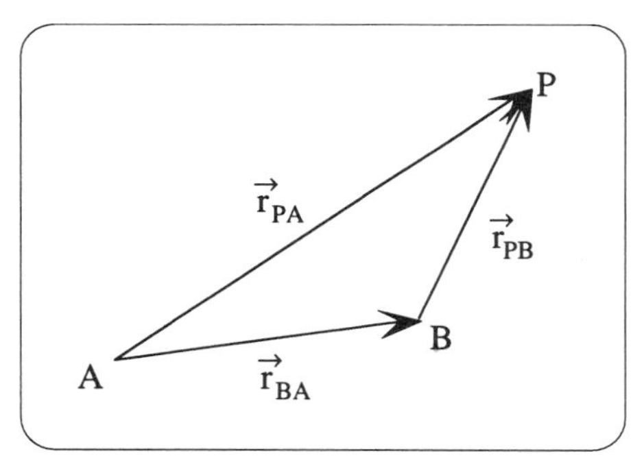

Fig. F-1. Position described relative to different reference frames.

New abilities

You should now be able to do the following:

(1) Find quantitatively the acceleration, or acceleration components, of a particle moving along a circular path.

(2) Relate the speed, period, and frequency of a particle moving with constant speed around a circle.

(3) Relate information about a particle's velocities or accelerations relative to different reference frames.

Grand summary about particle motion

The following table summarizes all the knowledge about motion discussed in the preceding chapters. (See also Appendix C.)

Basic descriptive concepts Position vector .... $\vec{r}$ Velocity .... $\vec{v} = d\vec{r}/dt$ Acceleration .... $\vec{a} = d\vec{v}/dt$	
Special motions ***Straight-line motion*** (acceleration a_x = constant) Rel (vel, t) .... $v_x - v_{xo} = a_x t$ Rel (disp, t)............. $D_x = x - x_o = v_{xo}t + \frac{1}{2}a_x t^2$ Rel (vel, disp)......... $v_x^2 - v_{xo}^2 = 2\, a_x D_x$	***Circular motion*** (speed v, radius r) Acceleration components Along $\vec{v}$ .... $a_v = dv/dt$ $\perp$ to $\vec{v}$ (to center)........... $a_c = v^2/r$ Period.................................... $T = 2\pi r/v$ Frequency .... $f = 1/T$
Motions relative to different frames Positions .... $\vec{r}_{PA} = \vec{r}_{PB} + \vec{r}_{BA}$ Velocities .... $\vec{v}_{PA} = \vec{v}_{PB} + \vec{v}_{BA}$ Accelerations................. $\vec{a}_{PA} = \vec{a}_{PB} + \vec{a}_{BA}$	Particle P, frames A and B. ($\vec{r}_{PA}$ = position of P relative to A, ...)

→ ***Go to Sec. 8F of the Workbook.***

Motion and Interaction of Particles

The preceding chapters introduced the concepts of velocity and acceleration, and showed how they provide useful ways of describing motion. With this preparation, we are now ready to address a far more ambitious and important question: Can one devise a theory capable of *predicting* how particles move when they are affected in various ways by other objects? The next few chapters will deal with this question and thus lead to some very far-ranging applications.

---

9 Newtonian Theory of Mechanics

The central goal of *mechanics* (the science of motion) is to invent a theory capable of explaining and predicting how objects move in a wide range of circumstances. To attain this goal, one needs to discover a few basic principles which can be used to explain and predict the motions of many diverse objects (such as baseballs, planets, electrons, etc.).

How can one discover such principles? The present chapter shows how a few simple observations and hypotheses lead to some principles, first formulated by Isaac Newton in the seventeenth century, which provide the basis for a highly successful theory of mechanics.

A. Introduction

Central question. To devise a theory of motion, we need to answer the following central question: How is the motion of an object influenced by other objects? Thus we need to be concerned with the *interaction* between objects.

Def: **Interaction:** An object A interacts with another object B if A is affected by the presence of B. (A-1)

The notion of interaction is familiar from everyday life. For example, one talks about an interaction between two people when the behavior of one affects the other. Similarly, in physics one says that a baseball interacts with a bat when its motion is affected by being hit with the bat.

Interaction is a *mutual* process. In other words, if A is influenced by the presence of B, then B is also influenced by the presence of A. However, the influence on one may be larger than that on the other. For example, in a collision between a car and a truck, both are affected — but the effect on the car may be far larger than the effect on the truck.

Thus we are centrally interested in answering the following question: *What precisely is the relation between motion and interaction?*

Approach

From simple to complex situations. The following strategy is useful to invent a theory or cope with other complex tasks: Start by considering some simple situations and exploit the resulting insights to deal with progressively more complex situations.

A consideration of simple situations can help one to formulate some *hypotheses* (i.e., tentative guesses suggested by a few observations or plausible arguments). If predictions based on these hypotheses are confirmed by many observations, then these hypotheses are adopted as the basic principles of a theory. (These principles remain, however, always subject to further revisions and refinement.)

Focus on particles. In accordance with the preceding strategy, we shall start by considering the motion of the simplest kinds of objects, namely *particles*.

Def: **Particle:** An object whose position can be adequately specified by that of a single point. (A-2)

For example, any object can be considered a particle if it is sufficiently small compared to all other distances of interest. An object can also be considered a particle if all its parts move with the *same* velocity and the *same* acceleration — since the motion of any of its points is then described in the same way. (For example, a sled sliding down along a hill may be considered a particle. On the other hand, a *rotating* object can *not* be considered a particle since different points on it move with different velocities.)

From few particles to many. We shall start by considering the simplest possible situation, that involving the motion of a single non-interacting particle. Then we shall try to understand the more complex situation of *two* particles interacting with each other. We shall then be prepared to deal with situations involving *three or more* interacting particles — and will thus be led to a theory of mechanics for *any* number of particles.

Such a theory should then be applicable to systems of any complexity since any object (including a liquid or gas) can be regarded as consisting of many small particles.

Complexities of everyday situations. Most situations familiar from everyday life are very complex because many particles commonly interact with many other particles. (For example, when a box slides along the floor, all the particles at the bottom of the box interact with many of the particles in the floor.) The familiarity with such common situations leads all of us to acquire various intuitive notions about them. These notions have some limited utility, but are often imprecise, inconsistent, and misleading in their predictions. Hence they do not provide a good starting point for disentangling the complexities and need to be replaced by more carefully formulated ideas.

➜ ***Go to Sec. 9A of the Workbook.***

B. Single non-interacting particle

Consider a single particle interacting with nothing else, e.g., a particle in outer space so very far from the sun and other stars that it is uninfluenced by their presence. What can one say about the motion of such a particle?

Suggestive observations. The answer to this question is suggested by some simple observations. For example, when a comet is very far from the sun, it moves relative to the stars with nearly constant velocity. Its velocity starts changing only when it comes close enough to the sun to interact appreciably with it.

On the surface of the earth, most ordinary objects interact appreciably with other objects. Hence special conditions are required to observe non-interacting particles. For example, if a hockey puck slides on a smooth horizontal surface of ice, interactions affecting the puck's horizontal motion are very small. As another example, consider a horizontal "air track" (a special steel track into which air is injected through many small holes). A "glider" on such a track then moves on a thin cushion of air so that interactions affecting the glider's motion are greatly reduced. Such a glider is then also observed to move with nearly constant velocity relative to the track.

Constant velocity relative to the stars. Astronomical observations, such as those of comets, suggest the following basic hypothesis:

Every non-interacting particle moves, relative to the stars, with constant velocity.	(B-1)

The motion of such a particle is thus very simple: If the particle is initially at rest, it will remain at rest. If it is moving, it will continue moving along a straight-line without acceleration.

Motion relative to other frames. Of course, there is no need to observe a particle's motion relative to the stars. For example, the particle might also be observed by someone sitting inside a rocket ship R moving relative to the stars with some velocity $\vec{v}_{RS}$. The velocity $\vec{v}_{PS}$ of the particle relative to the stars is then related to the velocity $\vec{v}_{PR}$ of the particle relative to the rocket ship by (8E-4). Thus

$$\vec{v}_{PS} = \vec{v}_{PR} + \vec{v}_{RS}$$

or

$$\vec{v}_{PR} = \vec{v}_{PS} - \vec{v}_{RS} \,. \tag{B-2}$$

Note that there is no real distinction between motion and rest. For example, a particle at rest relative to the stars would be observed to move relative to the rocket ship. Conversely, a particle at rest relative to the rocket ship would be moving relative to the stars.

Inertia principle. By (B-1), any non-interacting particle moves relative to the stars with a constant velocity $\vec{v}_{PS}$. Suppose that the rocket ship itself also travels relative to the stars with a *constant* velocity $\vec{v}_{PS}$. By (B-2), the velocity $\vec{v}_{PR}$ of the particle relative to the rocket ship would then also be constant. Hence (B-1) implies the following more general basic hypothesis well-confirmed by many observable implications:

Inertia principle: There exist reference frames relative to which every non-interacting particle moves with constant velocity.	(B-3)

Any such frame is called an *inertial frame.*

Def:	***Inertial frame:*** A reference frame relative to which every non-interacting particle moves with constant velocity.	(B-4)

According to the preceding comments, the stars constitute such an inertial reference frame, as do all other frames (like the preceding rocket ship) which move relative to the stars with *constant* velocity.

Example of a non-inertial frame

Consider a rocket ship traveling relative to the stars with a velocity $\vec{v}_{RS}$ which is *not* constant. As usual, any non-interacting particle would move with a constant velocity $\vec{v}_{PS}$ relative to the stars. But the velocity $\vec{v}_{PR}$ of such a particle relative to the rocket ship would, by (B-2), *not* be constant. This rocket ship would, therefore, *not* be an inertial frame.

Terrestrial reference frames. Suppose that one describes the motion of particles relative to a laboratory or other reference frame on the surface of the earth. Is such a reference frame an inertial frame (i.e., does every non-interacting particle move with constant velocity relative to such a frame)?

Strictly speaking, the answer is *no*. The reason is that the earth rotates around its axis and also revolves around the sun. Hence any laboratory on the surface of the earth really moves with a complicated accelerated motion relative to the stars.

> ***Evidence from observations of the night sky***
>
> A distant star is approximately a non-interacting particle remaining at rest relative to the sun and the other stars. Yet, all of us observe the whole firmament of stars rotating about the earth. This is because the earth spins around its axis and is thus somewhat like a merry-go-round rotating relative to the stars. The observed motion of the stars relative to the earth thus indicates clearly that the earth is not an inertial reference frame.

However, the earth rotates quite slowly, making only one revolution every 24 hours. During any time interval appreciably less than that, a laboratory on the earth moves thus with a *nearly constant velocity* of unchanged direction. If one is not interested in the greatest precision, or in predicting motions over time periods as long as hours, a reference frame fixed on the surface of the earth is, therefore, *approximately* an inertial frame. (This approximation is, however, *not* sufficient when one requires the precision needed for launching rockets or firing long-range artillery.)

Unless explicitly stated otherwise, we shall henceforth always describe the motion of particles relative to inertial (or approximately inertial) reference frames. This is usually simplest since every non-interacting particle is then known to move with constant velocity.

→ ***Go to Sec. 9B of the Workbook.***

C. Two interacting particles

As the next simplest case, consider two particles which may interact with each other, but which do not interact with any other objects. (For example, the two particles might be in outer space, fairly close to each other, but very far from any other objects.) As we shall see, an examination of two such particles, like those in Fig. C-1, leads to crucially important insights about motion and interaction.

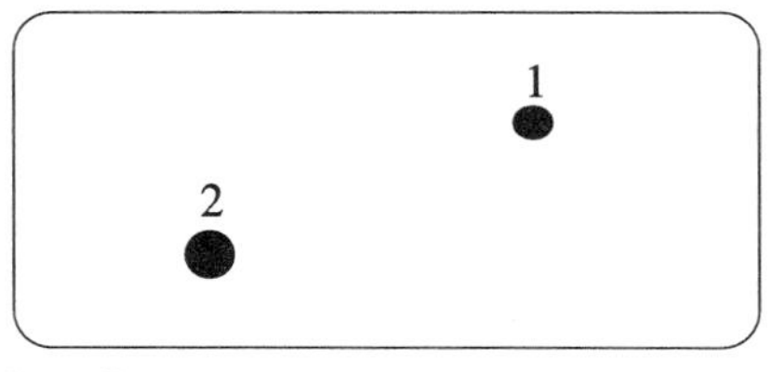

Fig. C-1. Two interacting particles.

Acceleration as indication of interaction

When the two particles are far apart, they don't interact with each other (i.e., each is unaffected by the presence of the other). In that case we know from the inertia principle (B-1) that each particle moves with constant velocity, i.e., with zero acceleration. On the other hand, if the particles are sufficiently close to each other that they *do* interact, then their accelerations are no longer zero. *The fact that a particle moves with non-zero acceleration indicates that the particle interacts with some other objects.*

Examples

(a) When a comet is very far from the sun, it moves with nearly constant velocity. But its velocity changes in direction and magnitude when the comet comes sufficiently close to the sun so as to interact appreciably with it.

(b) A hockey puck, sliding on smooth ice with nearly constant velocity, approaches another stationary puck. When the pucks come so close as to interact (i.e., when they hit each other), the velocity of each changes so that each is accelerated. Afterwards, when they have moved apart and no longer interact, each moves again with constant velocity.

Dependence of interaction on particle properties

Let us examine more closely the interaction between the two particles in Fig. C-1. The fact that the acceleration of a particle is affected by its interaction with the other particle implies the following:

> The acceleration of a particle depends on the properties of *both* interacting particles. (C-1)

Thus we need to answer the question: Just *how* does the interaction between the particles depend on their properties?

Position-dependent interactions. The simplest assumption is the following:

> The interaction between two particles at any instant depends only on their relative positions at that instant. (C-2)

Electromagnetic and other interactions can also depend on the velocities of the particles and on the past histories of their motions. Indeed, strictly speaking, the assumption (C-2) is never true, i.e., when a particle is displaced, the effect on another particle is only felt some time later. However, this time is often exceedingly short. (Such time delays are taken into account in Einstein's theory of relativity.)

This basic assumption describes many commonly observed interactions. It is widely useful and an excellent approximation, although of limited validity.

The hypothesis (C-2) implies the following about the accelerations of the particles:

> The acceleration of a particle, at any instant, depends only on the distance between the two interacting particles and is directed along the line joining them. (C-3)

(This must be true since no other direction is specified if the interaction depends solely on the relative positions of the particles.)

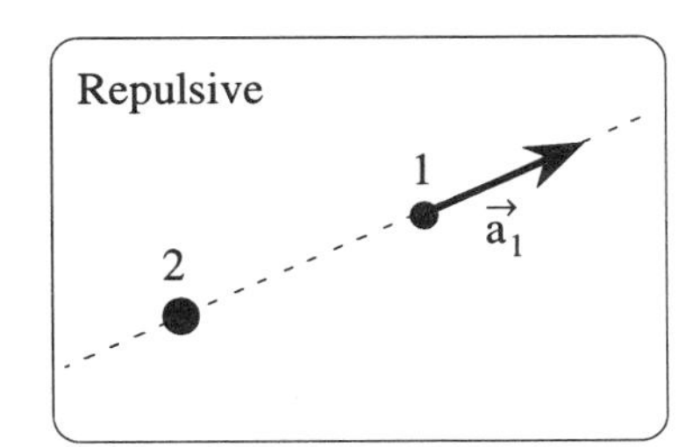

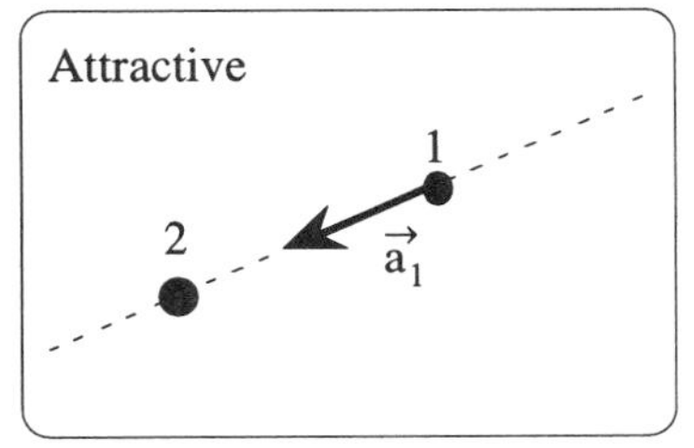

Fig. C-2. Repulsive and attractive interactions.

Repulsive and attractive interactions. The statement (C-3) implies that the acceleration of a particle (such as particle 1 in Fig. C-2) can only have two possible directions. (a) Either it can be directed *away* from the other particle, in which case the interaction is called *repulsive*. (b) Or it can be directed *toward* the other particle, in which case the interaction is called *attractive*.

Long-range and contact interactions. The interactions between two particles can be classified into two types, depending on how the interaction depends on the distance between the particles. (a) Some interactions are appreciable even if the distance between the particles is appreciable, and decrease only gradually as the distance between the particles increases. (This is schematically indicated in Fig. C-3a where the magnitude of the acceleration of one of the particles is denoted by a_1.) Such interactions are called *long-range interactions*. (b) Other interactions decrease abruptly to zero, as indicated schematically in Fig. C-3b, when the distance between the particles becomes much larger than atomic size. (In everyday life one would then say that the particles no longer "touch" each other.). Such interactions are called *contact interactions*.

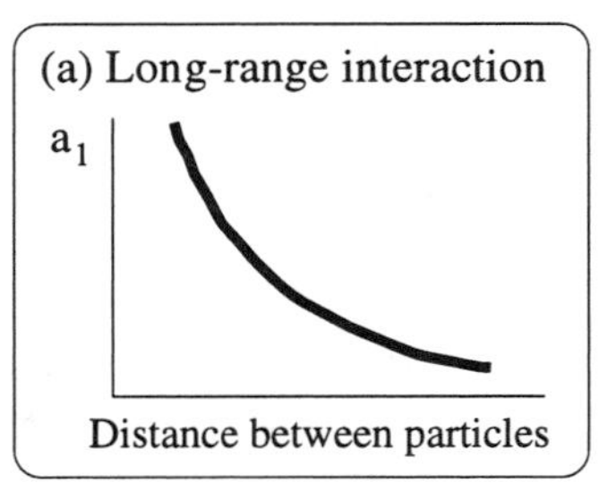

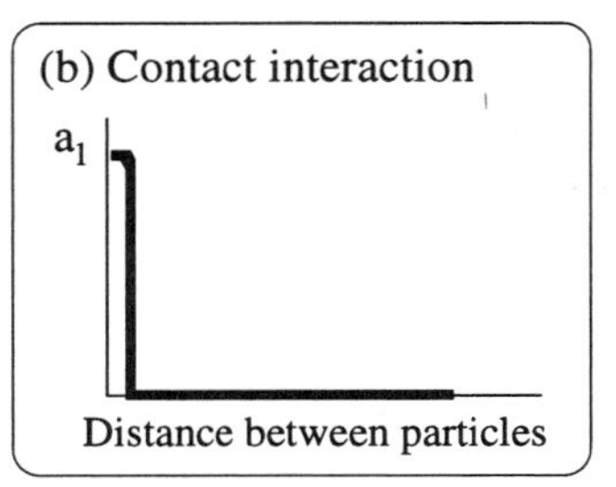

Fig. C-3. Long-range and contact interactions.

> ***Examples of interactions***
>
> A ball falls to the ground because it interacts with the earth, although it does not touch the earth. Similarly, a nail is attracted to a magnet, and thus interacts with it, although it does not touch the magnet. These are examples of long-range interactions.
>
> On the other hand, a ball interacts with a wall only when it actually hits the wall. This is an example of a contact interaction.

Reciprocity of interaction

The preceding discussion applies to the acceleration of either one of the particles. However, *each* of the particles experiences an acceleration as a result of its interaction with the other. Hence there arises the following question: How are the accelerations of the two particles related at any instant?

Special case of identical particles. This question is easily answered in the simple special case where the two particles are *identical* (e.g., two identical hockey pucks) so that all their properties are the same.

In this case the magnitudes a_1 and a_2 of the particles' accelerations must also be the same at any instant. In other words,

$$\text{for identical particles,} \qquad a_1 = a_2\,. \qquad \text{(C-4)}$$

Thus the ratio of these magnitudes $a_1/a_2 = 1$ at any instant.

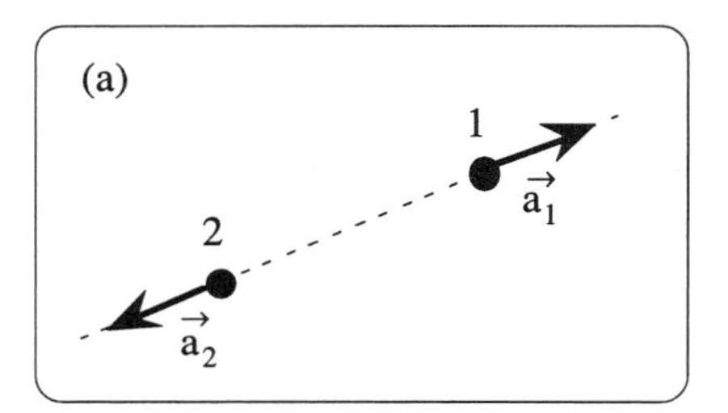

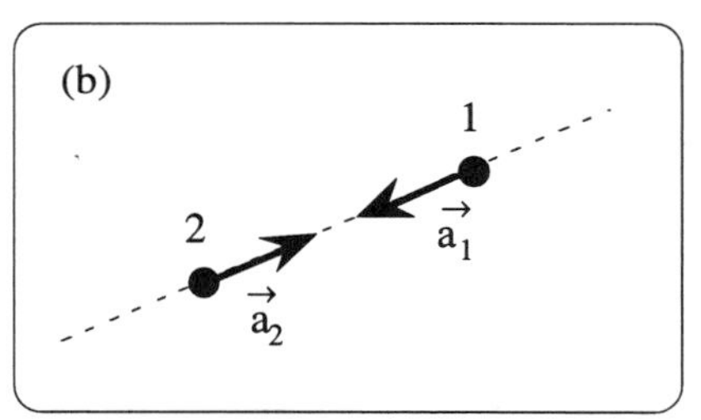

Fig. C-4. Relation between accelerations of identical particles.

Since identical particles have the same properties, the *directions* of their accelerations must be described in the same way. For example, if the acceleration of particle 1 is directed *away from* the other particle (as indicated in Fig. C-4a), the acceleration of particle 2 must also be directed *away from* the other particle. Similarly, if the acceleration of particle 1 is directed *toward* the other particle (as indicated in Fig. C-4b), the acceleration of particle 2 must also be directed *toward* the other particle. Thus it is apparent from Fig. C-4 that

$$\text{the accelerations of identical particles have } \textit{opposite} \text{ directions.} \qquad \text{(C-5)}$$

General case of any two particles. Consider now the general case where two particles are *not* necessarily identical. Then we may plausibly assume that the *directions* of the particles' accelerations are still opposite. However we no longer expect that the *magnitudes* of their accelerations are the same. (See Fig. C-5.) The simplest assumption then is that the ratio a_1/a_2 of these magnitudes, instead of being merely equal to 1, is some other *constant* number

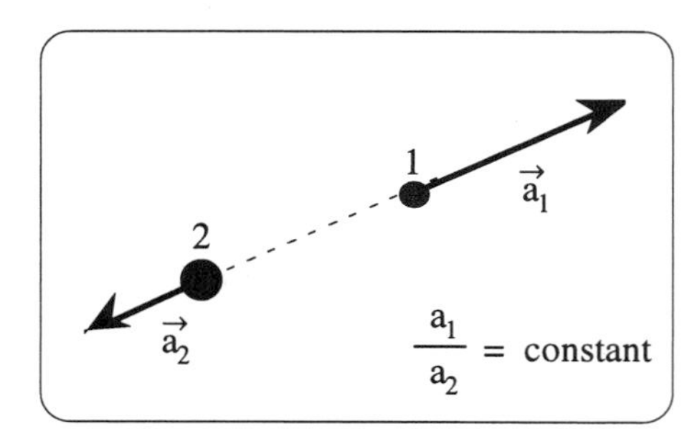

Fig. C-5. Relation between the accelerations of any two interacting particles.

(i.e., some number independent of the time, or the particles' positions, or their velocities, or the kind of interaction between them). In short, this number merely characterizes the particular particles under consideration.

Thus one is led to the following basic hypothesis well confirmed by observations of many predicted consequences:

Relation between accelerations: When two particles interact, their accelerations at any instant are oppositely directed. The ratio of their magnitudes is a constant depending only on the nature of these particles.	(C-6)

→ ***Go to Sec. 9C of the Workbook.***

D. Two particles: Mass and force

The preceding conclusions about the interaction of two particles lead to some important definitions.

Mass

Definition of mass. Since the ratio of the magnitudes of the accelerations depends only on the nature of these particles, one can characterize every particle by a property which is called its *mass* and denoted by the symbol m. Just like length or time, this property is defined comparatively (i.e., one compares the masses of two particles by comparing their accelerations). Thus one conventionally defines the ratio m_1/m_2 of the particles' masses to be equal to the constant ratio of the magnitudes of their accelerations.

Def: ***Mass ratio:*** $$\frac{m_1}{m_2} = \frac{a_2}{a_1}\,. \qquad \text{(D-1)}$$

Note that this definition relates the ratio of the masses to a_2/a_1 (rather than to a_1/a_2). The mass is defined in this way so that the particle which experiences the *smaller* acceleration is characterized by the *larger* mass.

> ***Example D-1.***
> When a ball hits a stationary baby carriage, the ball bounces back (thus experiencing an appreciable acceleration) while the baby carriage barely moves (thus experiencing only a very small acceleration). This observation indicates that the mass of the baby carriage is much larger than that of the ball.

Roughly speaking, the mass of a particle is a property describing the particle's tendency to maintain its velocity (i.e., to experience no acceleration) despite interactions with other objects. The property mass corresponds, therefore, roughly to the everyday notion of *inertia*.

Unit of mass. As in the case of other quantities which are defined comparatively (e.g., length and time), it is convenient to choose a particular standard with which all quantities can be compared. Thus one can choose a particular object as a standard of mass and then compare the masses of all other particles with it. By international convention, the SI standard of mass is a particular block of metal carefully preserved near Paris. The unit of mass (i.e., the algebraic symbol used to denote the mass of this standard) is *kilogram*, abbreviated as *kg*.

> ***Example D-2***
>
> Consider a hypothetical experiment where a particle collides with this kilogram standard. Suppose that, at some instant during this interaction, the particle's acceleration is a while that of the standard is a_S. According to the definition (D-1), the mass m of the particle would then be such that
>
> $$\frac{m}{\text{kilogram}} = \frac{a_S}{a}$$
>
> or $$m = \left(\frac{a_S}{a}\right) \text{kilogram} .$$
>
> For example, if the particle experiences an acceleration whose magnitude is 4.0 times larger than that of the standard kilogram, the mass of the particle would be 0.25 kg.

Typical values of mass. The mass of a typical book is about one kilogram. The mass of a typical person is about 75 kilogram.

Relation between accelerations. According to (D-1), a knowledge of the masses of two interacting particles determines the precise relation between the *magnitudes* of their accelerations at any instant. Thus (D-1) implies that

$$m_1\, a_1 = m_2\, a_2 . \tag{D-2}$$

Since the accelerations have opposite directions, the actual accelerations are then related by

$$\boxed{m_1 \vec{a}_1 = -m_2 \vec{a}_2\ .} \tag{D-3}$$

where the minus sign indicates the opposite directions of the vectors. The relation (D-3) thus summarizes compactly the relationship between the accelerations (magnitudes *and* directions) of two interacting particles.

Force

Description of interaction. As pointed out in (C-1), the existence of an interaction between two particles implies that the acceleration of any *one* of the particles (e.g., particle 1) depends on the properties of *both* particles. Thus one can write

$$\vec{a}_1 = \vec{I}_{12}$$

where $\vec{I}_{12}$ is a quantity which depends on the properties of *both* particles. This quantity $\vec{I}_{12}$ thus describes the interaction by specifying precisely how the acceleration of particle 1 is affected by its interaction with particle 2.

Definition of force. The relation (D-3) between accelerations can be expressed in simpler form by introducing, instead of $\vec{I}_{12}$, a slightly different quantity $\vec{F}_{12}$ which depends also on the mass m_1. Thus we can write

$$m_1\vec{a}_1 = \vec{F}_{12}\,. \qquad \text{(D-4)}$$

where the quantity $\vec{F}_{12}$ depends on the properties of both particles and thus specifies precisely how the acceleration of particle 1 is affected by its interaction with particle 2. The quantity $\vec{F}_{12}$ is called *the force on particle 1 by particle 2.* The general definition of force can then be stated as follows:

Def: ***Force:*** The force $\vec{F}$ on a particle by other objects is a quantity which depends on the properties of *all* the interacting objects and which is equal to $m\vec{a}$ where m is the mass of the particle and $\vec{a}$ is its acceleration. *(Force is thus a quantity which describes how the acceleration of a particle is affected by its interaction with other objects.)* (D-5)

Analogously to (D-4), the acceleration of particle 2 is related to the force on particle 2 by particle 1 so that

$$m_2\,\vec{a}_2 = \vec{F}_{21}\,. \qquad \text{(D-6)}$$

Relation between mutual forces. The particular definition (D-5) of force simplifies the description of interactions. For example, by (D-4) and (D-6), the relation (D-4) between the accelerations of the interacting particles leads to the following simple relation between the mutual forces $\vec{F}_{12}$ and $\vec{F}_{21}$ acting on them:

Relation between mutual forces: $\vec{F}_{12} = -\vec{F}_{21}\,.$ (D-7)

Historically, Newton called these mutual forces "action" and "reaction". These are very confusing words since they do not make clear which is which, or what acts on what. Do *not* use these unnecessary words.

In words, this relation between the mutual forces on two interacting particles can be stated as follows:

Mutual forces have equal magnitudes and opposite directions. (D-8)

The interaction between two particles can thus always be described by a *pair* of forces (a force $\vec{F}_{12}$ on particle 1 by particle 2 and a mutual force $\vec{F}_{21}$ on particle 2 by particle 1). These forces have opposite directions, but the same magnitude. The magnitude of the interaction can thus be described by the common magnitude of these forces.

Mutual forces have the equal magnitudes merely because of the particular *definition* (D-5) of force. It is only the corresponding relation (D-2) between the magnitudes of the accelerations which expresses a fundamental physical law.

Similarity to everyday notion of force. The concept "force" defined in physics has some similarity to the everyday notion that force is something which can affect the motion of an object. But in physics force is more precisely defined by (D-5), satisfies the relation (D-7 between mutual forces, and does not have various irrelevant connotations common in everyday life.

Units of force. The definition (D-4) or (D-5) of force implies that a force is specified in terms of the following units:

$$\text{SI units of force} = \text{kg}\left(\frac{\text{m}}{\text{s}^2}\right) \equiv \text{newton.} \tag{D-9}$$

Here the ≡ sign indicates that the unit *newton* (abbreviated by *N*) is merely a convenient abbreviation, defined to be equal to the combination of units to the left of this sign. In the British system the unit of force is the *pound* (abbreviated by *lb*). The approximate relation between the SI and British units is 1 newton ≈ 0.225 pound.

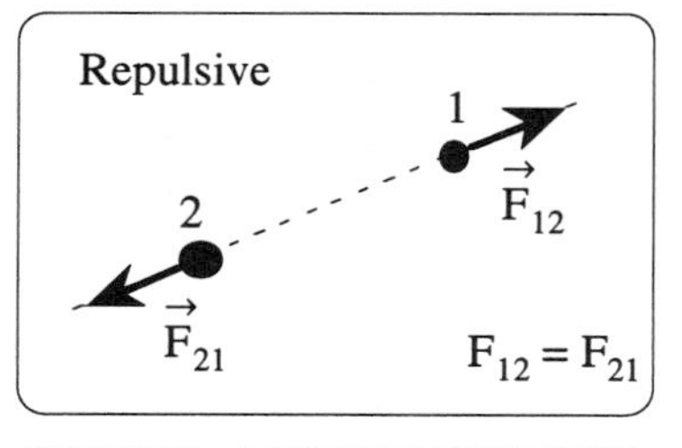

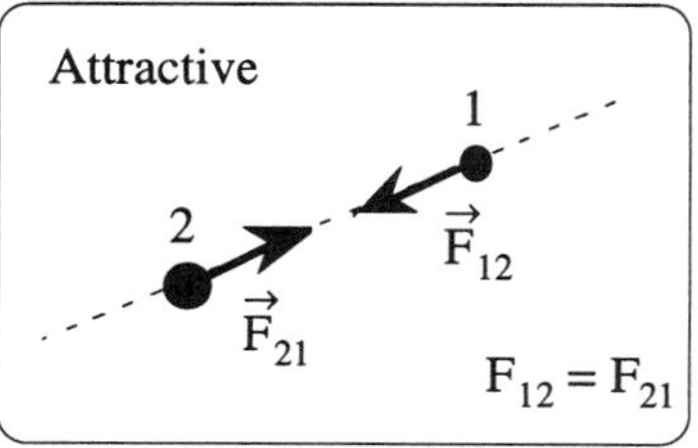

Fig. D-1. Forces on one particle by another.

Properties of two-particle forces. Section C discussed some basic properties of the accelerations of two interacting particles. Because of the definition (D-5) of force, these properties imply corresponding properties of the forces on these particles.

Thus (C-3) implies the following:

Central-force: At any instant, the force on a particle by another depends only on the distance between them, and is directed along the line joining them.	(D-10)

(Such a force is called *central* because its direction is along the line joining the centers of the particles.)

The force on each particle can thus be directed either away from the other particle or toward the other particle, i.e., it can be either repulsive or attractive. But in all cases the mutual forces must, at any instant, satisfy the relation (D-7) (i.e., they must have equal magnitudes and opposite directions, as illustrated in Fig. D-1).

As discussed in Sec. C, it is useful to distinguish between long-range and contact interactions. Correspondingly, *long-range forces* are those whose magnitude decreases gradually as the distance between the particles increases. (See Fig. D-2.) By contrast, *contact forces* are those whose magnitude decreases abruptly to zero when the distance between particles becomes much larger than atomic size (i.e., when the particles no longer "touch" each other).

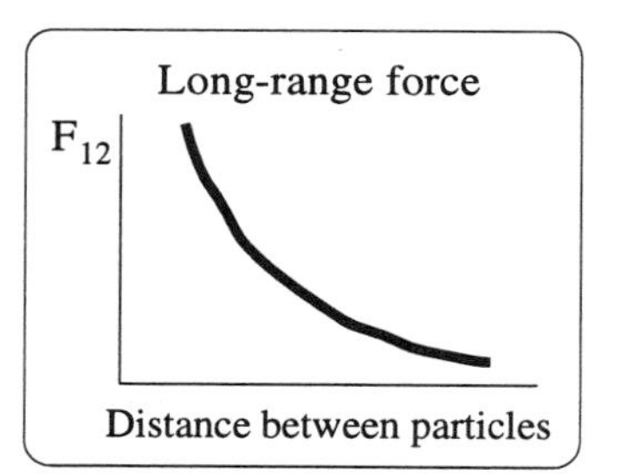

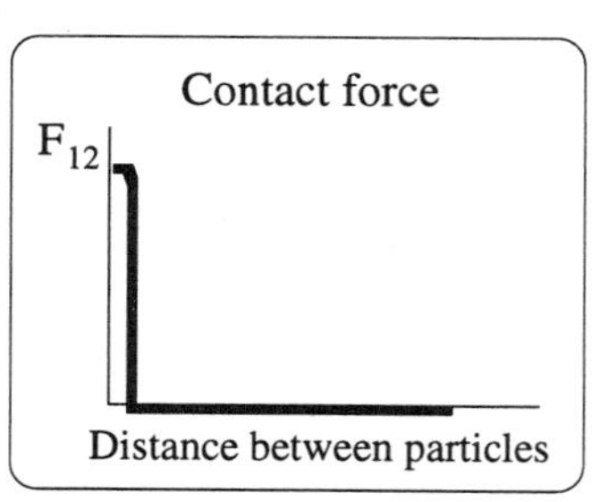

Fig. D-2. Long-range and contact forces.

→ ***Go to Sec. 9D of the Workbook.***

E. Three or more interacting particles

Three particles

We are now ready to consider the more complex situation of three particles, schematically illustrated in Fig. E-1. We suppose that these particles interact with each other, but are so far from all other objects that they interact with nothing else.

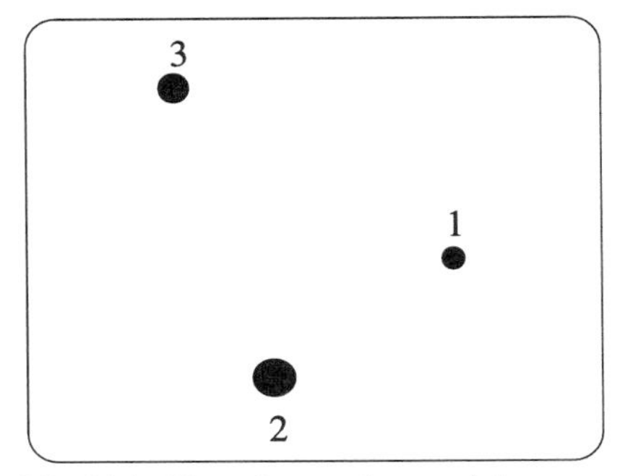

Fig. E-1. Three interacting particles.

Because of their mutual interaction, the particles don't move with constant velocity, but are accelerated. Thus we need to answer the following question: *How* does the acceleration of any one particle, say particle 1, depend on the properties of all the interacting particles?

Force describing the interactions. Analogously to (D-4) we can write for the acceleration of particle 1

due to 2 and 3, $$m_1 \vec{a}_1 = \vec{F}_{1,23} . \quad \text{(E-1)}$$

This relates the mass and acceleration of particle 1 to the quantity $\vec{F}_{1,23}$ which depends on the properties of *all* the particles and which can be called the *force on particle 1 by particles 2 and 3*. This force thus describes how the acceleration of particle 1 is affected by its interaction with the other two particles. However, can we specify in greater detail just *how* this force depends on the properties of all these particles?

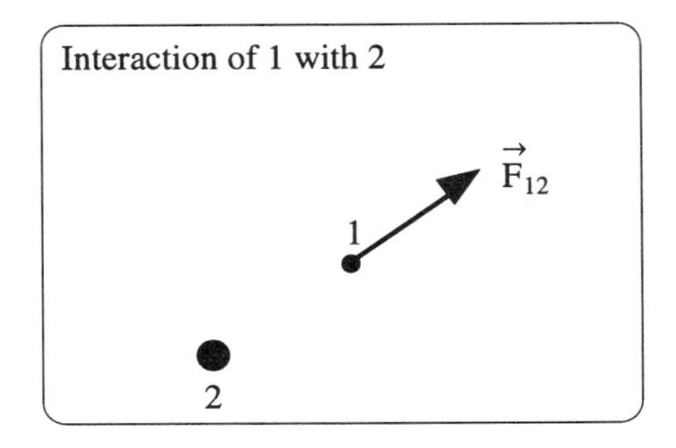

Simpler case of any two particles. To answer this question, imagine that particle 3 would not be there. Then we would merely face the familiar situation where particle 1 interacts solely with particle 2. The acceleration of 1 would then be simply related to the force on 1 by 2. Thus we could write, as in (D-4),

due to 2 alone, $$m_1 \vec{a}_1 = \vec{F}_{12} . \quad \text{(E-2)}$$

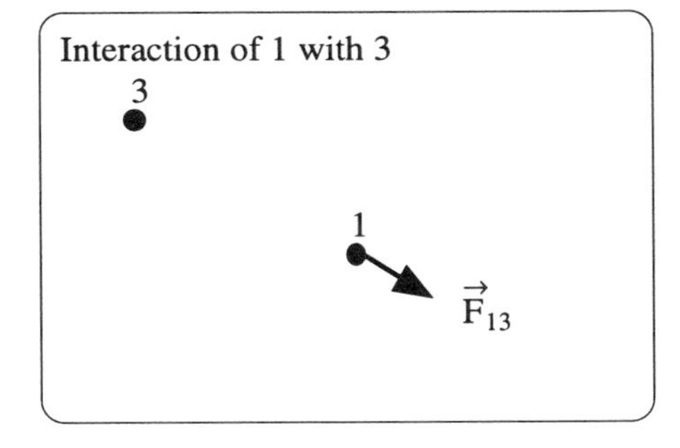

Similarly, imagine that particle 2 would not be there. Then we would merely face the situation where particle 1 interacts solely with particle 3. The acceleration of 1 would then be simply related to the force on 1 by 3. Analogously to (E-2) we could then write

due to 3 alone, $$m_1 \vec{a}_1 = \vec{F}_{13} . \quad \text{(E-3)}$$

Superposition principle. However, what happens when *both* particles 2 and 3 are present simultaneously? The simplest assumption answering this question (and well confirmed by many observable consequences) is expressed by the following principle:

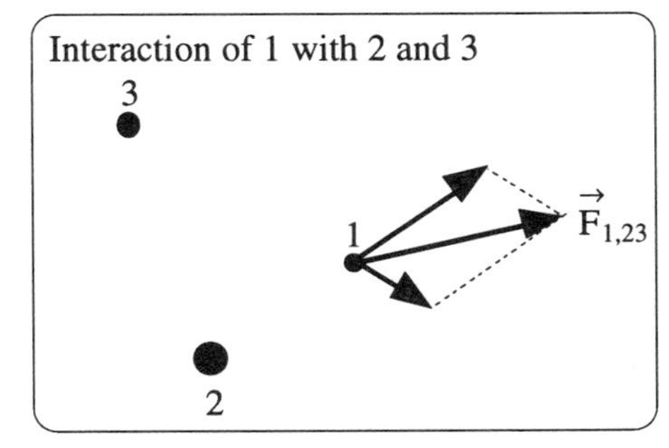

Fig. E-2. Superposition principle for three interacting particles.

> ***Superposition principle:*** The force produced by several particles present jointly is equal to the vector sum of the forces produced by these particles separately. (E-4)

Thus the force on particle 1 by particles 2 and 3 is simply specified by the following relation schematically illustrated in Fig. E-2:

$$\vec{F}_{1,23} = \vec{F}_{12} + \vec{F}_{13} . \qquad \text{(E-5)}$$

The superposition principle (E-4) provides an enormous simplification: It implies that the force on any one particle by *all* the others can be immediately found by simple vector addition if one knows the forces between individual pairs of particles!

Any number of particles

The superposition principle (E-4) is generally valid for *any* number of particles. Thus the force on any one particle (say, particle 1) by all the other particles is simply the vector sum of the individual forces on 1 by all the other particles.

$$\begin{aligned} \vec{F}_{1,\text{all}} &= \vec{F}_{12} + \vec{F}_{13} + \vec{F}_{14} + \ldots \\ &= \Sigma \vec{F}_{1s} \qquad \text{(E-6)} \end{aligned}$$

[Here the symbol Σ (the Greek letter *sigma*) has been used as an abbreviation to denote the previous sum ranging over all values of $s = 1,2,3 \ldots$]. The force on a particle by *all* the other particles can, therefore, be called the *total force* on this particle and can be denoted by $\vec{F}_{tot}$.

If a particle interacts with any other number of particles, the relation (E-1) can then be written in the following general form (first formulated by Isaac Newton):

Newton's law: $m\vec{a} = \vec{F}_{tot}$.	(E-7)

Here m is the mass of the particle and $\vec{a}$ is its acceleration. The total force $\vec{F}_{tot}$ on the particle (i.e., the force on this particle by *all* the other particles) is the vector sum of the individual forces on the particle by all the other particles.

Newton's basic law is of fundamental importance because it relates the motion any *one* particle (as described by its acceleration) to the total force which depends on the properties of *all* the interacting particles. This relationship can then be used to make predictions about the motion of each of the particles.

→ ***Go to Sec. 9E of the Workbook.***

F. Newton's laws of mechanics

The basic hypotheses discussed in the preceding sections constitute the theory of mechanics first formulated by Newton near the end of the seventeenth century. This theory has been remarkably successful in its ability to predict the motion of objects and is the basis of an enormous number of engineering applications.

Newton formulated these hypotheses in the form of the following three "laws of mechanics" (summarized here somewhat more clearly than Newton did):

> ***Law #1:*** There exist reference frames (called *inertial frames*) relative to which every non-interacting particle moves with constant velocity.

This first law is just the inertia principle (B-1).

> ***Law #2:*** The acceleration $\vec{a}$, relative to an inertial frame, of any particle of mass m is given by
>
> $$m\,\vec{a} = \vec{F}_{tot}$$
>
> where the quantity $\vec{F}_{tot}$ (called the *total force on the particle*) depends on the properties of all the interacting particles and is equal to the vector sum of the individual forces on this particle by all other particles.

This second law expresses Newton's basic law (E-7), defines the physics concept *force*, and incorporates the superposition principle (E-4).

> ***Law #3:*** The interaction between any two particles 1 and 2 can be described by a pair of mutual forces having the following properties:
> (a) The force $\vec{F}_{12}$ on particle 1 by particle 2, and the force $\vec{F}_{21}$ on particle 2 by particle 1, have equal magnitudes and opposite directions.
> (b) Each of these forces depends only on the distance between the particles and is directed along the line joining them.

Here statement *a* expresses the relation (D-7) between mutual forces and also permits one to define the concept *mass*. Statement *b* reflects the basic assumption (C-2) that the interactions depend only on the relative positions of the particles (an assumption which is often a good approximation, but not satisfied by all forces).

Specification of force. Newton's theory uses the concept *force* (a quantity depending on the properties of *all* the interacting particles) to describe how the acceleration of a particle is affected by its interaction with other particles. In specifying a force, it is essential to use the phrase

force ***on*** object 1 ***by*** object 2 (F-1)

to indicate clearly the relevant interacting objects (underlined above), i.e., the object *on* which the force acts and the object *by* which the force is exerted. Indeed, failure to specify *both* of these objects can lead to fatal confusions. (There is no ambiguity in talking about "the *total* force on an object" since this means "the force *on* this object *by* all other objects".)

Scope and limitations of Newton's laws. Despite their seeming simplicity, Newton's laws provide the basis of a theory with enormous predictive power. Indeed, if one knows how the forces between particles depends on their properties, Newton's law allows one to find the acceleration of each particle and thus to infer how each particle will move in the course of time.

Newton's laws deal only with particles. But, as we shall see later, these laws can be used to derive other laws applicable to complex systems consisting of many particles (e.g., rotating wheels, bicycles, cars, etc.).

Newton's theory leads to a tremendous range of practical applications in science and engineering. Indeed, the rest of this book will merely deal with some implications of this theory, implications which can be vastly extended in more advanced courses.

However, Newton's theory has some limitations. It implicitly makes some assumptions which are not valid when particles move with very high speeds (close to 3×10^8 m/s, the speed of light) or in some cases where particles are of atomic size. In these domains the theory must be refined and generalized (thus leading to the more complex theories of relativistic mechanics and quantum mechanics).

➜ ***Go to Sec. 9F of the Workbook.***

G. Exploiting Newton's laws

The cornerstone of Newton's theory of mechanics is his second law

$$m\,\vec{a} = \vec{F}_{tot}\,. \tag{G-1}$$

(We shall simply call this *Newton's law* unless we wish specifically to refer to another of his laws.)

Generality of Newton's law. Newton's law applies to *any* particle (whether it be a ball, a small portion of a liquid, a short portion of a string, or anything else). It establishes a fundamental relation between the *motion* of the particle (as described by its acceleration) and the *interaction* of the particle with other objects (as described by the total force on the particle).

Focus on a specific particle. Newton's law deals with a *specific* particle at any instant. Thus the quantities in (G-1) refer to the mass of the particle of interest, to the acceleration of *this* particle at any specific instant, and to the total force on *this* particle at *this same* instant.

Component form of Newton's law. Newton's law $m\vec{a} = \vec{F}_{tot}$ is an equation relating *vectors* (i.e., a particle's acceleration and the total force on it). It is usually much easier to deal with such a vector equation if the vectors are expressed in terms of their components along convenient directions. Newton's law is then equivalent to several numerical equations corresponding to these directions. These component equations can be readily obtained by exploiting the following properties of components [previously discussed in (7C-3) and (7C-4)]:

(a) If two vectors are equal, their components along any direction must be equal. Hence Newton's law (G-1) for any particle is equivalent to the following statement:

The component of $m\vec{a}$ along any direction must be equal to the component of the total force $\vec{F}_{tot}$ along this direction.	(G-2)

(b) Along any direction, the component of the sum of two vectors is equal to the sum of the components of these vectors. The corresponding implication for Newton's law is the following:

The component of the total force along any direction is the numerical sum of the components of the individual forces along this direction.	(G-3)

Knowledge needed to apply Newton's law. Newton's law is the most basic mechanics law relating the motion of a particle to its interactions with other particles. (See Fig. G-1.) To apply the law, one needs thus enough knowledge about the motions of particles and about their interactions. We have already learned a great deal about the *motion* of particles (e.g., about relations connecting accelerations to velocities and positions). But we still need to gain more knowledge about the *interactions* between particles. The next chapter will, therefore, begin to examine several such commonly occurring interactions.

➜ ***Go to Sec. 9G of the Workbook.***

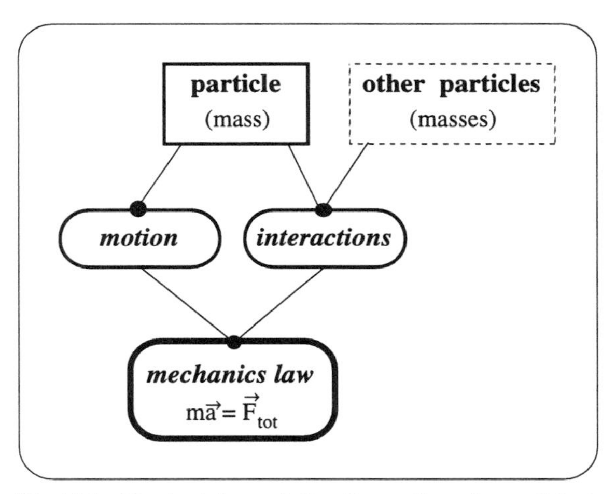

Fig. G-1. Newton's law relating the motion of a particle to its interactions with other particles.

H. Comment: Composite particles

When all parts of a system move with the *same* velocity and the *same* acceleration, the system can be regarded as a composite particle since it can be characterized by a single velocity and a single acceleration. For example, a sled sliding down along a hill can be considered a composite particle. A system consisting of a sliding sled, with a package fastened to its top, can also be regarded as a composite particle.

Suppose that two particles 1 and 2 form a composite particle C because they move with the same velocity $\vec{v}_C$ and the same acceleration $\vec{a}_C$. How then is the mass m_C of this composite particle related to the masses m_1 and m_2 of its constituents?

As is shown below, the definition of mass, together with the laws of mechanics summarized in Section F, implies the following simple answer to this question: *The mass of a composite particle is simply equal to the sum of the masses of its constituent particles.* Thus

$$\boxed{m_C = m_1 + m_2}\,. \tag{H-1}$$

Argument leading to the preceding result
(optional for interested readers)

The mass of the composite particle C can be determined by letting it interact with the standard particle S of mass m_S (e.g., the standard kilogram), as illustrated in Fig. H-1. By the definition (D-1), the mass m_C is then such that

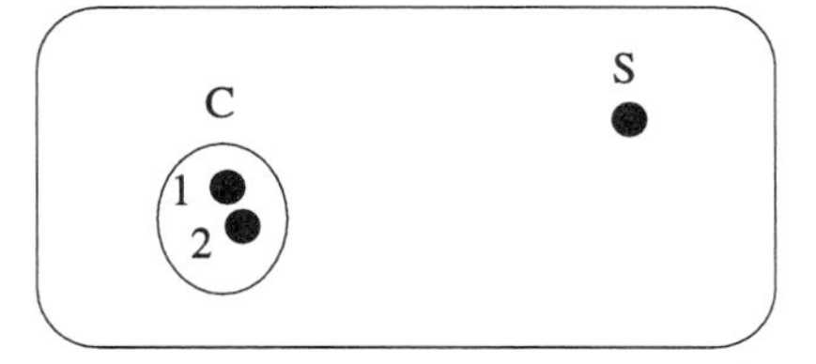

Fig. H-1. A composite particle C interacting with the standard of mass S.

$$\frac{m_C}{m_S} = \frac{a_S}{a_C}\,. \tag{H-2}$$

where a_C is the magnitude of the acceleration of C and a_S is the magnitude of the acceleration of S.

To examine how the motions of the particles in the diagram are related to their interactions, we can apply Newton's law to each one of them. Since particle 1 moves with the acceleration $\vec{a}_C$ of the composite particle, Newton's law applied to it yields

$$m_1\,\vec{a}_C = \vec{F}_{1S} + \vec{F}_{12} \tag{H-3}$$

where $\vec{F}_{1S}$ is the force on 1 by S, and $\vec{F}_{12}$ is the force on 1 by 2. Similarly, Newton's law applied to particle 2 yields

$$m_2\,\vec{a}_C = \vec{F}_{2S} + \vec{F}_{21} \tag{H-4}$$

where $\vec{F}_{2S}$ is the force on 2 by S, and $\vec{F}_{21}$ is the force on 2 by 1. Adding (H-3) and (H-4), one then gets

$$(m_1 + m_2)\,\vec{a}_C = \vec{F}_{1S} + \vec{F}_{2S} \tag{H-5}$$

since $\vec{F}_{12} = -\vec{F}_{21}$ because of the relation (D-7) between mutual forces.

Finally, Newton's law can also be applied to S and yields

$$m_S\,\vec{a}_S = \vec{F}_{S1} + \vec{F}_{S2} = -\,\vec{F}_{1S} - \vec{F}_{2S} \tag{H-6}$$

The right sides of (H-5) and (H-6) have equal magnitudes (although opposite signs). Hence the magnitudes of their left sides must also be equal, i.e.,

$$(m_1 + m_2)\,a_C = m_S\,a_S \tag{H-7}$$

or

$$\frac{(m_1 + m_2)}{m_S} = \frac{a_S}{a_C}\,. \tag{H-8}$$

Comparison of this relation with (H-2) then yields the result (H-1).

→ ***Go to Sec. 9H of the Workbook.***

I. Summary

Newton's law

For any *particle*, at any instant, relative to any *inertial frame*

$$\boxed{m\,\vec{a} = \vec{F}_{tot}}$$

where m = mass of particle,

$\vec{a}$ = acceleration of particle (describes its motion),

$\vec{F}_{tot}$ = total force on particle

= vector sum of the forces on the particle by all other particles.

(Describes interactions, depends on properties of *all* particles.)

Relation between mutual forces

Interaction between any two particles 1 and 2 is described by a pair of mutual forces (force $\vec{F}_{12}$ on 1 by 2, and force $\vec{F}_{21}$ on 2 by 1).

$$\boxed{\vec{F}_{12} = -\vec{F}_{21}}$$

Thus the mutual forces have equal magnitudes and opposite directions.

New abilities

You should now be able to do the following:

(1) Interpret properly the important concepts *inertial frame*, *mass*, and *force*.

(2) Interpret properly, and apply qualitatively in simple situations, Newton's law and the relation between mutual forces.

→ ***Go to Sec. 9I of the Workbook.***

10 Common Interactions

Newton's theory can be exploited to predict the motion of objects if one knows specifically how various interactions (and corresponding forces) depend on the properties of the interacting objects. The present chapter examines, therefore, some of the interactions commonly observed in everyday life.

Ultimately the properties of these interactions can be inferred from more basic knowledge about the properties of only very few fundamental interactions between atomic particles. Some of these fundamental interactions will be studied later in this book. However, for the time being we shall merely use some simple observations to elucidate the properties of some common interactions between objects.

Recent work has shown that fewer than four such fundamental interactions are sufficient to account for all interactions observed in the universe.

As pointed out in the preceding chapter, interactions can be classified into long-range interactions (which can be appreciable even if the interacting objects are far apart) and contact interactions (which are appreciable only if the objects are so close that they touch each other). The following sections examine the properties of one such long-range interaction (the gravitational interaction near the earth) and of several contact interactions.

Long-range interactions

A. Gravitational interaction near the earth

Consider a particle located near the surface of the earth and interacting only with the earth. As discussed in Sec. 5G and indicated in Fig. A-1, such a particle is accelerated downward (i.e., toward the center of the earth). The existence of this acceleration clearly indicates that the particle interacts with the earth. This interaction is called *gravitational*. It is a long-range interaction since it is manifest even if the particle is at an appreciable distance from the surface of the earth.

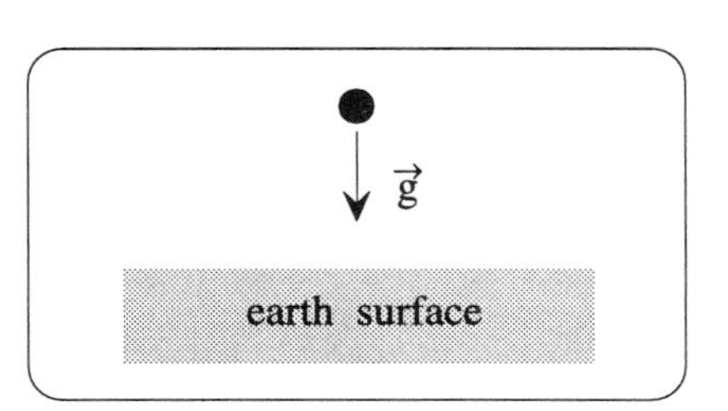

Fig. A-1. Gravitational acceleration due to the earth.

Properties of the gravitational acceleration. As discussed in Sec. 5G, it is remarkable that this gravitational acceleration is completely independent of the properties of the particle (e.g., independent of its size or the material of which it is made). Hence one can talk about the gravitational acceleration $\vec{g}$ at a given point (i.e., about the gravitational acceleration experienced by *any* particle located at this point).

The magnitude g of the gravitational acceleration is approximately 9.80 m/s^2 near the surface of the earth. It is nearly constant within any region of size much smaller than the radius of the earth, but becomes gradually less at larger heights (i.e., at points farther from the center of the earth).

Properties of the gravitational force. The preceding knowledge about the gravitational interaction allows one immediately to infer the corresponding properties of the gravitational force $\vec{F}_g$ exerted on a particle by the earth. To do this, one needs merely to apply to the particle Newton's law

$$m\vec{a} = \vec{F}_{tot} \,. \qquad \text{(A-1)}$$

If a particle interacts solely with the earth, its acceleration is $\vec{g}$. Furthermore, the total force $\vec{F}_{tot}$ on it is merely the gravitational force $\vec{F}_g$ by the earth. Newton's law (A-1) then implies that

$$m\,\vec{g} = \vec{F}_g \,.$$

Hence the gravitational force exerted by the earth on a particle of mass m is

Gravitational force: $\vec{F}_g = m\vec{g}\,.$	(A-2)

This gravitational force has the following properties: The *direction* of the force (like that of the gravitational acceleration) is downward, i.e., toward the center of the earth. The *magnitude* of this force at a particular location is simply proportional to the mass of the particle, but independent of any of its other properties (e.g., independent of its size, shape, or material).

Weight. The magnitude of the gravitational force on an object is called the *weight* of the object according to the following definition:

Def:	***Weight:*** The weight of an object is the magnitude of the gravitational force on it.	(A-3)

Hence (A-2) implies that the weight w of an object near the surface of the earth is simply

$$w = mg \,. \qquad \text{(A-4)}$$

The weight of an object depends thus both on its mass and on the magnitude of the gravitational acceleration at its location.

➜ ***Go to Sec. 10A of the Workbook.***

Contact interactions

B. Contact interactions and interatomic forces

Atoms (or molecules) interact appreciably with nearby atoms, although not with more distant ones. For example, the atoms in a solid are close to each other and thus exert strong forces on neighboring atoms. These forces maintain the atoms in their normal positions and prevent them from flying apart, i.e., prevent the solid from disintegrating.

Suppose that two objects are brought so close to each other that their separation becomes comparable to interatomic distances (about 10^{-10} m). From our large-scale point of view we would then say that the objects "touch" each other. The atoms near the touching surfaces then exert appreciable forces on each other. The result is an observed contact interaction between the objects (i.e., a contact force is exerted on each object by the other).

Example

As a simple example, consider a pot with an attached handle. Fig. B-1 illustrates such a pot maintained at rest by someone holding the handle. A force is then exerted on the pot by the handle. (If there were no such force, the pot would fall because of the gravitational force on it.)

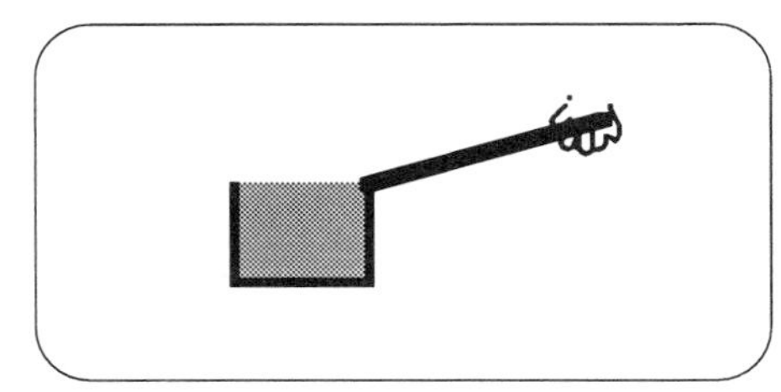

Fig. B-1. Pot held by an attached handle.

This is clearly a contact force, i.e., it is due to the interacting atoms adjacent to the touching surfaces. (Indeed, suppose that the handle were slighly separated from the pot so that there would be a small gap between them, as indicated in Fig. B-2. Then the handle would no longer exert a force on the pot and the latter would fall.)

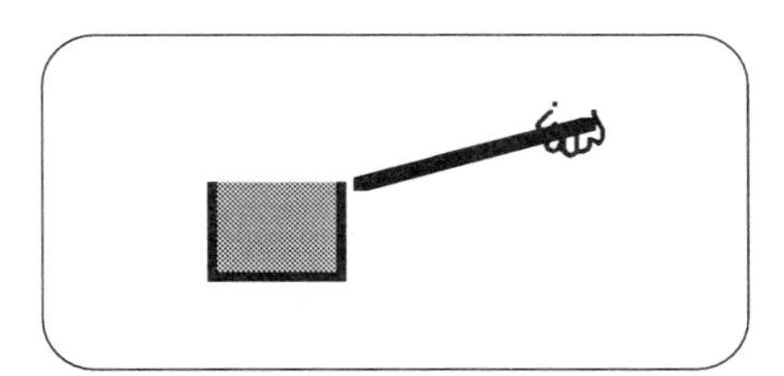

Fig. B-2. Handle slightly separated from the pot.

All the atoms in the handle, except those adjacent to the touching surfaces, have *no* direct effect on the interaction between the handle and the pot. However, they affect it indirectly. (For example, the person's hand exerts forces on those handle atoms immediately adjacent to the hand; these atoms, in turn, exert forces on their neighboring atoms; these, in turn, exert forces on their neighboring atoms; and so forth until forces are finally exerted on the atoms at the surface between the handle and the pot.) This is why the properties of the handle can affect the contact force which it exerts on the pot.

Note that the contact force exerted on the pot by the handle is *not* necessarily directed along the handle. For example, since the pot in Fig. B-1 is at rest, the total force on it must be zero. Hence the force exerted on the pot by the handle must be directed vertically upward (as indicated in Fig. B-3) so as to cancel the downward gravitational force on the pot.

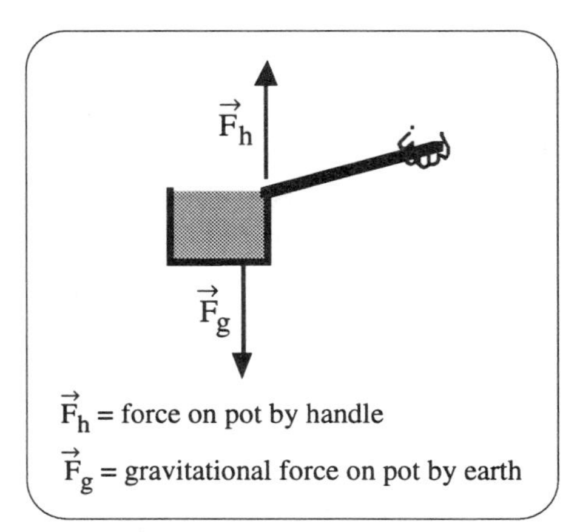

Fig. B-3. Forces on the pot.

The following sections examine the properties of some commonly occurring contact forces due to springs, due to strings, and due to non-adhering touching objects.

➜ ***Go to Sec. 10B of the Workbook.***

C. Interaction with a spring

Interactions due to deformation. When an object is deformed by changing its size or shape, the distances between neighboring atoms in the object are changed. The interactions between atoms are then correspondingly also changed and give rise to observable interactions.

A spring is a simple example of an object which can be readily deformed by increasing or decreasing its length (i.e., by stretching or compressing the spring). Consider a particle attached to the end of such a spring, as indicated in Fig. C-1. Whenever the spring is deformed from its normal state, the particle is accelerated. Thus the particle interacts with the spring. (This is clearly a contact interaction since the particle would be unaffected if it did not touch the spring.)

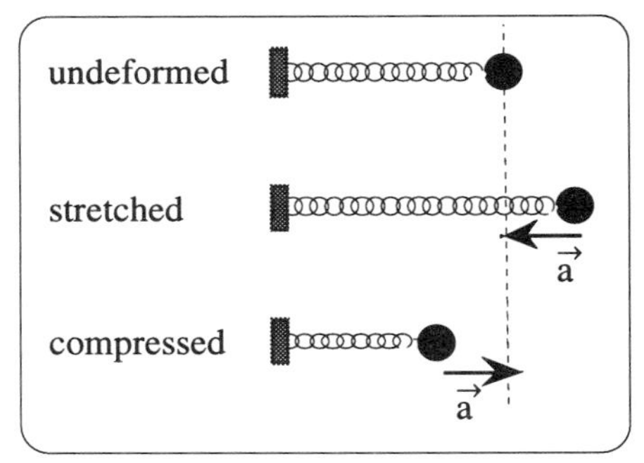

Fig. C-1. Interaction of a particle with a spring.

Fig. C-1 illustrates the interaction in greater detail. If the spring is stretched, the particle's acceleration is to the left. If it is compressed, the particle's acceleration is to the right. Thus the interaction can be attractive or repulsive, always opposing the deformation of the spring and tending to restore the spring to its undeformed state.

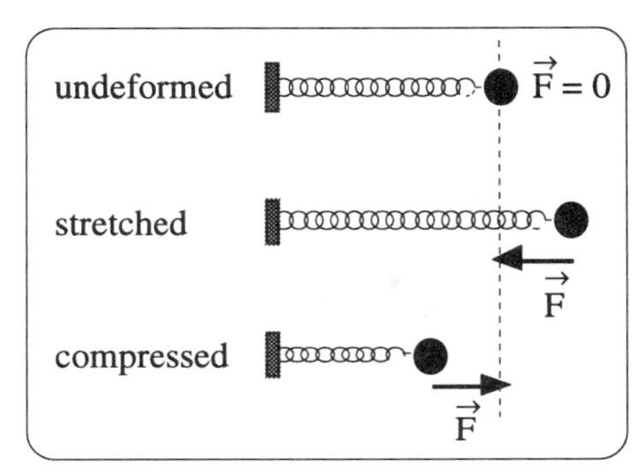

Fig. C-2. Force on a particle by a spring.

Properties of the force exerted by a spring. These properties of the interaction imply the following properties of the force on a particle by a spring:

> ***Force on a particle by a spring:***
> *Direction:* Along the spring, opposing its deformation.
> *Magnitude:* Increasing with deformation, zero if no deformation.

(C-1)

Thus the force has opposite directions, as indicated in Fig. C-2, depending on whether the spring is stretched or compressed. Furthermore, the magnitude of the force is larger if the magnitude of the spring's deformation is larger (i.e., either if the spring is stretched more or if it is compressed more).

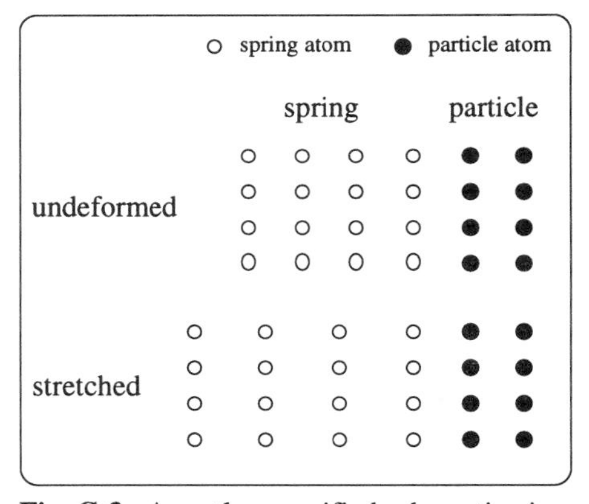

Fig. C-3. A vastly magnified schematic view of a few layers of atoms near the surface of contact between a particle and a spring.

Atomic origins of the force. When the spring is in its normal undeformed state, an atom in the spring experiences no net force due to neighboring atoms. But, if the spring is deformed, the separation between neighboring atoms is changed, as schematically illustrated in Fig. C-3. The forces between neighboring atoms then try to restore the atoms to their normal positions. For example, if the spring is stretched, the distance between neighboring atoms is increased. To restore the atoms to their normal positions, the forces between two neighboring atoms are then attractive, as indicated in Fig. C-4b. Conversely, if the spring is compressed, the distance between neighboring atoms is decreased. To restore the atoms to their normal positions, the forces between two neighboring atoms are then repulsive, as indicated in Fig. C-4c.

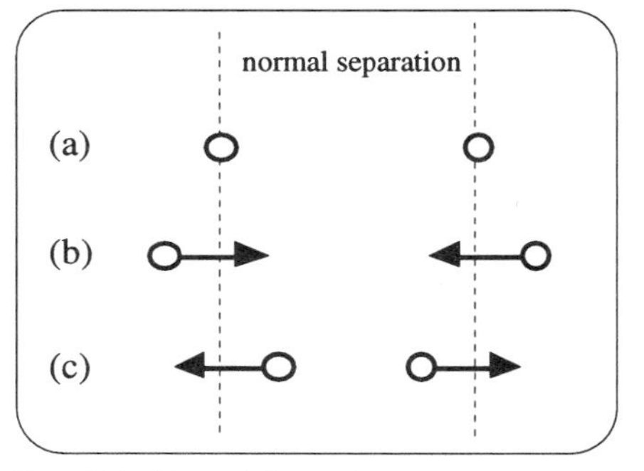

Fig. C-4. Mutual forces between neighboring atoms in a spring.

Each layer of atoms in the spring thus exerts on a neighboring layer a force which tends to restore the normal interatomic separation. Ultimately, these forces affect the layer of atoms adjacent to the particle attached to the spring and produce the observed force on this particle.

➜ ***Go to Sec. 10C of the Workbook.***

D. Interaction with a string

Force by a rubber band. Consider a particle attached to a rubber band. Such a rubber band acts somewhat like a spring. If it is stretched, it exerts on the particle a force tending to oppose the elongation of the rubber band (i.e., a force along the direction indicated in Fig. D-1).

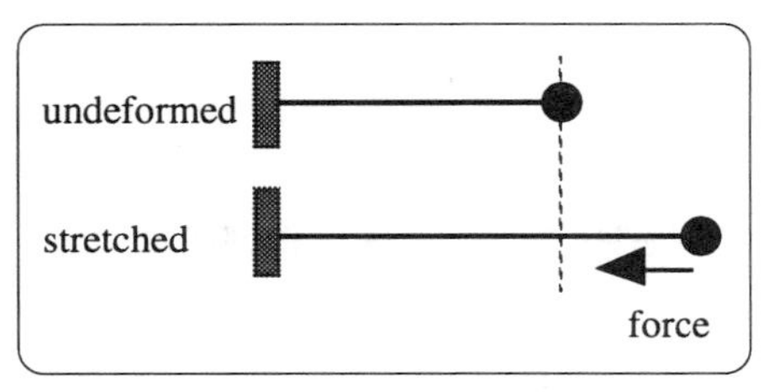

Fig. D-1. Particle attached to a rubber band.

However, unlike a spring, a rubber band lacks rigidity. It does *not* oppose compression, but merely buckles instead of remaining taut. Thus the rubber band exerts *no* force if it is not taut (i.e., if it is slack).

Force by a string. A string behaves just like a rubber band, opposing elongation but not compression The difference is that even a large force may cause only a negligibly small elongation of a string. (A sufficiently large force may, however, cause the string to break.) Hence it is often useful to approximate the situation by considering the length of a string as constant, irrespective of the magnitude of the force.

The preceding comments indicate that the forces between neighboring atoms in a string oppose the elongation of the string, but not its compression. The resultant force exerted on a particle by a string is thus directed along the string and *toward* it, i.e., it is an *attractive* force as indicated in Fig. D-2. (In other words, strings pull, but don't push.) A force having such a direction is called a *tension force* and is commonly denoted by the symbol $\vec{T}$.

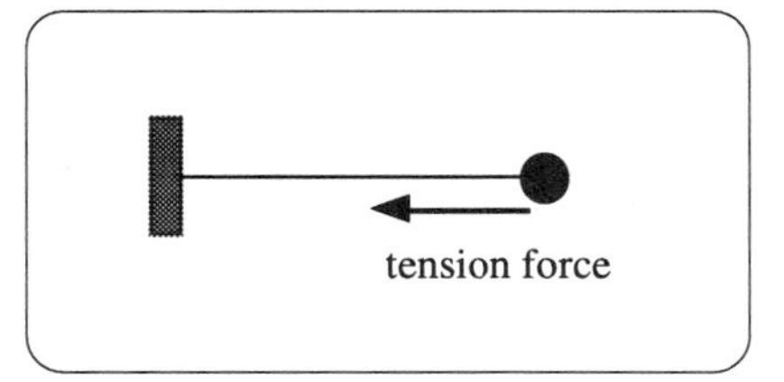

Fig. D-2. Force on a particle by a string.

The properties of the force exerted by a string can thus be summarized as follows:

Force on a particle by a string:		
Direction:	Attractive (opposing elongation of the string).	(D-1)
Magnitude:	Non-zero if the string is taut, zero if it is slack.	

Relations between forces. The interaction between the string and the attached particle can be described by the pair of mutual forces indicated in Fig. D-3 (i.e., the force on the particle by the string and the force on the string by the particle). As usual, these forces have the same magnitude and opposite directions.

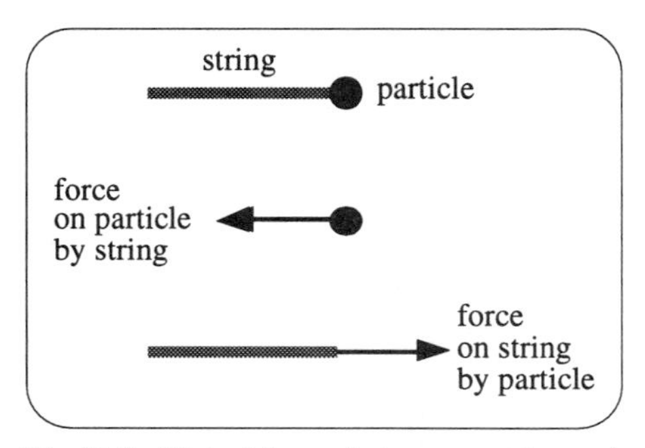

Fig. D-3. Mutual forces between a string and an attached particle.

Fig. D-4 indicates schematically the forces between neighboring atoms in a taut string. Each atom is attracted by its neighbors, thus opposing the elongation of the string.

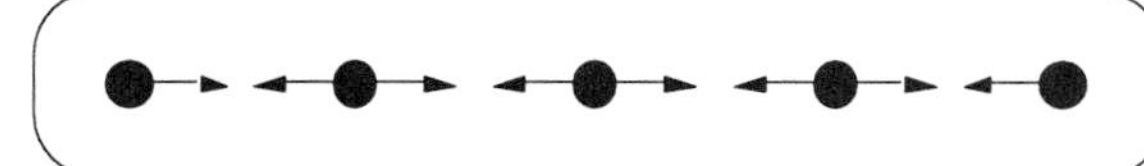

Fig. D-4. Schematic view, vastly magnified, of forces between neighboring atoms in a string.

➜ ***Go to Sec. 10D of the Workbook.***

E. Strings of negligible mass

The mass of a string is usually negligible compared to that of any other object with which it interacts. To good approximation, the mass of such a string can then be considered to be zero. The forces exerted by such a string have some specially simple properties.

Forces on a straight piece of string. Consider a straight piece of string of negligibly small mass. Suppose that two forces, of magnitudes F_1 and F_2, pull on the two ends of this piece of string, as indicated in Fig. E-1. What then is the relation between the magnitudes of these forces?

Application of Newton's law to this piece of string implies that

$$m\vec{a} = \vec{F}_{tot} \qquad \text{(E-1)}$$

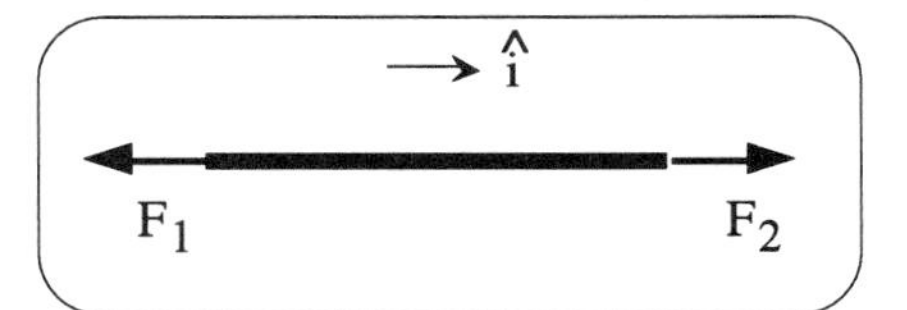

Fig. E-1. Forces on a piece of string.

where $\vec{a}$ is the acceleration of the string and $\vec{F}_{tot}$ is the total force on it. If the mass m of the string is negligible, m = 0 to excellent approximation. The left side of the equation (E-1) is then zero, even if the string is accelerating. Hence the total force on the string is negligible, i.e.,

$$\vec{F}_{tot} = 0 \, . \qquad \text{(E-2)}$$

Since the mass m of the string is negligible, the gravitational force on the string is also negligible. Hence the only forces acting on the string are the two oppositely directed forces acting on its ends. The relation (E-2), applied along the $\hat{i}$ direction parallel to the string, then implies that

$$F_2 - F_1 = 0$$

or

$$F_2 = F_1 \, . \qquad \text{(E-3)}$$

Note that this relation would also be true even if some other forces, *not* parallel to the string, were exerted on the string by some other touching objects.

Thus we arrive at the following conclusion:

If a piece of string has negligibly small mass, the magnitudes of the forces acting on its ends are equal (provided that no other forces parallel to the string act on it).	(E-4)

Forces at various points of a string. A string consists really of many small interacting pieces of string joined together (e.g., the pieces A, B, C, ... illustrated at the top of Fig. E-2). Focus attention on any small piece A of such a taut string of negligible mass, as indicated at the bottom of Fig. E-2. Suppose that a force of magnitude T is exerted on A by the string to the left of it. According to (E-4), a force of equal magnitude T must then be exerted on A by the adjacent string piece B to the right of it. But a mutual force to the left, of the same magnitude T, must then be exerted on B by A. According to (E-4), a force of equal magnitude T must then be exerted on B by the adjacent string piece C to

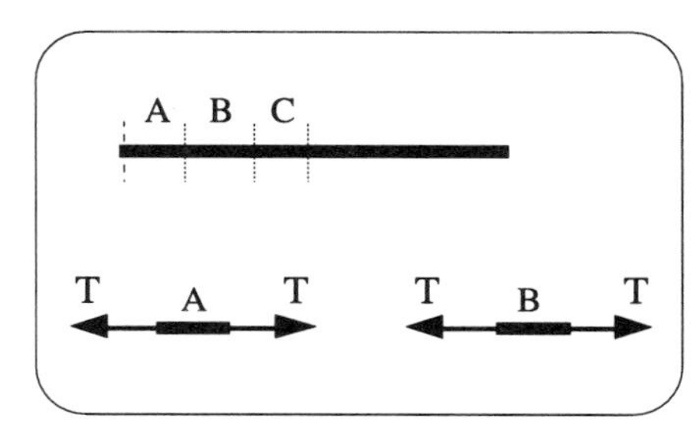

Fig. E-2. Adjacent small pieces of a string, and tension forces on them.

the right of it. By continuing this argument repeatedly for all adjacent pieces of the string, we arrive at the following conclusion:

> If a string has negligible mass, the magnitude of the tension force, exerted on any piece of string by an adjacent one, is the *same* at every point of the string (provided that no other forces parallel to the string act on it). (E-5)

Curved taut string. The conclusion (E-5) is also valid for a *curved* taut string (e.g., for a string wrapped around a post or pulley that exerts only forces perpendicular to the string). The reason is that every adjacent small string piece, if sufficiently small, is very nearly straight. Thus the preceding argument still holds (i.e., the tension forces acting on the ends of any such small string piece have equal magnitudes).

Example: Pulley and string with negligible masses

Consider the common situation, illustrated in Fig. E-3, where a string passes over a freely rotating pulley which has negligibly small mass. A negligibly small force, tangent to the pulley, is then sufficient to set the pulley rotating. The mutual force (parallel to the string), exerted on the string by the touching pulley, is then also negligibly small. Hence (E-5) implies that the magnitude of the tension force is the same at every point of this string.

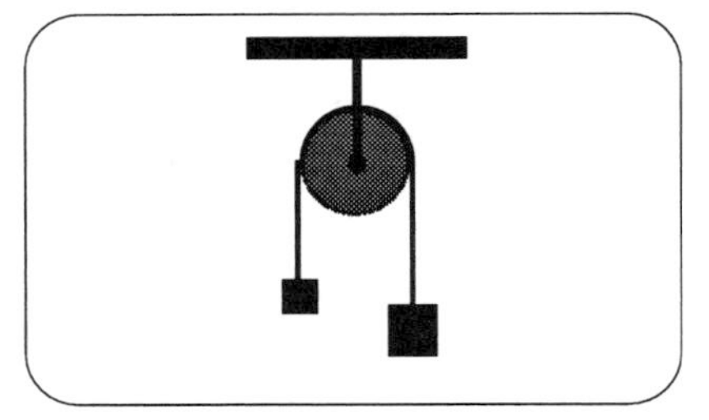

Fig. E-3. String passing over a pulley of negligible mass.

➜ ***Go to Sec. 10E of the Workbook.***

F. Interaction between non-adhering touching objects

Consider two solid objects which touch each other without adhering (i.e., which are not sticky or covered with glue). For example, the objects might be a box lying on a table, or a box sliding down along a chute. Because of interactions between the atoms near the surface of contact, a contact force is then exerted on one object by the other. (For example, Fig. F-1 illustrates such a contact force on a box sliding down along a chute.)

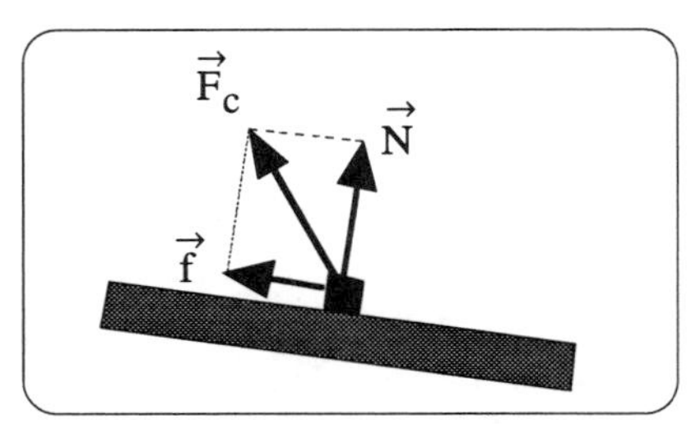

Fig. F-1. Contact force exerted on an object by another touching object (e.g., on a box by a chute).

The contact force $\vec{F}_c$ is ordinarily neither perpendicular nor parallel to the surface of contact. However, as indicated in Fig. F-1, it can always be expressed in terms of its component forces perpendicular and parallel to this surface. Thus one can write

$$\boxed{\vec{F}_c = \vec{N} + \vec{f}} \qquad \text{(F-1)}$$

where the component force $\vec{N}$ perpendicular to the surface is called the *normal force* and the component force $\vec{f}$ parallel to the surface is called the *friction force*. The simplifying advantage is that the properties of these two component forces can then be examined separately.

The word *normal* is here used with its meaning of *perpendicular.* (It is *not* the opposite of abnormal.)

Normal force

The atoms near the surface of contact of the touching objects interact so as to oppose being pushed closer together, i.e., so as to oppose compression of the objects. However, the atomic interactions do *not* oppose separation of the touching objects (since the contact surface is not sticky).

The touching surfaces may, however, stick together if they are extremely smooth and clean so that the atoms near them really approach each other very closely.

The normal force on an object by another touching object is thus directed *away* from this other object so as to oppose its compression, i.e., the force is *repulsive* as indicated in Fig. F-2. Furthermore, this force is clearly zero if the objects do not touch each other. The properties of the normal force can thus be summarized as follows:

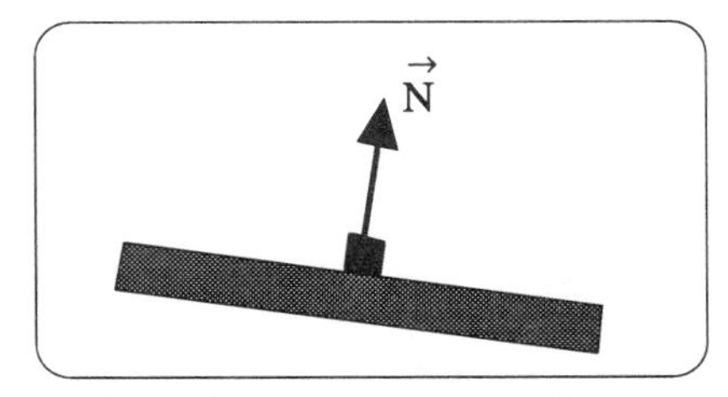

Fig. F-2. Normal force on an object by a another touching object (e.g., on a box by a chute).

Normal force		
Direction:	Perpendicular to contact surface, repulsive (opposing compression).	(F-2)
Magnitude:	Non-zero if objects touch, zero otherwise.	

As usual, the mutual normal forces exerted on each object by the other have the same magnitude, but opposite directions (as illustrated in Fig. F-3).

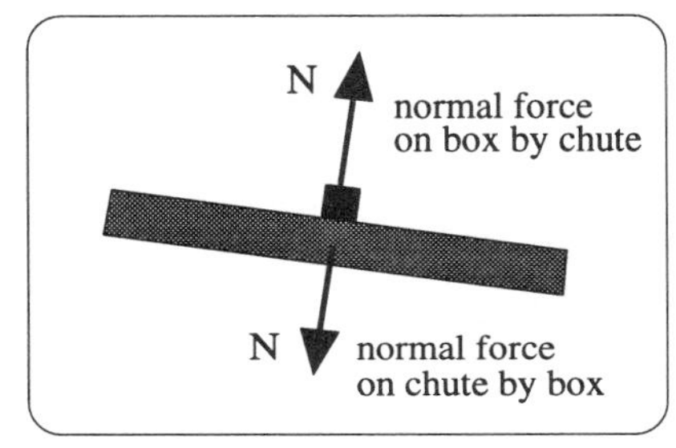

Fig. F-3. Mutual normal forces.

Magnitude of deformation. The deformation responsible for the normal force is readily apparent in some cases. For example, one can see that a rubber ball is compressed when it bounces against a floor. Similarly, Fig. F-4 illustrates a mattress deformed when a heavy object is placed on it. The mattress, opposing its compression, then exerts an upward normal force on the object.

However, in many cases the interacting objects are not so readily deformed. For example, if the heavy object of Fig. F-4 is placed on a thick slab of wood, like a table top, the deformation of the latter is so small as to be scarcely noticeable. To good approximation one can thus often consider objects to be so *rigid* that their deformation is negligible even if the forces exerted on them, or by them, are quite large.

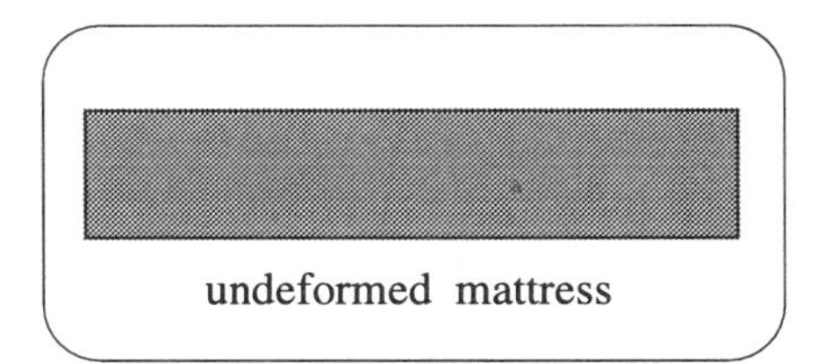

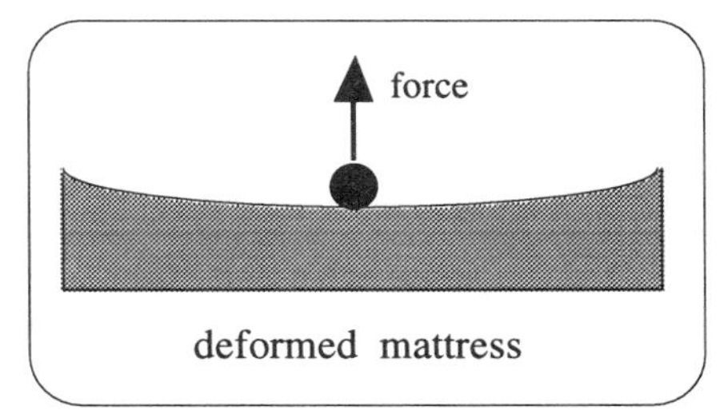

Fig. F-4. Force exerted on an object by a foam-rubber mattress.

Friction force

The friction force opposes sliding of the touching objects relative to each other. For example, Fig. F-5 indicates the friction force opposing the downward sliding motion of the box relative to the chute.

The friction force is particularly large if the surface of contact between the objects is rough. (When viewed through a microscope, the surfaces of touching objects look then somewhat like sheets of sandpaper whose bumps catch each other and thus resist sliding past each other.).

The direction of the friction force can be identified by the following method: (1) Determine how the objects would move relative to each other if there were no friction. (2) Then determine what direction the friction force must have so as to oppose this relative motion.

More specifically, it is useful to distinguish the following two cases.

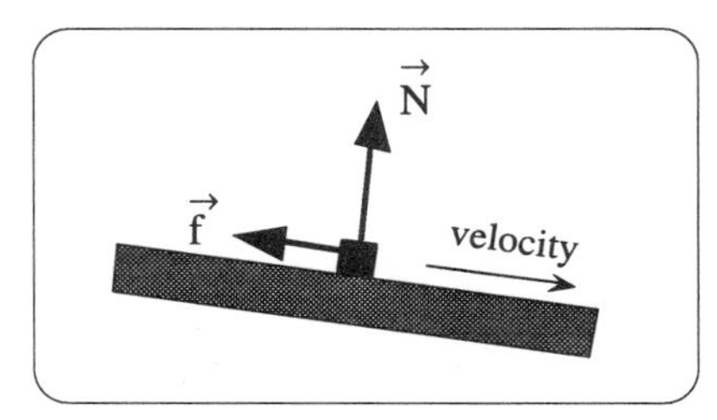

Fig. F-5. Normal and friction forces on a box by a chute.

* ***Kinetic case.*** (Objects *moving* relative to each other.) To oppose the relative motion, the friction force must then reduce the relative velocity of the objects. Accordingly, the friction force on an object is opposite to its velocity relative to the touching object.
* ***Static case.*** (Objects *at rest* relative to each other.) To oppose any relative motion, the friction force must then ensure that the relative velocity of the objects remains zero. Accordingly, the friction force on an object is such that the object's acceleration is zero relative to the touching object.

The properties of the friction force can thus be summarized as follows:

> ***Friction force***
> Parallel to surface, opposes *relative* sliding.
> If relative velocity $\neq 0$:
> Direction of friction force is opposite to relative velocity.
> If relative velocity $= 0$:
> Friction force is such that relative acceleration $= 0$.

(F-3)

The *magnitude* of the friction force will be examined more precisely in Chapter 12.

Example of kinetic friction

Consider a box sliding down along a chute. The friction force exerted on the box by the chute is then directed opposite to the velocity of the box relative to the chute (i.e., it is directed up along the chute, as indicated in Fig. F-5, so as to oppose the motion of the box sliding down along the chute.

Example of static friction

Consider a box resting on the horizontal floor, as illustrated in Fig. F-6a. The only horizontal force on the box is then the friction force exerted on it by the floor. This friction force must then be zero to ensure that the box is not accelerated, but remains at rest relative to the floor.

On the other hand, suppose that a person pulls on the box with a horizontal force $\vec{F}_0$ (as illustrated in Fig. F-6b), but does not pull hard enough to move the box. Then the friction force exerted on the box by the floor must ensure that the acceleration of the box relative to the floor remains zero. Accordingly, this friction force must have the same magnitude, but the opposite direction, as the applied force $\vec{F}_0$ (as illustrated in Fig. F-6b).

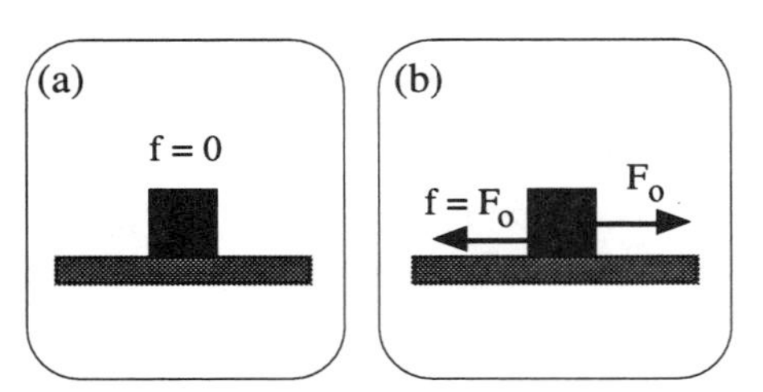

Fig. F-6. Friction force on a box at rest on the floor.

➜ ***Go to Sec. 10F of the Workbook.***

G. Summary

Long-range forces

Gravity near the earth: $\vec{F}_g = m\vec{g}$

Contact forces

Force by spring:

Direction: Along spring, opposing deformation.
Magnitude: Increasing with deformation, 0 if no deformation.

Force by string:

Direction: Attractive (opposing elongation).
Magnitude: $\neq 0$ if string is taut, 0 otherwise.

Force by non-adhering object: $\vec{F} = \vec{N} + \vec{f}$

Normal force $\vec{N}$:

Direction: $\perp$ to surface, repulsive (opposing compression).
Magnitude: $\neq 0$ if touching, $= 0$ otherwise.

Friction force $\vec{f}$: Parallel to surface, opposing relative sliding.

Relative motion: $\vec{f}$ is directed opposite to relative velocity.
Relative rest: $\vec{f}$ is such that relative acceleration = 0.

New abilities

You should now be able to determine the forces on a particle in any situation where it is interacting with other objects by any of the long-range or contact interactions discussed in this chapter.

➜ ***Go to Sec. 10G of the Workbook.***

11 Problem Solving in Mechanics

Newton's law provides the ability to predict or explain the motion of particles in a very wide range of situations. To do this, one needs to combine Newton's law with *all* of the knowledge acquired in the preceding chapters (i.e., knowledge about the description of motion and about the properties of various interactions). To deal with the resulting complexities, one also needs systematic methods for applying Newton's law and for solving diverse problems with it.

The present chapter discusses such methods and provides practice in applying them to solve various problems. By learning these methods and applying them consistently, you will achieve significant abilities to use physics knowledge for practical purposes. You will then also be well-prepared to deal with all the mechanics problems encountered in the rest of the book and beyond.

A. Systematic application of Newton's law

In order to apply Newton's law $m\vec{a} = \vec{F}_{tot}$, one must be able to express it correctly and conveniently in any particular situation. Specifically, application of Newton's law requires that one do the following:

(1) One must first describe in useful ways all the ingredients needed to apply Newton's law (i.e., all the information about a particle's mass, about its motion, and about all its interactions). This provides the requisite information about the particle's acceleration and the total force on it.

(2) One must then use this description to express Newton's laws in a readily usable form.

This section discusses useful methods for accomplishing these two important tasks.

System description by a system diagram

System. Any law is used by applying it to a particular *system* of interest.

Def: | ***System:*** Any object, or set of objects, which one wishes to consider. | (A-1)

In the case of Newton's law, which deals with particles, the system of interest must be a *particle* (i.e., either a simple or composite particle).

System diagram. A system can most usefully be described by drawing a diagram which makes all the relevant information about the system apparent in readily visualized form. (For example, the magnitudes and directions of vectors can more easily be described by arrows than by mere words.)

One needs a *separate* such diagram for every system of interest. Such a diagram is called a *system diagram* (or a *free-body diagram*). Such a separate diagram for a *single* system makes it easy to indicate all the forces exerted *on* this system.

A diagram displaying several systems together would be highly confusing because it makes it difficult to distinguish forces acting *on* a system from forces exerted *by* a system.

It is very important that the information in a system diagram is complete and correct. (Otherwise, all the equations and problem solutions based on it will be wrong.) Hence one needs a reliably effective method for drawing a system diagram.

Method for drawing a system diagram. An initial diagram, portraying the situation of interest, is essential before one can do anything else. To describe *any particular system* in this situation, one can then draw a system diagram by using the following method (also summarized in Fig. A-1):

Constructing a system diagram

(1) ***System.*** Indicate the system separately by itself.
(2) ***Mass.*** Indicate the system's mass.
(3) ***Motion.*** Draw arrows indicating the system's *velocity* and *acceleration* (including information about their magnitudes).
(4) ***All forces on the system.***
- ***Long-range.***
  - ***Interacting objects.*** Identify all the objects interacting with the system at long range (e.g., the earth interacting gravitationally).
  - ***Forces on system.*** Draw arrows indicating the corresponding forces *on* the system (including information about their magnitudes).
- ***Contact.***
  - ***Touching objects.*** Identify all the objects touching the system. (Mark and label contact points on the situation diagram.)
  - ***Forces on system.*** Draw arrows indicating the corresponding forces *on* the system (including information about their magnitudes).

(5) ***Components.*** Decompose oddly-directed vectors into components along convenient perpendicular directions.

(A-2)

System diagram

* **Separate system**
* **Mass**
* **Motion** (velocity, acceleration)
* **All forces *on* the system**
  * ***Long-range*** (e.g., gravity)
    * Interacting objects?
    * Forces *on* the system
  * ***Contact***
    * Touching objects? (Mark & label contacts.)
    * Forces *on* the system
* **Components**

Fig. A-1. Description of a system by a system diagram.

Example: Sled sliding down along a hill

Fig. A-2 shows a diagram illustrating a sled, of mass M, sliding down along a straight hill inclined at an angle θ from the horizontal. Fig. A-3 shows a system diagram describing the sled. The suggestions mentioned below were followed in constructing this diagram.

Suggestions for drawing system diagrams

The following suggestions are helpful to implement the preceding method for drawing system diagrams.

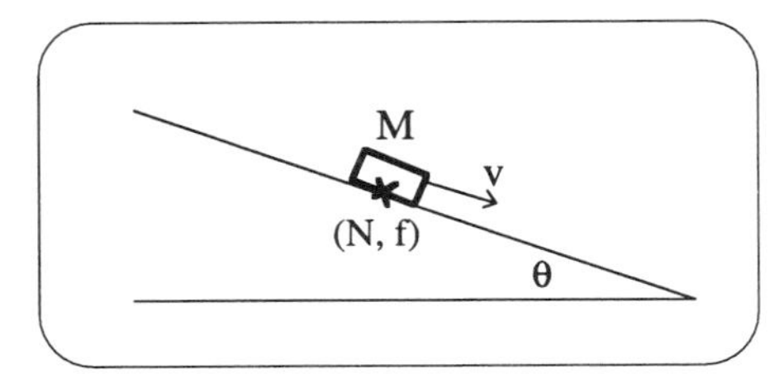

Fig. A-2. Situation diagram showing a sled sliding down along a hill.

Arrows indicating vectors. Indicate each vector by an arrow labeled by the magnitude of the vector (or by its component along this direction).

Distinguish vectors denoting motion quantities (like velocity and acceleration) from vectors denoting forces. As illustrated in Fig. A-3, this may be done by denoting motion quantities with open arrowheads and forces with solid arrowheads.

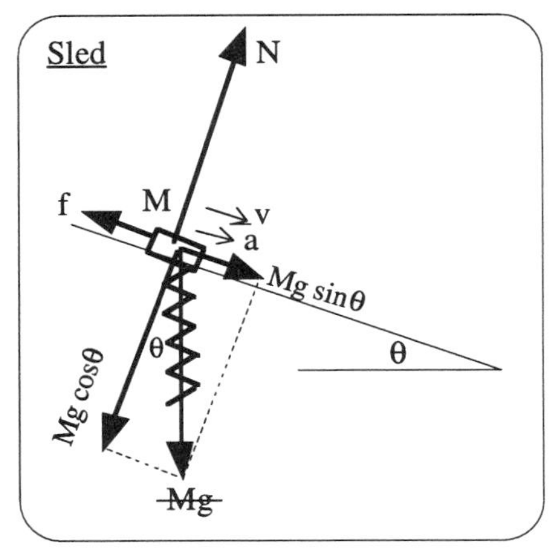

Fig. A-3. System diagram for the sled in Fig. A-2.

Since the system itself is supposed to be a particle, its size in the system diagram is irrelevant (e.g., it may be indicated by a point). It is then ordinarily most convenient to draw all arrows so that they start from the system, as indicated in Fig. A-3. (However, it may sometimes be useful to indicate an *attractive* force by an arrow starting from the system and pointing toward the attracting object, and to indicate a *repulsive* force by an arrow ending on the system and pointing away from the repelling object.)

Identifying contact forces. The following helps to identify reliably *all* contact interactions. (a) Examine the *situation diagram* and use a cross to mark on it every place where the system touches some other object. (b) Label this cross by the magnitude of the force (or the magnitudes of the component forces) describing the corresponding contact interaction. (c) Use this magnitude to label the corresponding force in the system diagram. (This makes it clear what particular interaction is described by this force.)

For example, in Fig. A-2 the contact interaction between the sled and the hill is indicated by the cross labeled by N and f, the magnitudes of the normal and friction forces. The corresponding forces *on* the sled (magnitudes and directions) are then shown indicated by the labeled arrows in the system diagram of Fig. A-3.

Convenient vector components. Decompose oddly-directed vectors into component vectors along directions parallel or perpendicular to the majority of vectors specifying forces or acceleration. (In this way one can attain a simple situation where *all* vectors are mutually perpendicular.)

To avoid confusion, cross out the original arrow (and associated label) denoting a vector which has been replaced by its component vectors. (For example, Fig. A-3 illustrates how the gravitational force on the sled has been replaced by its component forces parallel and perpendicular to the hill.)

Checking a system diagram. Check whether a system diagram is correct by verifying that the direction of the acceleration is consistent with that of the total force (as required by Newton's law).

For example, if the component of the acceleration along any direction is zero, the component of the total force along this direction must also be zero.

Similarly, if the component of the acceleration along any direction is positive, the component of the total force along this direction must also be positive.

Related system diagrams

A system diagram is always drawn for a *single separate* system. But there may be relations between the motions or interactions of various systems in a given situation. How can such relations among systems be indicated in their separate system diagrams?

Suppose that the motions of two systems are related. The directions of the velocities and accelerations of these systems should then be indicated in their system diagrams by arrows with the correct corresponding directions. Furthermore, if the magnitudes of velocities or accelerations are simply related, they should be expressed in terms of the same symbol.

Similarly, suppose that two systems are related by a mutual interaction. Then the corresponding mutual forces on the two systems must have opposite directions; these should be clearly indicated by the corresponding force arrows in the two system diagrams. Furthermore, these mutual forces have the same magnitude; this common magnitude should be indicated by the same symbol in both system diagrams.

Remember the previous suggestion that a contact interaction between two systems should be explicitly indicated by a labeled cross in the situation diagram. This greatly helps to identify the related mutual forces appearing in the separate system diagrams of these systems.

Example: Pulling a sled carrying a box

A sled, of mass M, is pulled along the horizontal snow-covered ground by means of a rope which applies a force of magnitude F_o at an angle θ from the horizontal. A box, of mass m, lies on the sled and remains at rest relative to it. Fig. A-4 illustrates the situation.

The labeled crosses in Fig. A-4 indicate the three contact interactions of the sled, i.e., its interaction with the touching rope (labeled by the magnitude F_o of the corresponding force), its interaction with the box (labeled by the magnitudes N and f of the corresponding normal and friction forces), and its interaction with the ground (labeled by the magnitudes N' and f' of the corresponding normal and friction forces).

Fig. A-5 shows the system diagram for the box and Fig. A-6 shows the system diagram for the sled. The box does not move relative to the sled. Hence the system diagrams indicate that the velocities (relative to the ground) of the box and of the sled are the same, and that their accelerations are also the same. The system diagrams also properly indicate that the mutual normal and friction forces between the sled and the box have opposite directions, but the same magnitude. (In Fig. A-6, the oddly directed force on the sled by the rope has also been decomposed into its component forces parallel and perpendicular to the ground.)

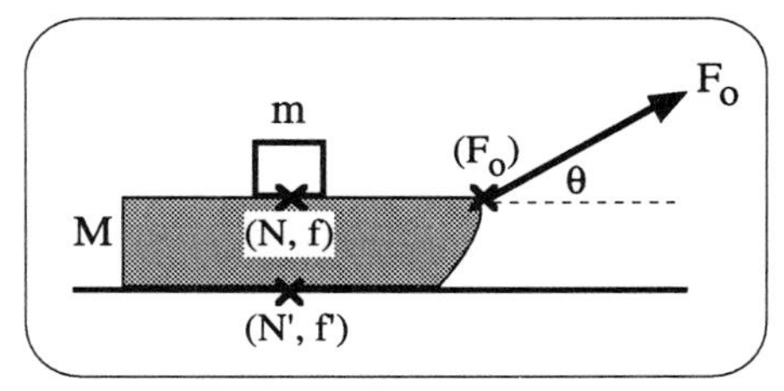

Fig. A-4. Situation diagram illustrating a box lying on a sled pulled by a rope.

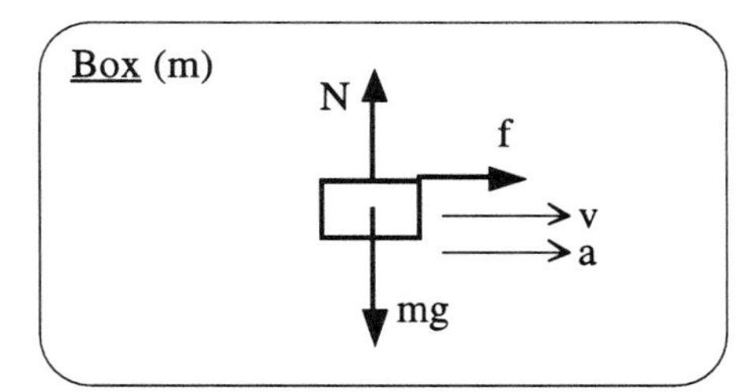

Fig. A-5. System diagram for the box.

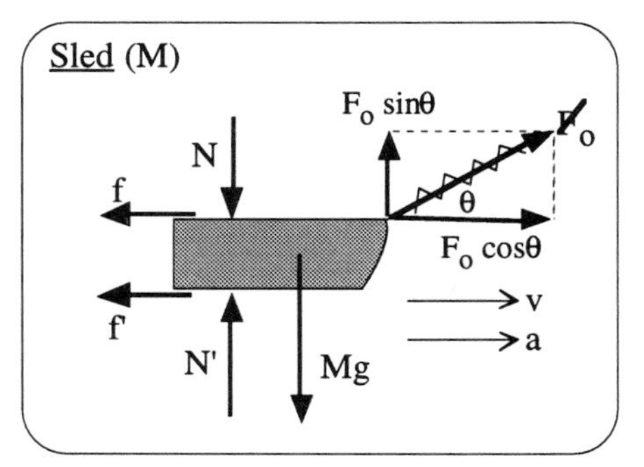

Fig. A-6. System diagram for the sled.

Applying Newton's law in component form

Newton's law

$$m\vec{a} = \vec{F}_{tot} \quad \text{(A-3)}$$

is a relation between two vectors, the acceleration $\vec{a}$ of a particle and the total force $\vec{F}_{tot}$ on it. The law implies, therefore, a corresponding relation between the components of these vectors along any direction. When the acceleration and all forces lie in a plane, Newton's law is then equivalent to two numerical relations for components along any two perpendicular (or other non-collinear) directions.

It is usually much easier to work with numbers than with vectors. To apply Newton's law, it is thus ordinarily best to express it in terms of its components along convenient directions. This can be easily done after one has constructed a system diagram for the system of interest. For then all the needed information, as well as the convenient directions, have already been identified and can be read off this diagram. One needs then merely to consider the components of the acceleration and of the individual forces along any such direction, and use Newton's law (A-3) to write the corresponding relation

$$\left.\begin{array}{l}\text{(mass)} \times \text{(acceleration component along this direction)} \\ = \text{(sum of all force components along this direction).}\end{array}\right] \quad \text{(A-4)}$$

Example: Applying Newton's law to a descending sled

Consider again Fig. A-2 illustrating a sled sliding down along a hill. The corresponding system diagram for the sled is shown in Fig. A-3. Newton's law, applied in the direction down along the hill, then yields the relation

$$\mathrm{Ma} = \mathrm{Mg}\,\sin\theta - \mathrm{f}\,. \quad \text{(A-5)}$$

Newton's law, applied along the direction along the normal force perpendicular to the hill, then yields

$$0 = \mathrm{N} - \mathrm{Mg}\,\cos\theta \quad \text{(A-6)}$$

since the sled's acceleration has no component perpendicular to the hill.

The preceding relations lead to the following conclusions. In (A-5), a denotes the positive magnitude of the sled's acceleration down along the hill. Hence the magnitude f of the friction force on the sled must be smaller than Mg $\sin\theta$ (where Mg is the weight of the sled). Similarly, (A-6) implies that the normal force exerted on the sled by the hill must be equal to Mg $\cos\theta$.

➜ ***Go to Sec. 11A of the Workbook.***

B. Problem-solving method

Now that we know how to apply Newton's law to any system, we are well prepared to solve a great variety of mechanics problems. However, a haphazard approach can easily lead to mistakes or cause one to get stuck. Thus one needs to use a systematic problem-solving method.

Major steps of the problem-solving method. As discussed in Sec. 6A, systematic problem solving involves the major steps outlined in Fig. B-1. In other words, it is always necessary to analyze the problem initially, then to

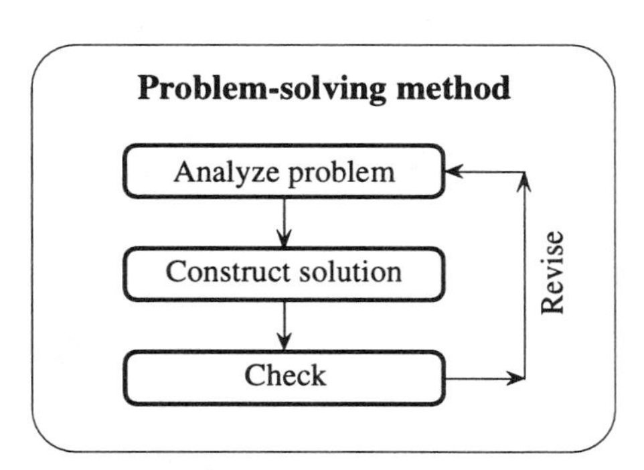

Fig. B-1. Major steps of the problem-solving method.

construct its solution, and finally to check and revise this solution. These steps are the same as the ones used previously to deal with the motion problems of Chapter 6. The following paragraphs, therefore, merely review these steps and indicate in what respects they need to be modified to deal with more complex mechanics problems.

Review and modification of the method

Initial analysis of a problem. The initial analysis of a problem involves the steps indicated in Fig. B-2. As previously discussed in Sec. 6B, this analysis begins with a basic problem description that specifies the situation and the goals (using predominantly one or more diagrams to summarize the relevant information).

Analysis of a problem

* **Basic description**
    * ***Situation***
        Known information
        Diagram & useful symbols
    * ***Goals***
* **Refined description**
    * ***Time sequence & intervals***
    * ***Physics description***
        (e.g., motion, interactions)

Fig. B-2. Initial analysis of a problem.

This description needs then to be refined by examining the time sequence of events and by redescribing the situation in terms of physics knowledge. In earlier problems, this physics description provided only some specific information about motion (e.g., about velocity and acceleration). But now this physics description involves ordinarily also information about interactions (e.g., about various forces).

Construction of solution. As previously discussed in Sec. 6C and indicated in Fig. B-3, the solution of a problem is constructed by decomposing the problem into successive subproblems and implementing the solutions of these. Success in this endeavor requires that one makes wise decisions in choosing useful subproblems. As indicated in Fig. B-3, a useful strategy for doing this involves the following: (a) Examining the status of the problem at any stage of the solution process (by identifying what information is known and unknown, and what obstacles hinder a solution); (b) identifying available options for subproblems that may help to overcome such obstacles; and (c) selecting a useful option among these.

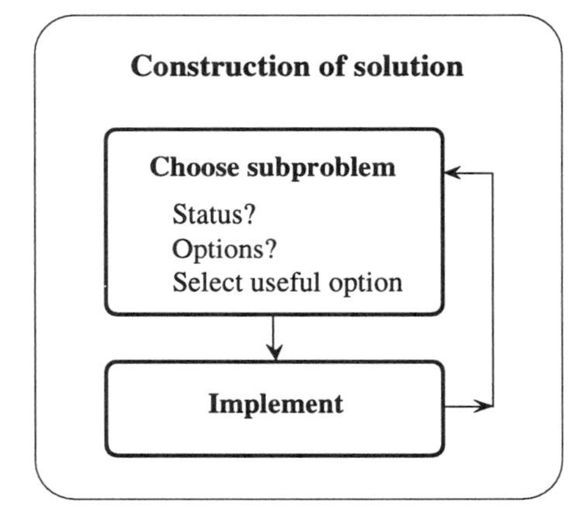

Fig. B-3. Constructing a solution by repeated decomposition into subproblems.

The two main obstacles hindering a solution are either a lack of useful information, or available information which is deficient because it contains unknown quantities. (See Fig. B-4.) If enough information is available, this second obstacle is readily overcome by combining available relations to eliminate the unknown quantities. The major obstacle which must be faced (and which looms particularly large whenever one starts a problem) is then the lack of enough useful information. How can this obstacle be overcome by finding useful relations?

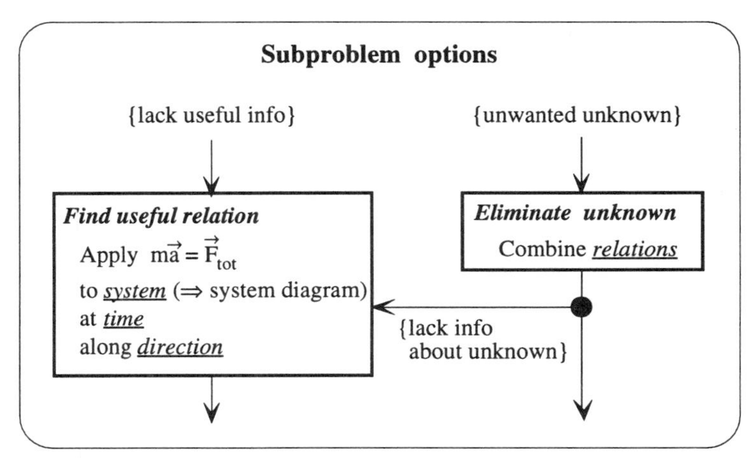

Fig. B-4. Main problem-solving obstacles and corresponding options for remedial subproblems.

In dealing with the simple problems of Chapter 6, useful information could be obtained merely from basic relations about motion. But, more generally, mechanics deals with the relation between motion and interactions. Correspondingly, the important information needed for mechanics problems is information relating the motion of systems to their interactions. The basic relations providing such information are mechanics laws — and the only such law which we have encountered up to now is Newton's law $m\vec{a} = \vec{F}_{tot}$. Hence we arrive at the following guideline: *Start any mechanics problem by applying Newton's law, and try to exploit all the implications of this law.*

Fig. B-4 (which is a modified version of Fig. C-2 in Chapter 6) makes this guideline explicit and indicates also what else one must do whenever one decides to apply Newton's law. (Needed choices are shown underlined.)

(1) One must choose to what system the law is to be applied, and at what time it is to be applied to this system. (The system is ordinarily one about which information is wanted, or one interacting with it. The time is usually a typical time or some time of special interest.)
(2) One must describe the chosen system at the chosen time by constructing a system diagram for it. (This provides the essential information needed for the actual application of Newton's law.)
(3) One must specify the particular direction along which Newton's law is to be applied.

When writing out the solution of a problem, it is then important to indicate the particular choices that have been made, as well as any relevant system diagrams and equations used to implement these choices.

Checking a solution. The solution of any problem must be checked to ensure that it is free of obvious deficiencies. In particular, one needs to assess whether all the goals of the problem have been attained, whether the answers are well-specified, whether they are self-consistent, whether they are consistent with other available information, and whether the solution is as good as possible. These standard checks, previously discussed in Sec. 6D and summarized in Fig. B-5, are applicable to *all* problems.

Checks

Goals attained?
- All wanted info found?

Well-specified?
- Answers in terms of knowns?
- Units?
- Vector magnitudes and directions?

Self-consistent?
- Units?
- Signs or directions?

Other-consistent?
- Sensible values?
- Special cases?
- Known dependence?
- Other solutions?

Optimal?
- Clear and simple?
- General?

Fig. B-5. Checks for a solution.

Prototype solution of a mechanics problem

The following illustrates application of the problem-solving method to a typical mechanics problem.

Problem statement *(banking of a road curve).* A road curve, in the shape of a circular arc of radius R, is to be designed so that a car can travel around it with a constant speed v, even if friction forces between the car and the road are negligibly small.

(a) What must be the "banking angle" (i.e., the angle between the road surface and the horizontal) so that this is possible?
(b) What must be the numerical value of this angle if R = 40. m and v = 9.0 m/s (i.e., about 20 mi/h)?

Solution. The following is the solution of the preceding problem.

Top view of car

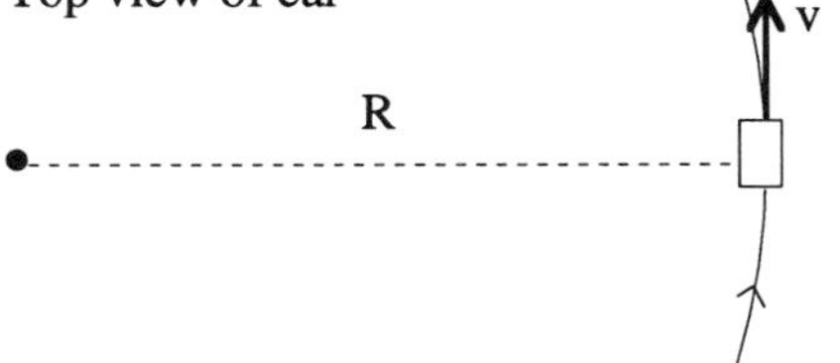

Side view of car

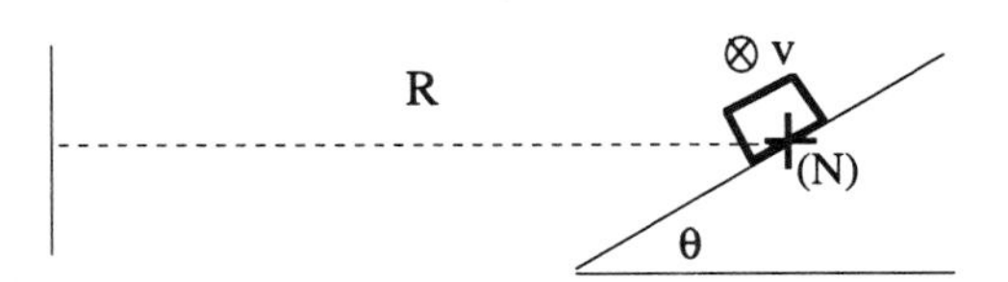

Radius R (40. m)
Speed v (9.0 m/s)

Known: R, v.

Goals: $\theta = ?$ (symbols)
$\theta = ?$ (number)

Analysis of problem

Situation diagram. (The side view is needed to show the banking angle θ.)

⊗ Symbol indicating a vector directed into the paper (so that the tail feathers of the arrow are visible).

The cross indicates the contact interaction between the car and the road. (It is labeled by the magnitude N of the corresponding normal force.)

Apply Newton's law, to car

$$m\vec{a} = \vec{F}_{tot}$$

along acceleration

$$\underline{m}\frac{v^2}{R} = \underline{N}\sin\underline{\underline{\theta}} \qquad (1)$$

vertically upward

$$0 = \underline{N}\cos\underline{\underline{\theta}} - \underline{m}g \qquad (2)$$

Car

N cosθ

N sinθ

θ

⊗v

a

m

θ

mg

$a = v^2/R$

Eliminate N by (1) and (2):

$$N = \frac{mg}{\cos\theta}$$

$$m\frac{v^2}{R} = mg\frac{\sin\theta}{\cos\theta} \qquad (3)$$

$$\frac{v^2}{R} = g\tan\theta$$

$$\tan\theta = \frac{v^2}{gR} \qquad (4)$$

or $$\boxed{\theta = \arctan\left(\frac{v^2}{gR}\right)} \qquad (5)$$

Construction of solution

{Lack info relating motion & interactions}

System diagram for car.

The mass of the car is unknown. Denote it by the symbol m.

Car's acceleration is horizontal, toward center of curve

Wanted unknowns have been underlined twice, unwanted unknowns once.

N and m are unwanted unknowns. Eliminate N by solving for it in (2), and then substituting the result into (1).

One can divide both sides of (3) by m. Hence the unknown mass m cancels.

arctan x denotes the angle whose tangent is x.

Numbers:

$$\tan\theta = \frac{(9.0\ \text{m/s})^2}{(9.80\ \text{m/s}^2)\ (40.\ \text{m})} = 0.207\,, \qquad \boxed{\theta = 12°} \qquad (6)$$

Checks:

Units of (4): Left side: no units (i.e., 1)

Right side: $\frac{(\text{m/s})^2}{(\text{m/s}^2)\ \text{m}} = 1$ OK

If v large: Expect large banking angle θ. Eq. (4) OK.

If R large (curve nearly straight): Expect small θ. Eq. (4) OK.

Checks (done in writing or mentally)

Check of unit consistency.

An angle has no units related to the standards of length, time, or mass.

Special cases.

The Workbook contains further examples of problem solutions and provides needed opportunities for practicing problem solving.

➜ ***Go to Sec. 11B of the Workbook.***

C. Further problem-solving suggestions

Starting from Newton's law. As already mentioned, it is usually best to start a problem by applying a basic mechanics law (like Newton's law), even if it is not immediately apparent how this will lead to the desired goal. Indeed, excessive fixation on the problem goal may sometimes fail to suggest any method of attaining it. On the other hand, one *does* know that practically any mechanics problem requires adequate information about the relation between motion and interactions — and application of Newton's law does provide such information. Furthermore, even if such information should ultimately prove to be unnecessary, it can only help and never do harm.

Introducing auxiliary unknown quantities. To apply physics knowledge to a given situation requires that this situation be analyzed in terms of specific quantities (like mass, acceleration, or forces). However, some of these quantities may not be mentioned in a problem. In such a situation, don't hesitate to introduce algebraic symbols to denote such unknown quantities. For example, they may ultimately cancel so that their actual values are irrelevant for the solution of the problem.

For example, in the prototype problem discussed in Sec. B, the mass of the car was not specified. However, this unknown mass could be simply specified by the symbol m, and this mass ultimately turned out to be irrelevant.

Solutions justified by basic physics principles. One must make sure that any problem is solved by logical reasoning from basic physics principles or definitions. After all, the central goal of science is to demonstrate the ability to predict or explain on the basis of a few such basic principles. Furthermore, it is only such well-justified reasoning that can ensure the correctness of a solution and guard one against misleading intuitions.

Of course, good intuitions (or even guesses) may sometimes be helpful in suggesting the solutions to some problems. But such solutions always need to be checked to verify that they are consistent with basic physics principles.

Working with symbols. As already pointed out in Sec. 6E, it is usually best to introduce useful symbols, to work with these symbols to obtain an algebraic result, and to substitute numbers only at the end. In this way, calculations are often simpler and less error-prone since one does not constantly need to work with cumbersome numbers and units. More important, the answers obtained are then generally applicable formulas which indicate useful qualitative implications which can be readily checked by their consistency with special cases.

Refinements of the problem-solving method

The main steps of the problem-solving method have been discussed in Sec. A. In the case of more complex problems, these steps may usefully be refined by the following additional actions.

Initial problem analysis. Some additional investigation of the problem may be useful at the end of the initial analysis. For example, some investigation of simple special cases can identify expected features of the solution and thus help one to avoid nonsense (during the solution process or in the final answers).

Construction of solution. A solution can be constructed sketchily and qualitatively, rather than in detail. In this way one can make an initial *plan* which outlines the major steps of the proposed solution. Particularly in the case of complex problems, such a plan can guide one in implementing the detailed actions ultimately required for a complete solution.

Checking and revising. It is clearly necessary to check and revise the solution of any problem. It is it is even more important to learn from what one has done so that one can check one's general knowledge and improve it for future purposes. Otherwise, all the work done to solve a problem is mostly wasted and without lasting benefits.

In particular, one can learn a great deal from one's past deficiencies by doing the following: (a) Identifying the difficulties or mistakes encountered in doing a problem, and diagnosing the reasons for them. (b) Specifying advice or warnings to help avoid such deficiencies in the future. (c) Making sure that one can now solve this problem, or similar ones, without any further difficulties or mistakes.

→ ***Go to Sec. 11C of the Workbook.***

D. Summary

Application of Newton's law

To apply Newton's law to a system, one must first describe the system separately by drawing a system diagram displaying the information summarized in Fig. D-1. Newton's law can then be expressed in terms of its components along convenient directions.

Problem-solving method

The problem-solving method discussed in Chapter 6 involves analyzing a problem, constructing its solution, then checking and revising it. These main steps of the method are equally applicable to more complex mechanics problems.

System diagram

* **Separate system**
* **Mass**
* **Motion** (velocity, acceleration)
* **All forces *on* the system**
    * ***Long-range*** (e.g., gravity)
        * Interacting objects?
        * Forces *on* the system
    * ***Contact***
        * Touching objects?
          (Mark & label contacts.)
        * Forces *on* the system
* **Components**

Fig. D-1. Description of a system by a system diagram.

To find useful relations between motion and interaction, Newton's law can be applied to some chosen systems, at some chosen times, along some chosen directions. Accordingly, the solution of any mechanics problem is ordinarily started by applying Newton's law.

New abilities

You should now be able to solve mechanics problems by using the preceding problem-solving method and your knowledge of Newton's law (together with your previous knowledge about motion and about the common interactions studied in Chapter 10).

➜ ***Go to Sec. 11D of the Workbook.***

12 More about Interactions

To deal with a larger range of commonly arising problems, it is necessary to examine more closely some of the interactions previously discussed in Chapter 10. Accordingly, the next two sections provide more quantitative details about the forces exerted by springs and the friction forces exerted by touching objects.

The subsequent sections then discuss two fundamentally important long-range interactions. One of these is the gravitational interaction between any two particles. (A knowledge of this interaction allows one not only to discuss particles interacting with the nearby earth, but to deal with a large range of astronomical problems.) The other long-range interaction is the electric interaction between charged particles. (This is the fundamental interaction between all atomic particles and is thus ultimately responsible for most properties of matter.)

A. Spring force for small deformations

As discussed in Sec. 10C, the force exerted on a particle by a spring is directed along the spring and opposes the deformation of the spring. The magnitude of this force, which is zero if the spring is undeformed, increases with increasing deformation of the spring.

To describe this force in more detail, consider a spring whose undeformed length is L_o. The spring has then some different length L when it is deformed. The quantity $x = L - L_o$ is called the *elongation* of the spring.

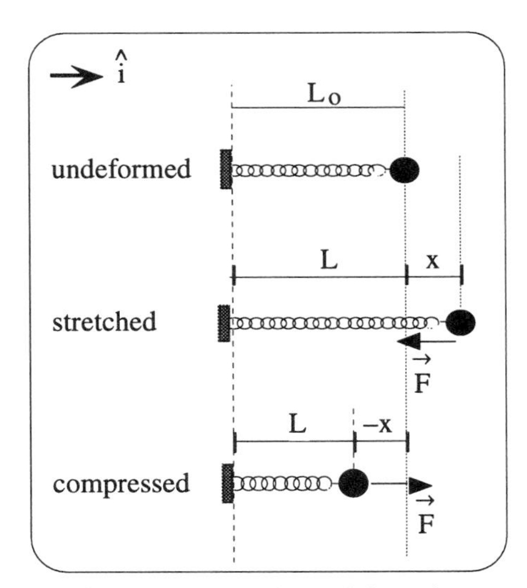

Fig. A-1. Force exerted by a deformed spring.

Component description of elongation and force. Fig. A-1 illustrates the situation. The indicated reference direction, specified by the unit vector $\hat{i}$, points from the spring toward the particle. The elongation x of the spring corresponds then to the component of the particle's displacement along this direction. Furthermore, the force $\vec{F}$ on the particle by the spring can be described by its component F_x along this direction.

Qualitative relation between force and elongation. As indicated in Fig. A-1, the force always opposes the deformation of the spring. If the elongation $x = 0$, the force component $F_x = 0$. If the spring is stretched, so that its elongation x is positive, F_x is negative. If the spring is compressed, so that its elongation x is negative, F_x is positive.

The *magnitude* of the force varies smoothly with the magnitude of the elongation. If the magnitude of the elongation x is larger, the magnitude of the force component F_x is correspondingly also larger.

The preceding qualitative characteristics of the spring force imply that a graph of F_x versus x should have the shape indicated in Fig. A-2.

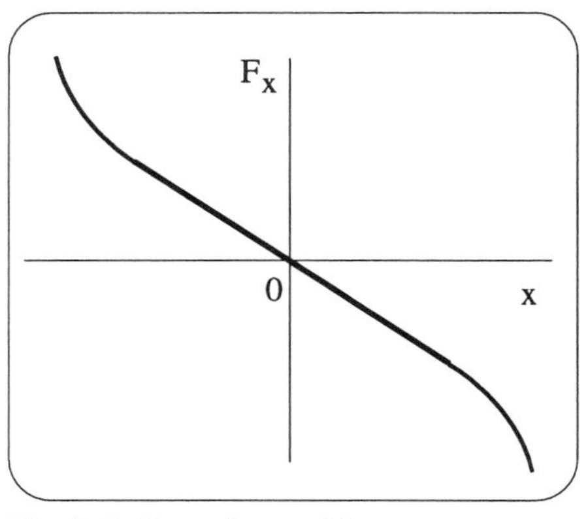

Fig. A-2. Dependence of force on elongation.

Proportionality of force and elongation. Any smooth curve is nearly straight in a sufficiently small region. Hence the curve in Fig. A-2 is nearly a straight line (with negative slope) in the region near x = 0 where the magnitude of the elongation is sufficiently small. In other words,

$$\text{if } |x| \text{ is sufficiently small,} \qquad F_x = -k\,x \tag{A-1}$$

where k is a positive constant.

This constant k is called the *spring constant.* It characterizes the particular spring by describing the relationship between its elongation and the force exerted by it. [A spring is called *ideal* if the proportional relationship (A-1) between the force and the elongation is valid for all elongations of interest.]

The proportionality expressed by (A-1) is sometimes called Hooke's law since it was first enunciated in 1678 by Robert Hooke, a contemporary of Newton.

➜ ***Go to Sec. 12A of the Workbook.***

B. Friction force

As discussed in Sec. 10F and indicated in Fig. B-1, the force exerted on an object by a non-adhering touching object can be described by two component forces, a *normal force* $\vec{N}$ perpendicular to the contact surface and a *friction force* $\vec{f}$ parallel to this surface.

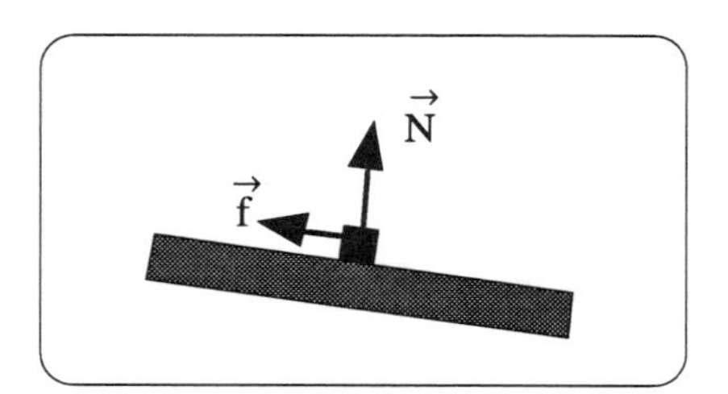

Fig. B-1. Normal and friction forces on an object by a touching solid surface.

The friction force always tends to oppose the relative motion of the touching objects. The direction of this friction force was discussed in Sec. 10F, but its magnitude was not fully specified. To specify it more quantitatively, it is again useful to consider separately the *kinetic* situation where the objects move relative to each other and the *static* situation where they do not move relative to each other.

Kinetic friction

Consider first the kinetic situation where the objects move relative to each other.

To oppose the relative motion, the friction force tends to reduce the relative velocity of the objects. Accordingly, *the direction of the friction force on an object is opposite to its velocity relative to the touching object.*

The magnitude of the friction force depends on the nature of the touching surfaces. It also depends on the normal force, being proportionately larger if this normal force is larger (i.e., if the touching surfaces are more tightly pressed against each other). Thus the magnitude f of the friction force can be described by the following relation:

for kinetic friction, $$f = \mu_k N \,. \tag{B-1}$$

Here μ_k is a constant, called the *coefficient of kinetic friction*, which depends on the nature of the touching surfaces of the objects. Observed values of this coefficient, for various kinds of touching surfaces, can be found compiled in handbooks. For example, the coefficient of kinetic friction for steel on steel is about 0.57, for brass on steel is 0.44, and for teflon on steel is 0.04.

Static friction

Consider now the static situation where the objects do *not* move relative to each other (i.e., where they are at rest relative to each other).

To oppose any relative motion, the friction force must then ensure that the relative velocity of the objects remains zero. Accordingly, *the friction force on an object is such that the object's acceleration relative to the touching object is zero.*

The preceding statement specifies both the direction and magnitude of the friction force, except for one additional restriction: The magnitude of the static friction force cannot exceed some maximum possible value. (Otherwise the objects start sliding relative to each other.) This maximum value depends again on the properties of the touching surfaces and on the magnitude of the normal force. This restriction on the magnitude f of the friction force can be described by the following inequality:

for static friction, $$f \le \mu_s N \,. \tag{B-2}$$

Here μ_s is a constant, called the *coefficient of static friction*, which depends on the nature of the touching surfaces of the objects. Observed values of this coefficient, for various kinds of touching surfaces, can be found compiled in handbooks. For example, the coefficient of static friction for steel on steel is about 0.74, for brass on steel is 0.51, and for teflon on steel is 0.04. (The coefficient of static friction μ_s is ordinarily larger than the coefficient of kinetic friction μ_k.)

Summary of friction properties

The preceding properties of the friction force $\vec{f}$ can then be summarized by the following statements:

Kinetic friction force: Direction opposite to relative velocity. $f = \mu_k N$	(B-3)

Static friction force:	Direction so that relative acceleration = 0. f so that relative acceleration = 0. $f \leq \mu_s N$

(B-4)

The relations (B-3) and (B-4) imply that the magnitude of the friction force does *not* depend on the size of the area of contact between the touching objects.

Reason for non-dependence on area

Fig. B-2a shows two touching objects B and C sliding relative to each other. Fig. B-2b shows the contact surface of B imagined subdivided into n identical small squares, each of area A_1. If the magnitude of the normal force on *one* of these squares is N_1, the magnitude of the friction force on this single square is

$$f_1 = \mu N_1 \tag{B-5}$$

where μ is the kinetic coefficient of friction.

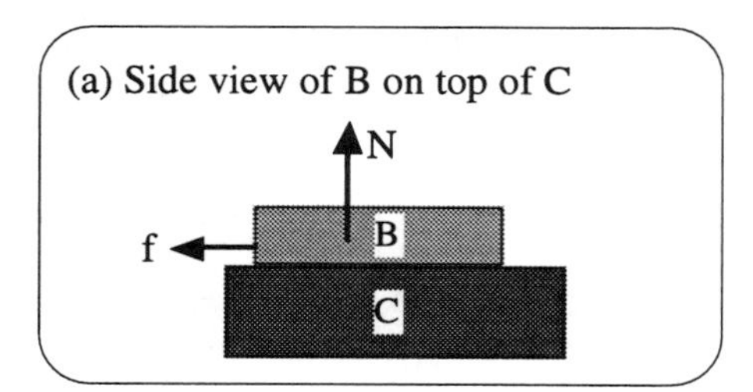

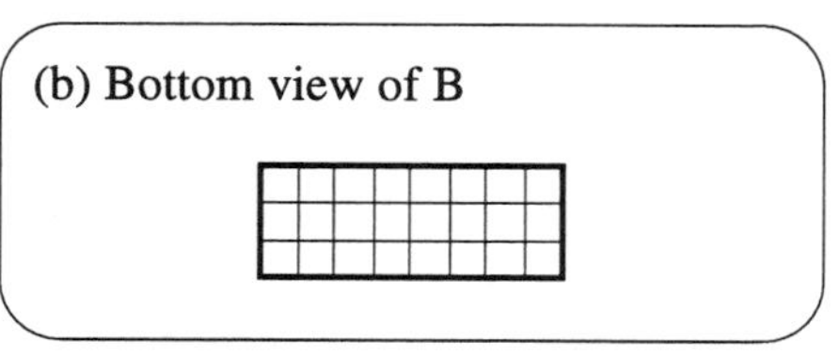

Fig. B-2. Contact between two objects B and C. (a) Normal and friction force on object B. (b) Contact surface responsible for these forces.

The magnitude f of the total friction force on B is n times larger than that on a single square, i.e.,

$$f = n f_1 = n \mu N_1 \, . \tag{B-6}$$

However, the magnitude N_1 of the normal force on a single square is n times smaller than that on the entire surface of B, i.e.,

$$N_1 = \frac{N}{n} \, . \tag{B-7}$$

The effects described by (B-5) and (B-6) compensate each other , thus yielding

$$f = n \mu \left(\frac{N}{n}\right) = \mu N \, . \tag{B-8}$$

Hence the resulting friction force f on B does not depend on the number n of touching squares (i.e., on the size of the contact surface between the objects).

Similar arguments apply to the case of static friction.

➜ ***Go to Sec. 12B of the Workbook.***

C. Gravitational Force

Sec. 10A discussed the gravitational force on a particle by the earth near its surface. This force is attractive (i.e., directed toward the center of the earth). The magnitude of this force is proportional to the mass m of the particle, but independent of any of its other properties.

Hypotheses about the gravitational force

Newton assumed that gravitational interaction exists not only between a particle and the earth, but more generally between *any* two particles or astronomical bodies. Let us then consider any two such particles (e.g., two astronomical bodies which are small compared to the distance R between them) and try to formulate plausible hypotheses about the properties of the gravitational force exerted on one by the other.

Direction of the force. The gravitational force on a particle by the earth is attractive. Hence one expects that the force between any two particles is attractive. Since mutual forces have opposite directions, the gravitational forces on the two particles should then have the directions indicated Fig. C-1.

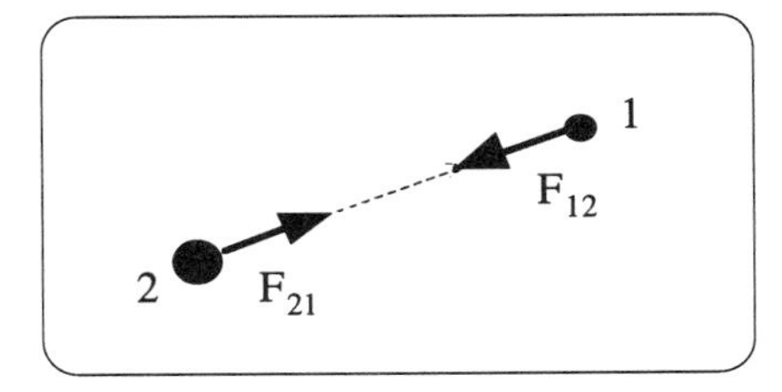

Fig. C-1. Mutual gravitational forces on two interacting particles.

Magnitude of the force. The gravitational force $m\vec{g}$ on a particle by the earth is proportional to the mass m of the particle. (For example, if the mass of the particle is 3 times larger, the magnitude of the gravitational force on it is also 3 times larger.) One would similarly expect that the magnitude F_{12} of the force on particle 1 by particle 2 should be proportional to the mass m_1 of particle 1, i.e., that

$$F_{12} \propto m_1 \,. \qquad \text{(C-1)}$$

(The symbol $\propto$ means *is proportional to*.) Similarly, the magnitude F_{21} of the force on particle 2 by particle 1 should be proportional to the mass m_2 of particle 2, i.e.,

$$F_{21} \propto m_2 \,. \qquad \text{(C-2)}$$

But the relation between mutual forces implies that the magnitudes of these forces are the same. Thus $F_{12} = F_{21} = F$, where F denotes this common magnitude. By (C-1) and (C-2), this common magnitude should then be proportional to *both* m_1 and m_2, i.e.,

$$F \propto m_1 m_2 \,. \qquad \text{(C-3)}$$

If the distance R between the particles is larger, the interaction between them (and thus also the magnitude F of the force) is expected to be smaller. The simplest assumption satisfying this requirement is that F depends on R so that

$$F \propto \frac{1}{R^n} \qquad \text{(C-4)}$$

where n is some constant greater than 1. Newton made the specific assumption that n = 2 (in order to fit Kepler's observations about the motions of the planets around the sun).

Gravitational force law

Thus Newton arrived at the following hypothesis about the magnitude and direction of the gravitational force $\vec{F}_g$ on any one particle by any other:

Gravitational force:		
Magnitude:	$F_g = G \dfrac{m_1 m_2}{R^2}$	(C-5)
Direction:	Attractive	

Here G is a constant independent of the properties or positions of the particles. This constant, called the *gravitational constant*, is thus a universal constant characteristic of our universe. [Do not confuse this constant G with the gravitational acceleration g (denoted by a lower-case letter g) which *does* depend on the particular position where a particle is located.] The numerical value of the gravitational constant G will be discussed later.

The implications of the hypothesis (C-5) have been confirmed by very many observations. Hence (C-5) is considered a fundamental law of nature and is called *Newton's universal law of gravitation*.

In summary, this law implies that every particle exerts on any other particle an attractive gravitational force which depends on the masses of both particles (being proportionately larger if the mass of either of these particles is larger). This force decreases with increasing separation of the particles, being inversely proportional to the square of the distance between them. (For example, if the distance between the particles is twice as large, the magnitude of the force is *four* times smaller.)

Gravitational force by (or on) a spherical body

Newton's gravitational law (C-5) specifies the gravitational force on a particle P by another particle. The force on a particle P by any *system*, consisting of many particles, can then readily be found. By the superposition principle, it is simply the vector sum of the force exerted on P by each of the other particles.

For example, suppose that one wants to use the gravitational force law (C-5) to find the force exerted on a particle P by the earth. This law cannot be directly applied since the whole earth (unless very far away) can certainly *not* be considered a particle. However, the earth *can* be regarded as consisting of a very large number of very small particles, as schematically indicated in Fig. C-2. The force on P can then be obtained by adding vectorially the forces exerted on P by each of these particles. This calculation can readily be done (by using integral calculus) and yields the following simple result:

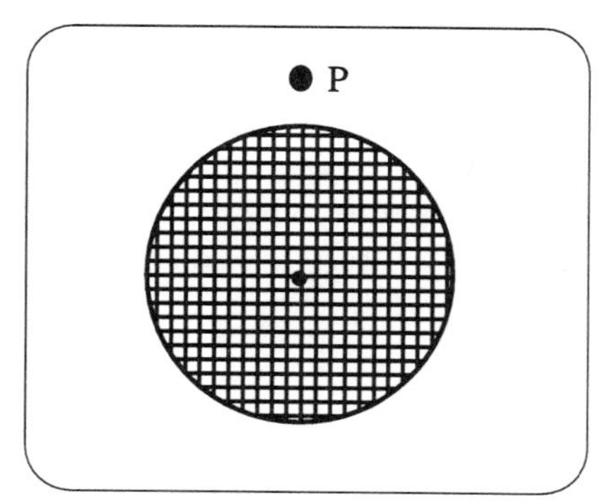

Fig. C-2. Gravitational interaction between a particle and all the particles in the earth.

> ***Gravitational force by a spherically symmetric body:*** If a particle is located outside a spherically symmetric body, the force on the particle by the body is the same as if the entire mass of the body were concentrated at its center. (C-6)

Despite its simplicity, this result is far from obvious. [For example, it would *not* be true if the exponent of R in the denominator of the gravitational force law (C-5) would not be exactly equal to 2.]

A similar result holds for the force *on* a body. Thus the gravitational force exerted *on* a spherically symmetric body by a particle outside it is the same as if the entire mass of the body were concentrated at its center.

Magnitude of the gravitational constant

Smallness of gravitational forces. According to (C-5), the gravitational force on an object by another depends on the product of their masses. If at least one of these objects is of astronomical size and has a correspondingly large mass (e.g., if it is the earth), then the gravitational force can be appreciable. However, if *both* objects are of everyday size and have correspondingly small masses, then the gravitational force is very much smaller. Indeed, gravitational forces between everyday objects (e.g., between two billiard balls) are utterly negligible compared to most other forces.

Determination of G. How can one determine the magnitude of the gravitational constant G? In principle, one can do this by using the gravitational force law (C-5). One needs only measure the gravitational force between two particles (or two spherical bodies) of known masses and measure the distance R between their centers. All the quantities in (C-5), except G, would then be known; hence G itself could easily be calculated.

The preceding method of determining G is, however, difficult to implement for the following reasons: (a) If the bodies are of everyday size, the gravitational force between them is so very small that it is very difficult to measure. (b) If at least one of the bodies has a large mass because it is of astronomical size (e.g., the earth or the moon), the gravitational force is large and could readily be measured. However, the mass of such an astronomical body is not known in terms of kilograms (since there is no apparent way of comparing its mass with the standard kilogram stored in Paris).

The only way of overcoming these difficulties is to pursue the first option by constructing a very sensitive instrument capable of measuring the very small gravitational force between two objects of everyday size. In 1798 Henry Cavendish constructed such an instrument (a *torsion balance*) and thus succeeded in measuring the tiny gravitational force exerted on one lead ball by another. This measurement was highly important for two reasons: (a) It confirmed experimentally Newton's hypothesis that the gravitational force exists not only between massive astronomical bodies, but between *any* two objects. (b) It determined the actual numerical value of the gravitational constant. The value thus found is

$$\boxed{G = 6.672 \times 10^{-11} \text{ N m}^2/\text{kg}^2 .} \qquad \text{(C-7)}$$

Applications of the gravitational force law

Many useful results can be obtained by using Newton's law of mechanics, together with the gravitational force law (C-5) and the known value of G. For example, it is possible to find the gravitational acceleration g at any distance from the earth, to determine the mass of the earth, and to make quantitative predictions about the motions of the planets around the sun (or the motions of artificial satellites around the earth). The following example illustrates this last application. You yourself can readily explore other applications by doing the problems in the Workbook.

Example: Planetary motion

Fig. C-3 illustrates a planet (e.g., the earth or Mars) revolving around the sun with a speed v in a circular orbit of radius R. The planet's acceleration $\vec{a}$ is then directed toward the sun. The attractive gravitational force $\vec{F}_g$ on the planet by the sun is also directed toward the sun. Indeed, this is the *total* force on the planet since no other forces act on it. (The forces by the other planets are negligible compared to the gravitational force exerted by the much more massive sun.)

Newton's law $m\vec{a} = \vec{F}_{tot}$, applied to the planet along the direction toward the sun, implies that

$$m\left(\frac{v^2}{R}\right) = G\frac{Mm}{R^2}\ .$$

Hence

$$v^2 = \frac{GM}{R}\ . \qquad \text{(C-8)}$$

The planet's speed v is thus independent of its mass m and is smaller if the planet's distance from the sun is larger.

The planet's speed v is related to its period T, i.e., to the time required to go once around the circumference $2\pi R$ of its orbit. Thus

$$v = \frac{2\pi R}{T}\ . \qquad \text{(C-9)}$$

Hence (C-8) implies that

$$\frac{4\pi^2 R^2}{T^2} = G\frac{M}{R}$$

so that

$$\frac{R^3}{T^2} = \frac{GM}{4\pi^2}\ . \qquad \text{(C-10)}$$

The right side of (C-10), and hence its left side too, is the same for every planet moving around the sun. Thus this relation implies that the planet at a larger distance from the sun has a longer period, but in such a way that the ratio R^3/T^2 is the *same* for every planet. This precise relationship was actually discovered by Johann Kepler in 1619 (well before Newton's birth) on the basis of extensive astronomical observations.

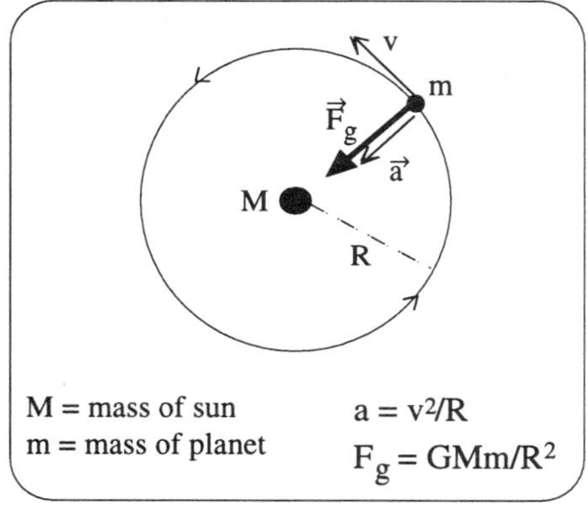

Fig. C-3. Motion of a planet around the sun.

➜ ***Go to Sec. 12C of the Workbook.***

D. Electric force

Electric interactions. All matter is ultimately composed of various kinds of atoms. Any such atom consists of a *nucleus* (having a size of about 10^{-15} m) built of smaller fundamental particles called *protons* and *neutrons*. This nucleus is surrounded by a cloud of other fundamental particles, called *electrons*, so that the entire atom has a size of about 10^{-10} m. (The mass of the electron is 9.11×10^{-31} kg, that of the proton is 1.67×10^{-27} kg, and that of the neutron is almost the same as that of the proton.)

Electrons and protons can be characterized by a property called electric *charge*, and they interact with each other by an interaction which is called *electric*. This electric interaction is of the utmost importance since it is responsible for holding together all atoms and molecules. It is also the basis for the innumerable electrical devices and practical applications that pervade our modern technological society.

Electric interactions can be observed in their most primitive form when one rubs a glass rod with a silk cloth or separates two pieces of Scotch tape. In such processes electrons get transferred from one object to the other. As a result, the objects become charged so that they exert electric forces on each other.

Electric charge. The electric charge q of a particle is defined in terms of the electric force exerted on it. For example, Fig. D-1 shows a point P near some charged object. One can then measure the electric force exerted by this object on any charged particle placed at P. Suppose that $\vec{F}_1$ is the electric force on a particle 1 placed at this point, and that $\vec{F}_2$ is the electric force on a particle 2 placed at this point. Then the ratio q_1/q_2 of the charges of these particles is *defined* so that

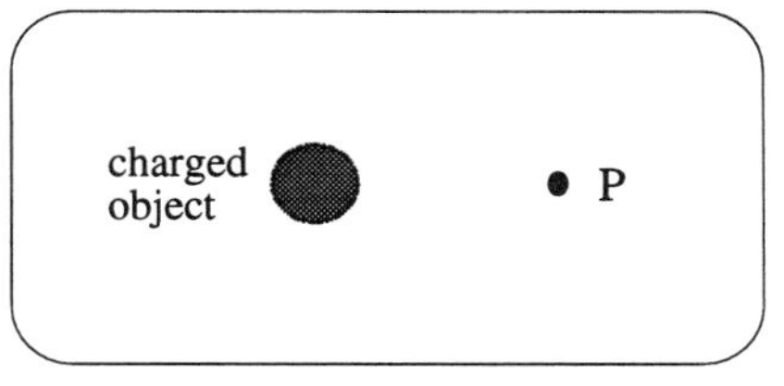

Fig. D-1. Point P near a charged object.

$$\frac{q_1}{q_2} = \pm \frac{F_1}{F_2} \qquad \text{(D-1)}$$

where F_1 and F_2 are the magnitudes of the respective forces. Here the plus sign is used if the forces have the same direction, and the minus sign if they have opposite directions.

The two possible signs in (D-1) indicate that the charge of a particle can be positive or negative (or zero if no electric force acts on it). For example, the proton and electron have charges of opposite signs, and the neutron has zero charge.

Note that only the *ratio* of two charges is defined by (D-1). As is done in the case of other quantities (like length, time, or mass), it is then convenient to introduce a standard of charge and to compare all charges against it. There is no need here to discuss the particular standard of charge adopted by international convention, except to say that the unit assigned to it is called *coulomb* (abbreviated by *C*). In terms of this unit, the electron has a charge of $-e$, and the proton a charge of $+e$, where the magnitude e of the electronic charge is equal to

$$e = 1.60 \times 10^{-19}\ \text{C}\,. \qquad \text{(D-2)}$$

Force between two charged particles. Consider two charged particles (whether atomic particles, or large-scale particles like charged glass spheres). As usual, the mutual electric forces on these particles must have the same magnitude and opposite directions (as indicated in Fig. D-2). The common

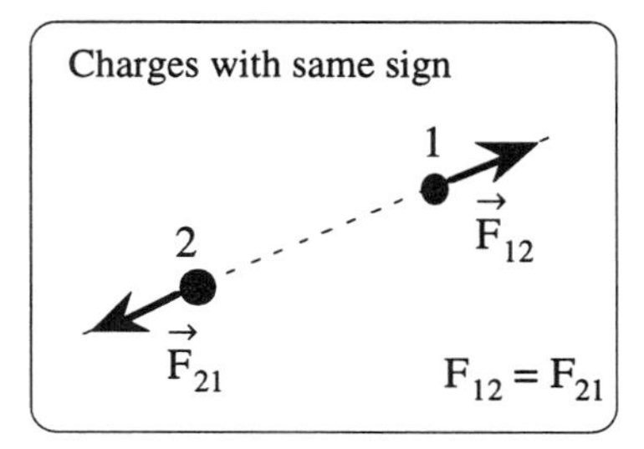

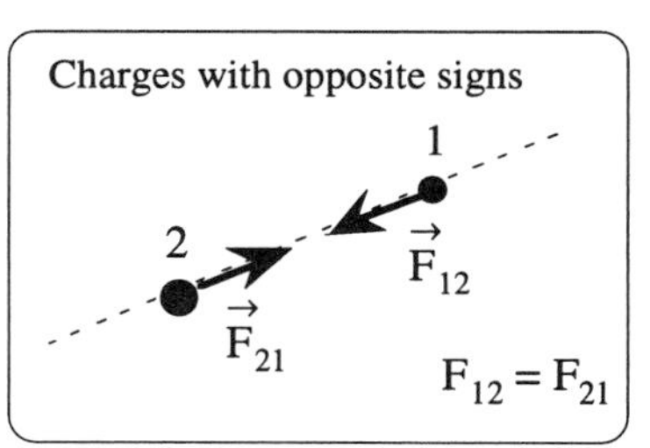

Fig. D-2. Electric forces on two charged particles.

magnitude of the electric force on one particle by the other depends on their charges q_1 and q_2, and on the distance R between them. Charles A. Coulomb (the person in whose honor the unit of charge is named) first determined experimentally in 1785 how the electric force $\vec{F}_e$ depends on these quantities. This dependence is summarized by the following statement (called *Coulomb's law*):

Electric force:

Magnitude: $F_e = k_e \frac{|q_1 q_2|}{R^2}$ (D-3)

Direction: Repulsive if charges have same signs, attractive if they have opposite signs.

The vertical bars on the right side of (D-3) indicate that only the magnitudes of the charges are relevant in specifying the magnitude of the force.

Here the electric force constant k_e is approximately equal to 9×10^9 Nm^2/C^2. Somewhat more precisely, its value is

$$k_e = 8.988 \times 10^9 \; Nm^2/C^2 \, . \quad \text{(D-4)}$$

Note that the electric force, just like the gravitational force specified in (C-5), is inversely proportional to the square of the distance between the particles. However, the electric interaction is otherwise totally different from the gravitational interaction.

The preceding paragraphs summarize the basic properties of the electric interaction and are sufficient to deal with some of its basic applications. However, electric interactions are so very important that they will be discussed at much greater length in your later study of physics.

➜ ***Go to Sec. 12D of the Workbook.***

E. Predictive power of mechanics

Suppose that one knows how particles interact (i.e., that one knows how the force on any particle by any other depends on their positions and other properties). What predictions can then be made by using Newton's law $m\vec{a} = \vec{F}_{tot}$?

To be specific, consider any number of interacting particles with known masses and known interactions. (For example, the particles might be the sun and the planets, all interacting with each other by gravitational forces.) Suppose that one knows the position $\vec{r}$ and velocity $\vec{v}$ of each particle at a particular instant of time t_0. How well can one then predict the positions and velocities of these particles at some future time?

Predictions about a slightly later time

Let us first answer a simpler question: What predictions can one make about the situation at a *slightly later time*, e.g., one millisecond after t_o?

Predictions from velocity. Knowing the velocity $\vec{v}$ of a particle at the time t_o, one can use the following reasoning (outlined in Fig. E-1):

* One can determine the displacement of the particle during the next millisecond. (From the definition of velocity, this displacement is $d\vec{r} = \vec{v}\,dt$, where dt = 1 millisecond.)
* One can then determine the position of the particle one millisecond after t_o. (This position is just its original position plus its displacement, i.e., $\vec{r} + d\vec{r}$.)

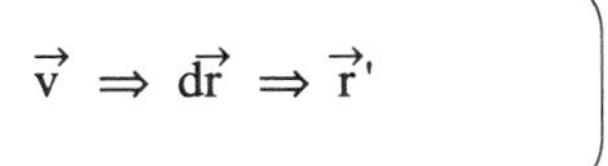

Fig. E-1. Knowledge of velocity allows prediction of position at a slightly later time.

Predictions from positions. Knowing the *positions* of all particles at the initial time t_o, one can use the following reasoning (outlined in Fig. E-2):

* One can determine the force on any particle by any other particle (since we know how the interactions of the particles depend on their positions).
* One can then determine the *total* force on any such particle (by merely adding vectorially all the individual forces just determined).
* One can then determine the acceleration of any such particle (by using Newton's law $m\vec{a} = \vec{F}_{tot}$).
* One can then determine the change of velocity of the particle during the next millisecond. (From the definition of acceleration, this change of velocity is $d\vec{v} = \vec{a}\,dt$, where dt = 1 millisecond.)
* One can then determine the velocity of the particle one millisecond after t_o. (This velocity is just its old velocity plus its change of velocity, i.e., $\vec{v} + d\vec{v}$.)

$$\vec{r}, \vec{r}_1, \vec{r}_2, \ldots \Rightarrow \vec{F}_{tot} \Rightarrow \vec{a} \Rightarrow d\vec{v} \Rightarrow \vec{v}'$$

Fig. E-2. Knowledge of all positions allows prediction of velocity at a slightly later time.

Resultant prediction. Thus we arrive at the following conclusion: *Knowing the position and velocity of every particle at any time t_o, one can determine the position and velocity of every particle one millisecond later.*

Predictions about any other time

Repetition of infinitesimal predictions. The preceding prediction process can now be repeated in the following way: (a) Knowing the positions and velocities of all the particles one millisecond after t_o, one can use this information (as before) to determine their positions and velocities one millisecond later, i.e., 2 milliseconds after t_o. (b) One can then, in turn, use this information to determine the positions and velocities of the particles one millisecond later, i.e., 3 milliseconds after t_o. (c) One can then, in turn, use this information to determine the positions and velocities of the particles one millisecond later, i.e., 4 milliseconds after t_o. By continuing this process (as schematically indicated in Fig. E-3), one can thus determine the positions and velocities of the particles a year after t_o, or a century after t_o, or at any other time after the initial time t_o.

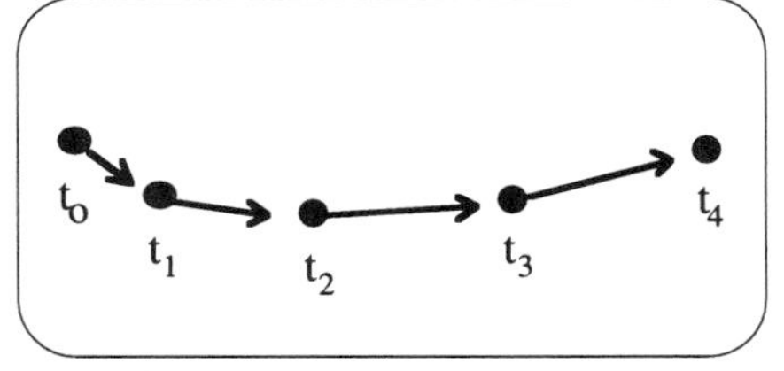

Fig. E-3. Successive predictions of position and velocity at times one millisecond apart.

Thus we arrive at the following conclusion:

A knowledge of the positions and velocities of particles at any time allows one to predict their positions and velocities at any other time.	(E-1)

Implications. Newton's law of mechanics, used together with a knowledge of interactions, should then allow one to predict the entire future history of any system of particles, no matter how complex. This is why Newton's theory of mechanics had a profound influence, ushering in the "Age of Enlightenment" with its confidence in the human ability to understand nature. More realistically, this is also why Newton's theory is of such tremendous practical importance in all of science and engineering.

One might object that the predictive power implied by (E-1) is illusory since it requires so very many repetitive calculations. But this objection can often be overcome. (a) There are mathematical techniques (such as calculus) which can express the required reasoning in more easily implemented forms. (b) All the many calculations are individually very simple, involving merely additions and multiplications. Hence they are particularly well suited to modern electronic computers which can perform such operations with enormous speed. The prediction process outlined in the preceding paragraphs is, therefore, often practically feasible. For example, it is used to predict precisely the motion of planets or to navigate spacecrafts along desired trajectories.

The conclusion (E-1) has, however, the following limitations. (a) Its implied predictive power cannot be achieved in practice because of excessive complexity. For example, systems consisting of very many highly interacting particles (such as liquids or the earth's atmosphere) are so complex that predictions are very hard to achieve. Furthermore, even in seemingly simpler systems, very slight changes in the situation at some initial time t_0 can result in drastically different outcomes; predictions in such cases are very difficult. (b) In the realm of atomic particles, some of the basic assumptions of Newtonian mechanics are not valid. The extraordinarily great predictive power implied by (E-1) is then, correspondingly, no longer realized.

➜ ***Go to Sec. 12E of the Workbook.***

F. Summary

Interactions and corresponding forces

The following table summarizes the properties of all the interactions discussed in this chapter and the earlier Chapter 10.

Interaction	Force	
	Direction	Magnitude
Long-range interactions		
Gravitational		
Due to earth, near surface	$\vec{F}$ vertically down (to center of earth)	$F = mg$ $\{g = 9.80\ m/s^2\}$
Due to any particle	$\vec{F}$ attractive	$F = Gm_1m_2/R^2$ $\{G = 6.67 \times 10^{-11}\ N\ m^2/kg^2\}$
Electric		
Due to charged particle	$\vec{F}$ repulsive for charges of same sign, attractive for opposite signs.	$F = k_e\,\lvert q_1q_2\rvert/R^2$ $\{k_e = 8.99 \times 10^9\ N\ m^2/C^2\}$
Contact interactions		
Due to spring	$\vec{F}$ along spring, to undeformed position (opposes deformation)	F increases with deformation x. ($F = 0$ if $x = 0$.)
For small deformation	....	$F_x = -kx$
Due to string	$\vec{F}$ attractive {tension force} (opposes elongation)	$F \neq 0$ if string is taut, $= 0$ if slack.
Due to touching object	$\vec{F} = \vec{N} + \vec{f}$	
Normal force	$\vec{N}$ perpendicular to surface, repulsive (opposes compression)	$N \neq 0$ if touching, $= 0$ otherwise.
Friction force	$\vec{f}$ parallel to surface (opposes *relative* sliding)	
Kinetic (relative motion)	Direction opposite to relative velocity	$f = \mu_k N$
Static (no relative motion)	Direction so that relative accel = 0	f so that relative accel = 0 $f \leq \mu_s N$

Predictive power of mechanics

Knowledge of positions and velocities at any one time permits predictions of positions and velocities at any other time.

New abilities

You should now be able to apply Newton's law (and the problem-solving method) to solve mechanics problems involving any of the long-range and contact interactions discussed in Chapter 10 and the present chapter.

➜ ***Go to Sec. 12F of the Workbook.***

13 Kinetic Energy and Work

Newton's law relates the acceleration of a particle to the total force on it. It thus allows one to infer how a particle's velocity or position changes in the course of time.

However, instead of being centrally interested in the time, one may merely want to know how a particle's velocity is related to its position. (For example, one may not be interested in how long a *time* is required for a ball to fall a certain distance. Instead, one may merely want to know the ball's velocity after traversing this distance.) How then can Newton's law be used to relate directly a particle's velocity and position, without an explicit mention of time?

By addressing this question, the present chapter will lead to some new concepts and relations extremely useful for explaining or predicting the motion of objects.

A. Kinetic-energy law for small displacements

Consider a particle moving under the influence of forces due to other objects. (For example, the particle might be the baseball, illustrated in Fig. A-1, which moves along a curved path while affected by the gravitational force due to the earth.)

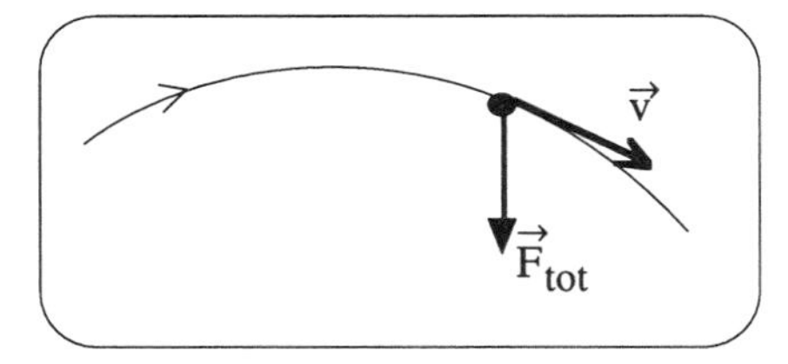

Fig. A-1. Motion of a ball.

Newton's law $m\vec{a} = \vec{F}_{tot}$ relates the acceleration of any such particle to the total force on it. Since the acceleration $\vec{a}$ is just the rate of change of velocity with time t, Newton's law asserts that

$$m\frac{d\vec{v}}{dt} = \vec{F}_{tot}\,. \qquad \text{(A-1)}$$

But, by its definition, the velocity $\vec{v}$ is just the rate of change of position with time, i.e.,

$$\vec{v} = \frac{d\vec{r}}{dt} \qquad \text{(A-2)}$$

where $d\vec{r}$ is the particle's small displacement during the small time interval dt.

The relations (A-1) and (A-2) both involve the small time interval dt. Hence it should be possible to combine these relations so as to eliminate dt, and thus to find an equation relating directly the velocity $\vec{v}$ and position $\vec{r}$. The only difficulty is the algebra required to deal with vector relations such as (A-1) and (A-2). But this difficulty can readily be overcome by expressing these relations in terms of their numerical components along convenient directions.

Relating velocity and position

Let us then express all vectors in terms of their components along the total force and perpendicular to it.

Components of Newton's law. *Along* the direction of this total force, Newton's law (A-1) implies that

$$m\frac{dv_F}{dt} = F_{tot}\,. \qquad \text{(A-3)}$$

Here v_F denotes the component of the particle's velocity along the total force. Furthermore, the component of the total force along itself is simply its magnitude F_{tot}.

In a direction *perpendicular* to the total force, Newton's law (A-1) implies that

$$\frac{dv_\perp}{dt} = 0$$

since the component of the total force, perpendicular to itself, is just zero. This relation shows that the particle has no acceleration perpendicular to the total force. Hence the particle's velocity component $v_\perp$, perpendicular to the total force, remains unchanged during any small time dt, i.e.,

$$v_\perp = \text{constant (during dt).} \qquad \text{(A-4)}$$

Components of the definition of velocity. Similarly, the definition of velocity (A-2) implies a corresponding relation for the components of the vectors along the total force. Thus

$$v_F = \frac{dr_F}{dt} \qquad \text{(A-5)}$$

where dr_F is the component (along the total force) of the particle's small displacement during the small time interval dt.

Elimination of the time. The time interval dt can now be readily eliminated by combining the component equations (A-3) and (A-5). For example, this can be done by multiplying the corresponding sides of these equations. Thus one gets

$$m\,v_F\frac{dv_F}{dt} = F_{tot}\frac{dr_F}{dt}$$

or, multiplying both sides by dt,

$$m\, v_F\, dv_F = F_{tot}\, dr_F\, . \qquad \text{(A-6)}$$

The time interval dt has now been eliminated. Hence (A-6) relates directly the component v_F of the particle's velocity and component dr_F of its displacement.

As shown below, this relation can be further simplified by expressing it in terms of the speed v (i.e., the *magnitude* of the velocity). Thus (A-6) is equivalent to

$$\boxed{d(\tfrac{1}{2}mv^2) = F_{tot}\, dr_F\, .} \qquad \text{(A-7)}$$

This equation is the desired relation between a particle's speed and position.

Simplifications leading to (A-7)

The product $v_F\, dv_F$ in (A-6) can be expressed in terms of v_F^2. According to (5C-9), an infinitesimal change of v_F^2 is equal to

$$d(v_F^2) = 2\, v_F\, dv_F\, .$$

Furthermore, the magnitude v of the velocity is related to its components so that

$$v^2 = v_F^2 + v_\perp^2\, .$$

Hence $$d(v^2) = d(v_F^2) + d(v_\perp^2) = d(v_F^2)$$

since $d(v_\perp^2) = 0$ [i.e., since we know from (A-4) that the perpendicular component of the velocity remains unchanged]. By combining all this information, the left side of (A-6) can be expressed in the form

$$m\, v_F\, dv_F = \frac{1}{2} m\, d(v_F^2) = \frac{1}{2} m\, d(v^2) = d(\tfrac{1}{2}mv^2)$$

and thus yields the left side of (A-7).

Definitions of kinetic energy and work

The relation (A-7) can be written in simpler form by introducing convenient abbreviations for the quantities appearing on both of its sides. Accordingly, we define a quantity K, called *kinetic energy*.

Def: ***Kinetic energy:*** $K = \frac{1}{2} mv^2\, .$ (A-8)

Furthermore, consider *any* force $\vec{F}$ and an infinitesimal component dr_F of the particle's displacement along this force. Then we can define a quantity d'W called the infinitesimal *work* done on the particle by this force.

Def: ***Infinitesimal work:*** $d'W = F\, dr_F\, .$ (A-9)

Here the work d'W is infinitesimal since dr_F is infinitesimal. But, because of the extra factor F in (A-9), it is not merely defined as a small difference. [To make this distinction explicit, the symbol d' (with a prime) is used to denote any infinitesimal quantity, even if this quantity is not a difference.]

With these definitions, the relation (A-7) can be written in the following simple form

Kinetic-energy law *(for small displacement):* $dK = d'W_{tot}$	(A-10)

where $d'W_{tot}$ is the infinitesimal work done on the particle by the *total* force.

The kinetic-energy law (A-10) relates the change of a particle's kinetic energy to the work done on it along a small displacement. To understand the significance of this law, we must now examine the meanings of the newly introduced concepts of *kinetic energy* and *work*.

Kinetic energy

The kinetic energy (A-8) of a particle describes its motion in terms of its mass m and speed v.

Non-negative number. The kinetic energy $K = \frac{1}{2} mv^2$ is a *number* (rather than a vector) since both the mass m and speed v are numbers. It must be positive since both the mass and speed are positive, but can be zero if the particle's speed is zero. The kinetic energy is larger if the particle's mass is larger or its speed is larger.

Units of kinetic energy. By (A-8), the SI units of kinetic energy are:

$$\text{Units of kinetic energy} = \text{kg (m/s)}^2 = \text{kg m}^2/\text{s}^2 \equiv \text{joule} \tag{A-11}$$

where *joule* (abbreviated by *J*) is merely a convenient abbreviation for the preceding combination of units.

The unit joule is named after James Prescott Joule (1818-1889), an English physicist who carried out fundamental studies relating different forms of energy.

Infinitesimal work

The infinitesimal work $d'W = F\, dr_F$ depends on the force $\vec{F}$ exerted on the particle by some other object and thus describes the interaction of the particle with this object.

Defining method. According to the definition (A-9), the work done on a particle by a force $\vec{F}$, when the particle moves through a small displacement $d\vec{r}$, is found by doing the following (as illustrated in Fig. A-2).

Finding infinitesimal work	(A-12)
(1) Identify the magnitude F of the force acting on the particle. (2) Multiply this magnitude by the component dr_F of the particle's small displacement along the force. (The component of the displacement *perpendicular* to the force is irrelevant.)	

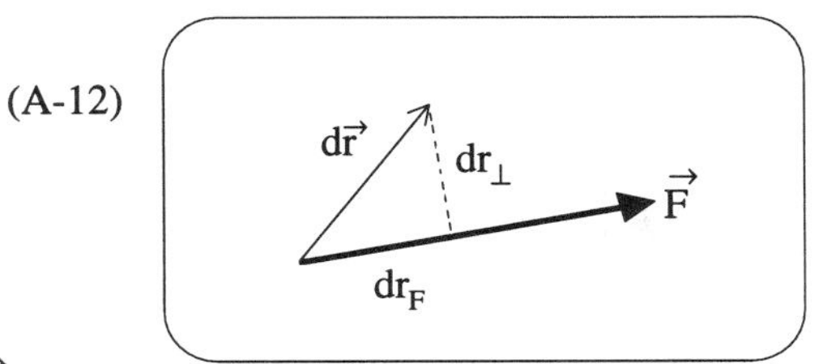

Fig. A-2. Method for finding infinitesimal work.

(The displacement $d\vec{r}$ is supposed to be sufficiently small that the force $\vec{F}$ has the same value throughout this displacement.)

Number with any sign. The work is a number (rather than a vector) since the magnitude F of the force is a number and the numerical component dr_F of the small displacement is also a number. The magnitude F of the force must be positive or zero; but the component of the displacement along the force can be positive, negative, or zero (depending on the direction of the displacement relative to the force). Hence the work itself can also be positive, negative, or zero.

Units of work. By (A-9), the SI units of work are:

$$\text{Units of work} = \text{N m} = (\text{kg m/s}^2)\ \text{m} = \text{kg m}^2/\text{s}^2 = \text{joule} \qquad \text{(A-13)}$$

i.e., they are the *same* as the units of kinetic energy. [Of course, this must be so to ensure the consistency of units in the relation (A-10) between kinetic energy and work.]

> ***Example: Work done along some small displacements***
>
> The downward gravitational force on a ball near the surface of the earth has a magnitude F = 2 newton. This ball can move a small distance s = 0.01 meter along each of the several directions indicated in Fig. A-3. In each of these cases, some work d'W is done on the ball by the gravitational force.
>
> (a) If the ball moves vertically downward (so that its displacement is along the force), d'W = Fs = (2 N) (0.01 m) = 0.02 J.
>
> (b) If the ball moves vertically upward (so that its displacement is opposite to the force), d'W = F (–s) = – 0.02 J.
>
> (c) If the ball moves horizontally (so that its displacement is perpendicular to the force), d'W = 0 (since the component of the displacement along the force is zero).
>
> (d) If the ball moves downward at an angle $\theta = 30°$ from the horizontal, d'W = F (s sin30°) = 0.01 J (since sin30° = 0.50).

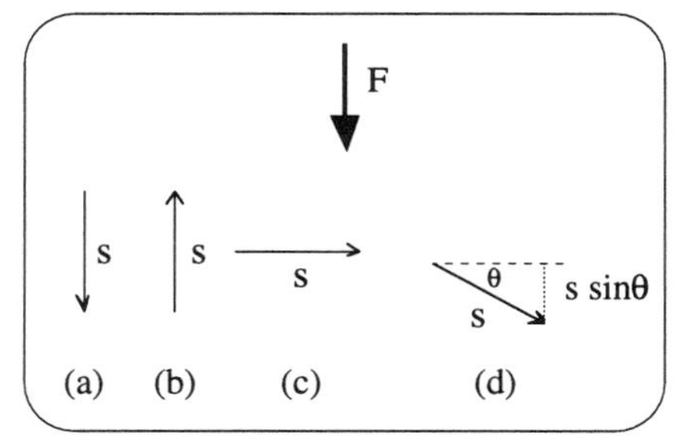

Fig. A-3. Work done along various small displacements.

Relation between kinetic energy and work

The kinetic-energy law (A-10) specifies a fundamental relation between a particle's *change* of kinetic energy and the work done on it. The kinetic energy involves the particle's speed; the work is that done by the *total* force acting on the particle. Hence the kinetic-energy law (just like Newton's law) relates the motion of a particle to its interactions. But it describes motion and interactions in terms of kinetic energy and work (instead of acceleration and forces).

The kinetic-energy law (A-10) also indicates the physical significance of the concept *work* defined in (A-9). *Work is a quantity describing how a particle's speed is affected by its interaction with other objects.* It is thus closely related to the concept *force* (since force is a quantity describing how a particle's *acceleration* is affected by its interaction with other objects.)

> ***Example relating speed and work***
>
> While falling under the influence of gravity, a ball descends vertically through some small distance. The work done on the ball by the gravitational force is then positive since the ball's displacement is along the force. According to the kinetic-energy law (A-10), the change of the ball's kinetic energy is then also positive. Hence the ball's kinetic energy, and thus also its speed, increases.

Conversely, suppose that the ball, after being thrown up, is moving upward through some small distance. Then the work done on the ball by the gravitational force is negative since the ball's displacement is opposite to the force. According to the kinetic-energy law (A-10), the change of the ball's kinetic energy is then also negative. Hence the ball's kinetic energy, and thus also its speed, decreases.

→ ***Go to Sec. 13A of the Workbook.***

B. General kinetic-energy law

Kinetic-energy law for any path

The kinetic-energy law stated in (A-10) is only valid for very small displacements of a particle. However, the law can be easily extended to so as to apply to *any* motion of a particle.

Particle moving along some path. Consider a particle which moves along some path from a point A to another point B, as illustrated in Fig. B-1. The motion of the particle along this path consists then of a sequence of successive infinitesimal displacements. The kinetic-energy law (A-10) can be applied to each of these. Thus it must be true that

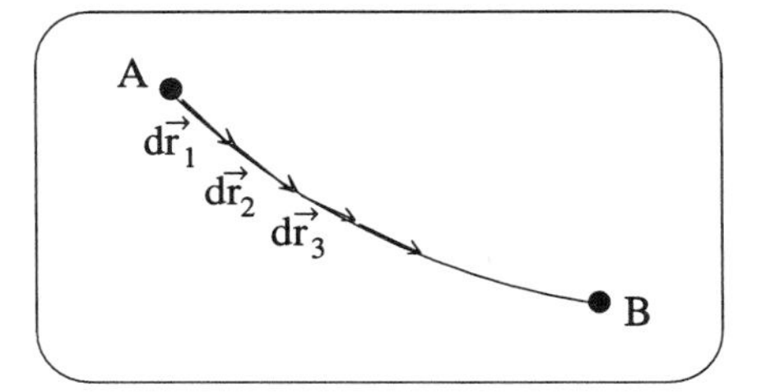

Fig. B-1. Infinitesimal displacements (much exaggerated) of a particle moving along some path.

$$(dK)_1 = (d'W_{tot})_1, \quad (dK)_2 = (d'W_{tot})_2, \quad (dK)_3 = (d'W_{tot})_3, \ldots$$

Here $(dK)_1$ is the change of the particle's kinetic energy during the first small displacement, and $(d'W_{tot})_1$ is the work done on it by the total force during this displacement. Similarly, $(dK)_2$ is the change of the particle's kinetic energy during the second small displacement, and $(d'W_{tot})_2$ is the work done on it by the total force during this displacement. And so forth. Adding the contributions from all the successive displacements, one then gets

$$(dK)_1 + (dK)_2 + (dK)_3 + \ldots = (d'W_{tot})_1 + (d'W_{tot})_2 + (d'W_{tot})_3 \ldots \quad \text{(B-1)}$$

Entire change of kinetic energy. The sum of all the successive small changes of kinetic energy on the left side of (B-1) is just equal to the entire change ΔK of the particle's kinetic energy as it moves from A to B. In other words

$$(dK)_1 + (dK)_2 + (dK)_3 + \ldots = \Delta K = K_B - K_A \quad \text{(B-2)}$$

where K_A is the particle's initial kinetic energy at the point A and K_B is its final kinetic energy at the point B.

Work along a path. The right side of (B-1) can be expressed in more compact form by introducing a convenient abbreviation. Let us then introduce a quantity W_{AB} called "the work done on a particle by a force $\vec{F}$ along the entire path from A to B". This work is defined to be the sum of the infinitesimal works along the successive displacements along this path. This definition can be summarized by writing

$$W_{AB} = \int_A^B d'W \qquad \text{(B-3)}$$

where the symbol $\int$ denotes a sum of infinitesimal quantities along the path from A to B. (Mathematicians call this a *line integral.*) Even more compactly, one can omit explicit mention of the endpoints of the path and simply write

The symbol $\int$ (called the integral sign) is merely an elongated form of the letter S (indicating a sum).

The sum approaches a well-defined limiting value if the works d'W are chosen small enough.

Def: **Work:** $W = \int d'W = \int F\, dr_F$ (B-4)

where we have recalled the definition (A-9) of the infinitesimal work d'W.

Kinetic-energy law. In (B-1) the work along the particle's path is that done by the *total* force on the particle. With the preceding definition of work, (B-1) can then be summarized to yield the following general form of the kinetic-energy law:

Kinetic-energy law: $\Delta K = W_{tot}$. (B-5)

Stated in words, this law says the following: *The change of a particle's kinetic energy is equal to the work done on the particle by the total force acting on it.* Unlike the earlier statement (A-10), the form (B-5) of the law is applicable to *any* displacement of a particle, no matter how large.

The kinetic-energy law (B-5) is very useful. It will be discussed further throughout the rest of this chapter and frequently applied afterwards.

Work along a path

Specification of work. A complete specification of work requires the phrase

work done on <u>particle</u> by <u>force</u> along <u>path</u> (B-6)

where one needs to specify each of the underlined entities.

The work W done on the particle by some force $\vec{F}$ can be calculated from its definition (B-4)

$$W = \int d'W, \quad \text{where } d'W = F\, dr_F\,. \qquad \text{(B-7)}$$

In other words, this work is just the sum of the infinitesimal works done on the particle along all the successive small segments of its path.

Method for calculating work. The method for calculating this work thus involves the following main steps:

Finding work along a path	
(1) Calculate the infinitesimal work done on the particle for each of its successive infinitesimal displacements along the path (by multiplying the magnitude of the force by the component of the particle's displacement along this force). (2) Add up all these infinitesimal works.	(B-8)

As indicated in Fig. B-2, the force on a particle is ordinarily *different* at different positions. The work done on the particle, along any small segment of its path, must thus be calculated by using the force existing at the particular position of this segment.

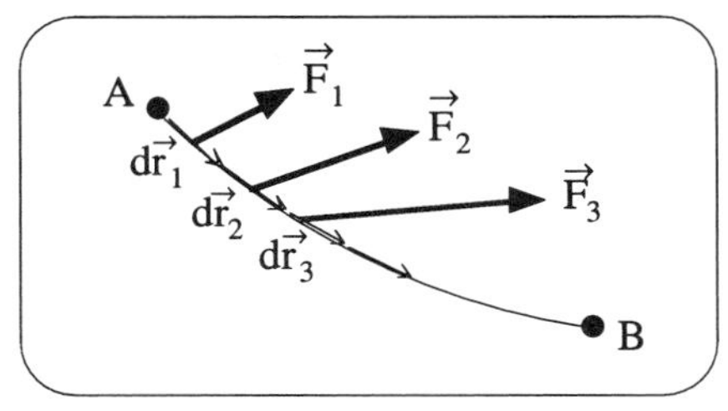

Fig. B-2. Forces at different positions of a particle's path.

The calculation of work can thus be fairly complex. However, it is quite simple in many important special cases, like those examined in the next section.

➜ ***Go to Sec. 13B of the Workbook.***

C. Work done by some common forces

Work done by a constant force

Consider the special case where the force $\vec{F}$ on a particle is *constant* (so that it has everywhere the same magnitude and the same direction). For example, the force might be the downward gravitational force due to the earth near its surface.

Work done along a small displacement. The work $d'W = F\, dr_F$ done on the particle along any small displacement depends then only on the component dr_F of the displacement along the fixed direction of the force (e.g., only on the downward component of the displacement if this is the gravitational force). For example, in Fig. C-1 the components (along the force) of all the small displacements have the *same* magnitude s. Since the constant force has the same magnitude F everywhere, the *same* work Fs is then done by this force along each of the different small displacements 1, 2, and 3. (A work –Fs, of the same magnitude but opposite sign, is done by this force along the displacements 4 or 5 since the component of the displacement along the force is then negative.)

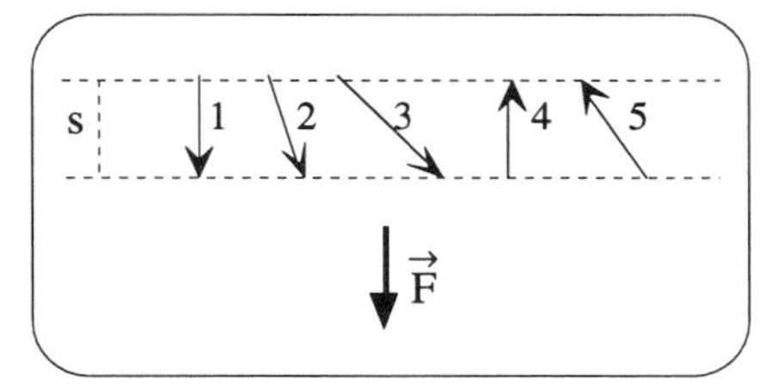

Fig. C-1. Different infinitesimal displacements in the presence of a constant force.

Work done along a large displacement. The work done along any path, such as the path from A to B in Fig. C-2, is then simply the sum of the works done along all the successive small displacements along this path. Thus

$$W = F\,(dr_F)_1 + F\,(dr_F)_2 + \ldots = F\,[(dr_F)_1 + (dr_F)_2 + \ldots]$$

or, in more compact notation,

$$W = \int F dr_F = F \int dr_F$$

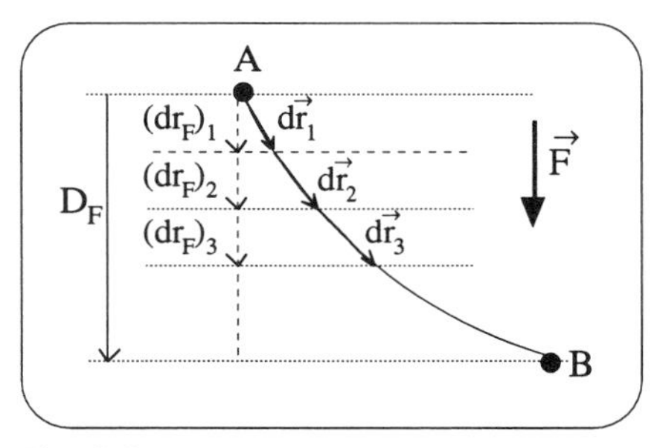

Fig. C-2. Path of a particle in the presence of a constant force.

since the force has the same magnitude F everywhere and is thus a common factor. But the sum of the components of all the successive small displacements is just the component D_F of the particle's *total* displacement along the force (as illustrated in Fig. C-2). Hence one obtains the following simple result

$$\text{if } \vec{F} \text{ is constant,} \qquad W = F\,D_F\,. \tag{C-1}$$

Path-independence of the work. The work in (C-1) depends on the particle's total displacement from one point to the other. However, it does *not* depend on the path traversed by the particle between these points. Thus the work done by a constant force is the same for *any* path joining the same two points.

The reason for this conclusion is apparent from Fig. C-3 which shows two different paths leading from a point A to another point B. The first path consists of successive small segments labeled 1, 2, 3, ...; the second path consists of successive small segments 1', 2', 3', ... As indicated in Fig. C-3, the components, along the force, of the small displacements along corresponding segments are the same. Hence the work done on the particle along segment 1 is the same as the work done on it along segment 1', the work done along segment 2 is the same as the work done along segment 2'; and so forth. Hence the work done along the entire first path is the same as the work done along the entire second path.

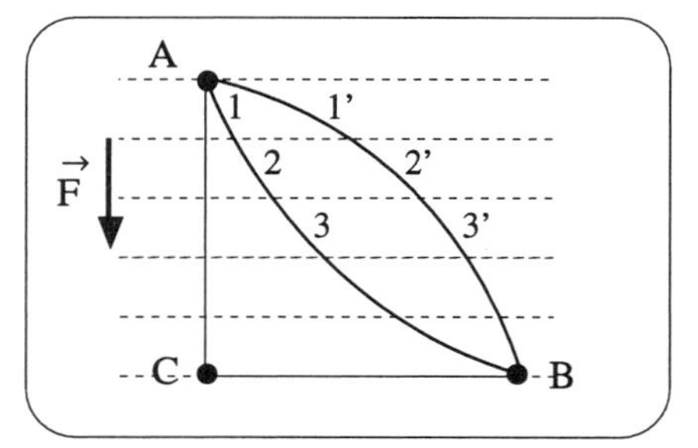

Fig. C-3. Different paths between the two points A and B.

The preceding conclusion is equally true for more complicated paths, like the path APQRSB shown in Fig. C-4. Here the work done from Q to R has the same magnitude, but opposite sign, as the work done from R to S. Hence these works cancel so that the entire work done along PQRS is just the work done along PQ, and this is again the same as the work done along segment 2 of the first path.

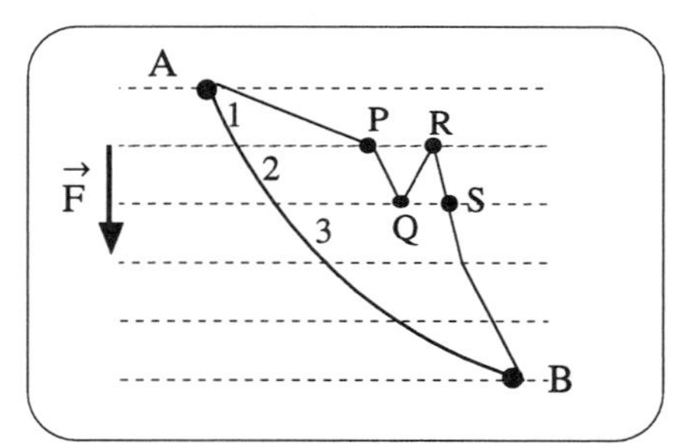

Fig. C-4. A complicated path from A to B.

The preceding remarks can be summarized by the following statement:

The work done by a *constant* force is independent of the path.	(C-2)

Simple paths for calculating work. If the force is constant, one may thus choose *any* path to calculate the work done on a particle moving between two points. In particular, one can simplify the calculation by choosing a very simple path. For example, in Fig. C-3, the work done along any path from A to B is the same as the work done along the path ACB, i.e., it is simply equal to the work along the straight path AC parallel to the force (since the work along the remaining path CB, perpendicular to the force, is zero).

Example: Speed of a projectile

A projectile is launched with a speed v_O from a point O on the ground, as illustrated in Fig. C-5. Air resistance is negligible. What then is the projectile's speed v_P when it is at some other point P at a height h above the ground?

Let us apply the kinetic-energy law to the projectile between the instant of launch and the instant when the projectile is at P. Then

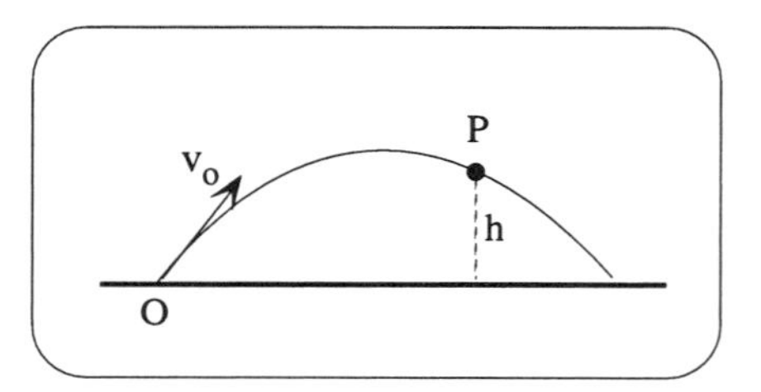

Fig. C-5. Speed of a projectile.

$$\Delta K = K_P - K_o = W_g \qquad \text{(C-3)}$$

since the work done by the total force is here merely the work W_g done by the gravitational force. But, if the projectile has a mass m, the downward gravitational force on it has a constant magnitude mg. According to (C-1), an amount of work mg(–h).is then done by this force on the projectile when it moves up through a vertical distance h. Hence (C-3) implies that

$$\frac{1}{2}mv_P{}^2 - \frac{1}{2}mv_o{}^2 = mg(-h)\,.$$

A little algebra then yields the speed at any point P. Thus

$$v_P = \sqrt{v_o{}^2 - 2gh.} \qquad \text{(C-4)}$$

Note that it would have been much more difficult to obtain this result by using arguments based on forces and accelerations.

Work done by a friction force

Suppose that a particle moves along the horizontal floor. The *magnitude* f of the friction force exerted on the particle by the floor is then constant. However, the *direction* of the friction force is *not* constant since this direction is always opposite to the particle's displacement, as indicated in Fig. C-6 (i.e., this direction is always such as to oppose the particle's motion relative to the floor).

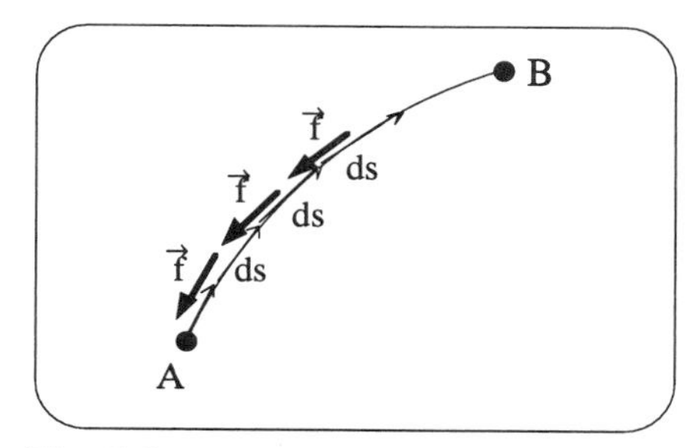

Fig. C-6. Friction force on a particle moving along a horizontal floor.

When the particle moves an infinitesimal distance ds along its path, the work d'W done on it by the friction force is equal to

$$d'W = f\,(-ds) = -\,f\,ds$$

because the component of the particle's displacement along the force is –ds. (This work is negative since the displacement is directed *opposite* to the friction force.) By the definition (B-4), the work done by the friction force along some entire path, like that indicated in Fig. C-6, is then

$$W = \int(-f\,ds) = -\,f\int ds \qquad \text{(C-5)}$$

since f is merely a constant factor multiplying every term in the sum. But the last sum in (C-5) is just the sum of all the successive infinitesimal distances moved along the path (i.e., it is the total length L of the path). Hence the work done by a friction force of constant magnitude f is simply

$$\boxed{\text{if } f = \text{constant}, \qquad W = -\,fL\,.} \qquad \text{(C-6)}$$

Path-dependence of frictional work. Note that the work done by the friction force depends on the length of the path and thus *does* depend crucially on the nature of this path. For example, if a particle moves from a point A to another point B along the longer path 1 in Fig. C-7, the magnitude of the work done by the friction force is larger than if the particle moves between the same two points along the shorter path 2.

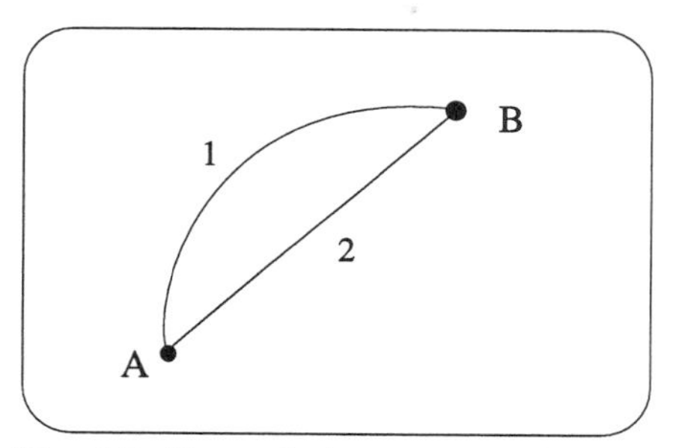

Fig. C-7. Two paths between the same two points.

➜ ***Go to Sec. 13C of the Workbook.***

D. Work and vector dot product

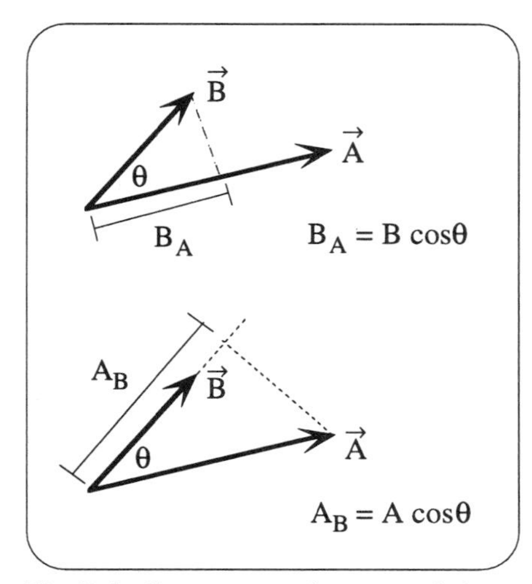

Fig. D-1. Components of vectors relative to each other.

Alternative calculations of work. According to the definition (A-9), the work d'W done by a force $\vec{F}$ along a small displacement $d\vec{r}$ is equal to

$$d'W = F\, dr_F\,. \qquad \text{(D-1)}$$

To calculate this work, one thus needs to multiply the magnitude of one vector $\vec{A}$ by the component B_A of a second vector $\vec{B}$ along the first. Thus one needs to find the product

$$A\, B_A\,. \qquad \text{(D-2)}$$

In the case of the work (D-1), the vector $\vec{A}$ is the force and $\vec{B}$ is the displacement. But the product (D-2) occurs also in many other situations and can be calculated in various useful ways.

As is apparent from the top part of Fig. D-1, the component of $\vec{B}$ along $\vec{A}$ is just $B_A = B \cos\theta$ if θ is the angle between the two vectors. Thus (D-2) is equal to

$$A\, B_A = A\, B \cos\theta\,. \qquad \text{(D-3)}$$

The result (D-3) is generally valid if the cosine is considered negative when θ is larger than 90°.

The result (D-3) can also be expressed in another way by noting that $A \cos\theta = A_B$, the component of $\vec{A}$ along $\vec{B}$. (This is apparent from the bottom part of Fig. D-1.) Thus (D-3) can be equivalently expressed in any one of the following ways

$$A\, B_A = A\, B \cos\theta = B\, A_B\,. \qquad \text{(D-4)}$$

Definition of dot product. Accordingly, it is useful to denote (D-4) by the following convenient abbreviation

Def: ***Dot product:*** $\vec{A}\cdot\vec{B} = A\, B \cos\theta = A\, B_A = B\, A_B$ (D-5)

The quantity $\vec{A}\cdot\vec{B}$ thus defined introduces a particular way of multiplying two vectors so as to find their so-called *dot product* (denoted by a dot between the vector symbols).

According to its definition, the dot product of two vectors is obtained by multiplying the *magnitudes* of these vectors by the cosine of the angle between them. Equivalently, it is obtained by multiplying the magnitude of either vector by the component of the other vector along it.

Simple properties of the dot product. According to the definition (D-4), the dot product of two vectors is a *number* (i.e., *not* a vector). This number can be positive or negative (depending on whether the angle between the two vectors is less than or greater than 90°). The dot product is zero if the vectors are perpendicular (since $\cos 90° = 0$).

The *square* of a vector is defined as the dot product of the vector with itself, i.e.,

$$\vec{A}^2 = \vec{A}\cdot\vec{A} = A\,A\cos 0° = A^2 \tag{D-6}$$

and is thus merely equal to the square of the *magnitude* of this vector.

Commutative property. The definition (D-4) of the dot product implies that the order of multiplication of the vectors is irrelevant, i.e.,

$$\vec{A}\cdot\vec{B} = \vec{B}\cdot\vec{A}\,. \tag{D-7}$$

Distributive property. The relation

$$(\vec{A}+\vec{B})\cdot\vec{C} = \vec{A}\cdot\vec{C} + \vec{B}\cdot\vec{C} \tag{D-8}$$

would be true for the products of ordinary numbers. Is it also true for the dot products of vectors? By the definition (D-4) of the dot product, the question is equivalent to asking whether

$$(\vec{A}+\vec{B})_C = A_C + B_C\,,$$

i.e., whether the component of a sum of two vectors, along some other vector $\vec{C}$, is equal to the sum of their components. But we already saw in (7C-4) that this is true. Hence the relation (D-8) is true.

Applications to work. The definition (D-4) of the dot product allows one to express the infinitesimal work (D-1) in the compact form

$$d'W = \vec{F}\cdot d\vec{r}\,. \tag{D-9}$$

This work can then be calculated in several equivalent ways, corresponding to the several equivalent expressions (D-5) of the dot product. For example, this work can be calculated (a) by multiplying the magnitude of the force, the magnitude of the small displacement, and the cosine of the angle between them; or (b) by multiplying the magnitude of the force by the component of the small displacement along this force; or (c) by multiplying the magnitude of the small displacement by the component of the force along it.

Using the notation of the dot product, the general definition (B-4) of work can be summarized in the form

Def: **Work:** $W = \int d'W = \int \vec{F}\cdot d\vec{r}\,.$ (D-10)

The previously discussed properties of the dot product can then often be exploited to simplify the calculation of work.

➜ ***Go to Sec. 13D of the Workbook.***

E. Superposition principle for work

Work done by several forces. Suppose that the force $\vec{F}$ on a particle is equal to the sum of two other forces $\vec{F}_1$ and $\vec{F}_2$ so that

$$\vec{F} = \vec{F}_1 + \vec{F}_2 . \tag{E-1}$$

(For example, the force $\vec{F}$ might be the total force on a particle acted on by the forces due to two other objects.) How then is the work done by the total force $\vec{F}$ related to the works done by the individual forces?

If the particle is displaced by a small amount $d\vec{r}$, the corresponding small work done on the particle by the total force $\vec{F}$ is then

$$d'W = \vec{F} \cdot d\vec{r} = (\vec{F}_1 + \vec{F}_2) \cdot d\vec{r} = \vec{F}_1 \cdot d\vec{r} + \vec{F}_2 \cdot d\vec{r}$$

where the last step has exploited the property (D-8) of the dot product. But the last two terms in this relation are merely the work $d'W_1$ and $d'W_2$ done by the two individual forces. Thus

$$d'W = d'W_1 + d'W_2 . \tag{E-2}$$

Since this relation is true for the works done along any small displacement, it must also be true for the works done along some entire path so that

$$W = W_1 + W_2 . \tag{E-3}$$

Superposition principle. Thus we arrive at the following general conclusion:

Superposition principle for work: The work done by a sum of forces is equal to the sum of the works done by the individual forces.	(E-4)

In particular, *the work done by a total force is equal to the total work* (i.e., to the sum of the works done by all the individual forces).

Note that the sum of forces is a *vector* sum, but that the sum of the corresponding works is an ordinary *numerical* sum.

Utility of the superposition principle. To appreciate the great utility of the superposition principle (E-4), consider two different methods which can be used to calculate the work W_{tot} done on a particle by the total force due to all other objects.

Method 1. Calculate the total force on the particle by adding the forces exerted by all the other objects. (Since forces are vectors, this requires one to add vectors.) Then use this total force to calculate the work.

Method 2. Calculate separately the work done by each of the individual forces. Then add all these works. (Since works are just numbers, this requires one merely to add ordinary numbers.)

Since the addition of numbers is much simpler than that of vectors, the second method is ordinarily far simpler — and often allows one easily to solve problems which would otherwise be difficult or impossible. The following example provides an illustration.

Example: Speed of a pendulum bob

The bob of a pendulum is released from rest at a height h above the lowest point of its swing. What then is the speed of the bob at this lowest point?

Fig. E-1 illustrates this situation where the bob is released at the point A and reaches its lowest position at B. Fig. E-2 describes the bob at any typical time between these. To find the bob's speed at B, we can apply the kinetic-energy law to the bob moving from A to B. Thus

$$\Delta K = W_{tot}$$

or

$$K_B - K_A = W_g + W_T \,. \qquad \text{(E-5)}$$

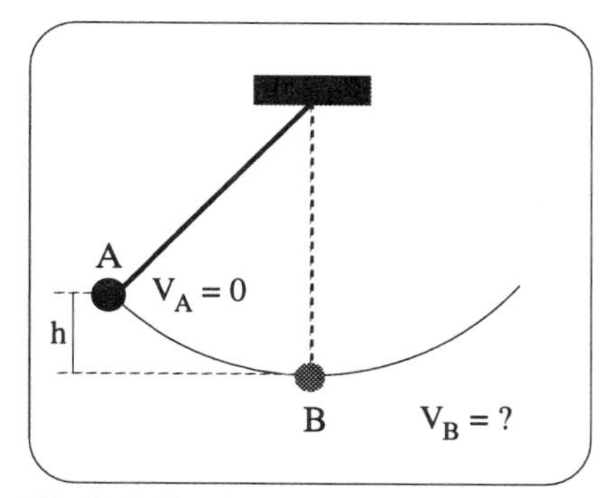

Fig. E-1. Pendulum released from rest.

Here we have exploited (E-4) to claim that the total work done on the bob is the work W_g done by the gravitational force plus the work W_T done by the tension force due to the string.

The work W_g done by the constant downward gravitational force depends merely on the bob's vertical displacement h along this force. Thus $W_g = mgh$ if m is the mass of the bob.

Any infinitesimal displacement of the bob is perpendicular to the tension force since the tension force is everywhere perpendicular to the bob's circular path. Hence the tension force does *zero* work on the bob along any small displacement. Thus it also does *zero* work along the entire path of the bob.

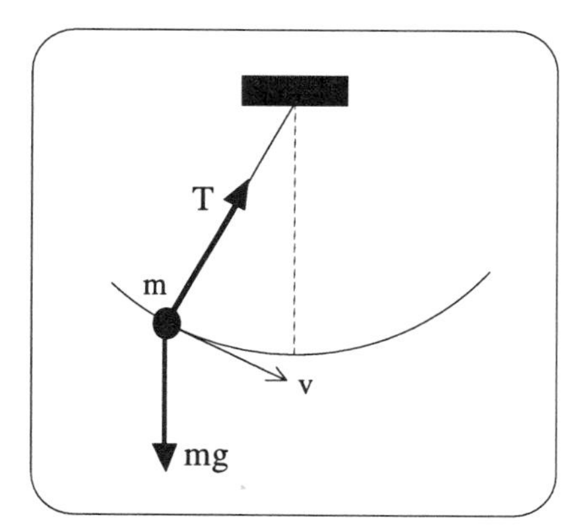

Fig. E-2. Forces on the pendulum bob.

The kinetic-energy law (E-5) applied to the bob from A to B thus yields

$$\frac{1}{2} m v_B{}^2 - 0 = mgh + 0 \,.$$

Hence

$$v_B{}^2 = 2gh$$

or

$$v_B = \sqrt{2gh} \,. \qquad \text{(E-6)}$$

Note that the speed attained by the ball is exactly the same as if the string did not exist and the ball had merely fallen vertically through the distance h. In other words, the radial force exerted by the string affects the acceleration of the bob so as to deflect it into a circular path, but it does *not* affect the bob's speed.

[The preceding problem would be extremely difficult to solve if, instead of adding the two simple works in (E-5), one tried to calculate the total work by finding the total force on the bob.]

➜ ***Go to Sec. 13E of the Workbook.***

F. Power

The time required to do some work may sometimes be important. (For example, it may be more difficult to carry a box up a flight of stairs rapidly rather than slowly, although the work done would be the same.) Thus one may also be interested in specifying the *rate* of doing work.

Definition of power. Suppose that a small amount of work d'W is done during a short time dt. Then the rate of doing work, called *power* and denoted by $\mathcal{P}$, is defined as follows:

Def: **Power:** $$\mathcal{P} = \frac{d'W}{dt}\,. \qquad \text{(F-1)}$$

This definition properly implies that, if the same work is done in a shorter period of time, the power (i.e., the rate of doing work) is larger.

Units of power. The definition (F-1) implies that the SI units of power are

$$\text{units of power} = \frac{\text{joule}}{\text{second}} \equiv \text{watt} \qquad \text{(F-2)}$$

The unit *watt* is named in honor of James Watt (1736-1819), the inventor of the steam engine.

where the unit *watt* (abbreviated by *W*) is merely a convenient abbreviation for the preceding combination of units.

For example, a 100-watt light bulb is one which requires that 100 joules of work are done per second by electric forces (in order to cause electrons to move through the filament of the light bulb and thus to heat the filament sufficiently to produce light).

Power delivered to a moving particle. The work d'W done on a particle by a force $\vec{F}$, when the particle moves through a small displacement $d\vec{r}$, is

$$d'W = \vec{F} \cdot d\vec{r}\,.$$

By the definition (F-1), the power delivered to the particle by this force is obtained by dividing this work by the small time dt during which it is done. Thus

$$\mathcal{P} = \frac{d'W}{dt} = \vec{F} \cdot \frac{d\vec{r}}{dt}$$

or

$$\mathcal{P} = \vec{F} \cdot \vec{v} \qquad \text{(F-3)}$$

since $d\vec{r}/dt$ is just the velocity $\vec{v}$ of the particle.

➜ ***Go to Sec. 13F of the Workbook.***

G. Summary

Kinetic-energy law: $\Delta K = W_{tot}$ (W_{tot} = total work)

Definitions:

Kinetic energy: $K = \frac{1}{2} mv^2$ (m = mass, v = speed)

Work: $W = \int d'W = \int \vec{F} \cdot d\vec{r}$ (sum of small works along path)

Power: $\mathcal{P} = \frac{d'W}{dt}$

Dot product of vectors:

Definition: $\vec{A} \cdot \vec{B} = A\, B \cos\theta = A\, B_A = B\, A_B$

Properties: $\vec{A} \cdot \vec{B} = \vec{B} \cdot \vec{A}$

$(\vec{A} + \vec{B}) \cdot \vec{C} = \vec{A} \cdot \vec{C} + \vec{B} \cdot \vec{C}$

New abilities

You should now be able to do the following:

(1) Calculate the work done by a force along any small displacement, and find the power delivered by such a force.

(2) Calculate the work done along any path by a constant force or by a friction force of constant magnitude.

→ ***Go to Sec. 13G of the Workbook.***

14 Potential Energy and the Energy Law

As discussed in the preceding chapter, the work done on a particle by a *constant* force (like the gravitational force near the earth) is independent of the particle's path. In the present chapter we shall see that the work done by all fundamental forces (e.g., gravitational or electric forces) shares the same simple property. The calculation of work can then be greatly simplified by introducing the concept of *potential energy*. Furthermore, the kinetic-energy law can be expressed in the form of a general *energy law* which is very widely useful and important.

A. Potential energy

Conservative forces. A *constant* force (like the gravitational force near the surface of the earth) is independent of the position of the particle on which it acts. As discussed in Sec. 13C, the work done on a particle by such a force is independent of the particle's path or speed (i.e., it depends solely on the particle's initial and final positions). A force having this simple property is called *conservative* in accordance with the following definition:

Def: **_Conservative force:_** A force which does the same work independent of the path. (A-1)

The force is called *conservative* because (as shown in Sec. B) it can lead to the constancy of a quantity called "energy".

For example, the constant gravitational force near the earth is conservative, but a friction force is not.

Potential energy due to conservative forces

As we shall see later, most important forces are conservative even though they are not constant. How can one exploit the resulting simplicity?

An analogy from daily life

When hiking in mountainous terrain, one is often interested in one's *descent*, i.e., in the vertical distance descended between two points. (For example, one's descent is positive when walking downhill, and negative when walking uphill.) Note that the descent between any two points A and B is the *same* no matter what path is chosen to walk between them.

To find the descent between these points, one could walk between them along some path — and laboriously add up the vertical components of the displacements along all the successive steps along this path.

However, the task is much simpler if one knows the vertical heights of the point A and B above some standard level S (e.g., sea level). For then the descent between these points can be immediately obtained by merely subtracting their heights.

Standard position for calculating work. The preceding analogy suggests the following question: If a force is conservative, how can the work done by it, along any path, be found from mere information about the endpoints A and B of this path?

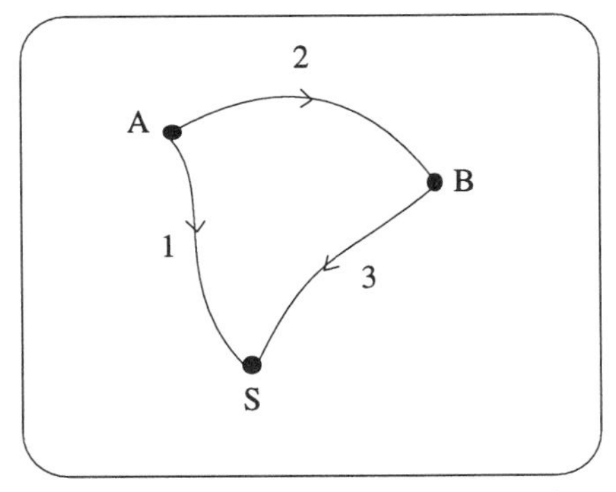

Fig. A-1. Alternate paths from a point A to a standard position S.

Let us choose some point S as a standard position. One can then go from the point A to the standard position S either directly (as indicated by the path 1 in Fig. A-1), or by passing through the point B (as indicated by the successive paths 2 and 3 in Fig. A-1). Some work W_{AS} is done in the first case, and some work $(W_{AB} + W_{BS})$ is done in the second case. But, since the work is independent of the path, the work along each of these two paths from A to S must be the same. Thus

$$W_{AS} = W_{AB} + W_{BS}$$

or

$$W_{AB} = W_{AS} - W_{BS} \,. \tag{A-2}$$

This shows that the work between any two points A and B can be found by simple subtraction if one knows the work done from each of these points to the standard position S.

Definition of potential energy. The result (A-2) can be simplified by introducing the following definition of a quantity U_A

$$\boxed{U_A = W_{AS} \,.} \tag{A-3}$$

This quantity is called the *potential energy* of the particle at the point A (relative to the standard position S). This definition can be stated in words as follows:

Def: ***Potential energy:*** The potential energy U_A (due to specified conservative forces) of a particle at a point A, relative to some standard position S, is the work W_{AS} done on the particle by these forces along any path from A to S. (A-4)

Relation between work and potential energy. In terms of this definition, the result (A-2) can be written as

$$W_{AB} = U_A - U_B = -(U_B - U_A) \,. \tag{A-5}$$

Thus the work is simply related to the change ΔU of the particle's potential energy. This change is defined as the final potential energy U_B at the point B minus the original potential energy U_A at the point A. Hence the relation (A-5) can be written more compactly as

$$\boxed{W = -\Delta U\,.} \tag{A-6}$$

The work done on a particle by a conservative force can thus be calculated very simply from a knowledge of the particle's potential energy due to this force. The resulting simplification is particularly great in the case of conservative forces more complex than the constant gravitational force near the earth. For, once the potential energy has been determined, any work can be found merely be subtracting potential energies, without needing to make any further complex calculations.

Properties of the potential energy. Because of its definition (A-3), the potential energy has the same properties as work. Thus it is a *number* (which can be positive, negative, or zero) expressed in terms of the SI unit *joule*.

The potential energy U_S at the standard position S is just the work W_{SS} done in going from S back to S along any path — and thus also the work done in just remaining at S. Since this work is obviously zero, it is generally true that

$$\boxed{\text{at the standard position,} \qquad U_S = 0\,.} \tag{A-7}$$

The standard position S can be chosen in any convenient way. The *value* of the potential energy depends, correspondingly, on the choice of this standard position. However, the *difference* between the potential energies at any two points does *not* depend on the choice of standard position. Indeed, by (A-6), it is just equal to the work done between these points.

If a particle moves along a force, the work done by this force is positive. By (A-5) or (A-6) the particle's potential energy correspondingly decreases. Thus one arrives at the following qualitative conclusion:

> A particle's potential energy decreases if the particle moves along the direction of the corresponding force, and increases if the particle moves opposite to this force. (A-8)

Gravitational potential energy near the earth

The preceding ideas are readily applied to the simple case of a particle, of mass m, acted on by the constant downward gravitational force $m\vec{g}$ due to the neighboring earth. Suppose that this particle is located at some point A at a vertical height h above some chosen standard position S, as illustrated in Fig. A-2. By the general definition (A-3), the potential energy U of the particle at this point is then simply equal to the work W_{AS} done on the particle by the gravitational force when the particle moves along any path from A to S. Since the component of the particle's displacement along the downward gravitational force is merely h, (13C-1) implies that $W_{AS} = (mg)\,h$. Hence

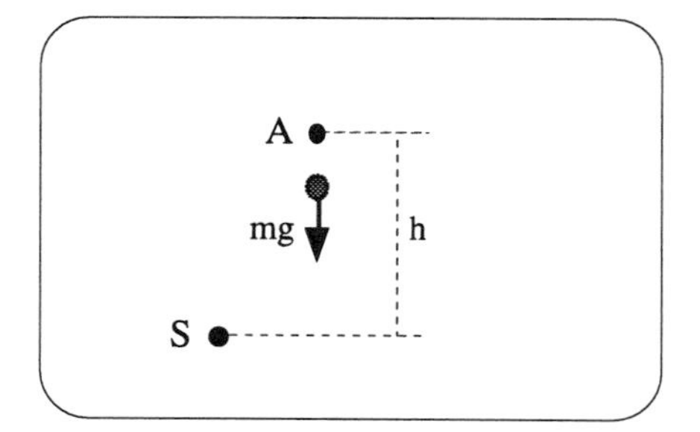

Fig. A-2. Position A of a particle relative to a standard position S.

$$\boxed{U = mgh\,.} \tag{A-9}$$

Here h is positive if the position of the particle is above the standard position S (so that the component of the particle's displacement from A to S is positive). Conversely, h is negative if the position of the particle is below the standard position S (so that the component of the particle's displacement from A to S is negative).

Example: Gravitational work on a stone falling into a lake

Fig. A-3 indicates a stone, having a weight 2 newton, which is dropped from a point A located 3 meters above the surface of a lake. The stone lands at a point B at the bottom of the lake 4 meters below the surface . What is the work W done on the stone by the gravitational force during this fall?

Since the stone falls vertically through a total distance of 7 m, direct calculation of the work yields

$$W = (2\text{ N})\,(7\text{ m}) = 14\text{ J}. \qquad \text{(A-10)}$$

Alternatively, one can first find the potential energies of the stone at A and B. By (A-9), these potential energies, relative to a standard position S chosen at the surface of the lake, are

$$U_A = (2\text{ N})\,(3\text{ m}) = 6\text{ J}\,; \quad U_B = (2\text{N})\,(-4\text{ m}) = -8\text{ J}\,.$$

By (A-5), the work W is then

$$W = U_A - U_B = 6\text{ J} - (-8\text{ J}) = 14\text{ J}$$

which agrees properly with (A-10).

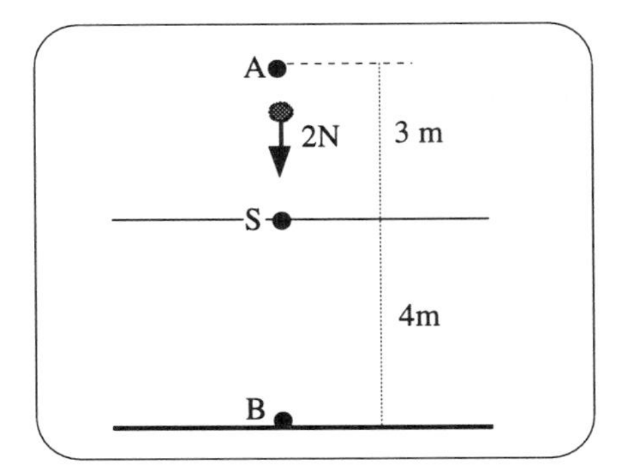

Fig. A-3. Stone falling into a lake.

➜ ***Go to Sec. 14A of the Workbook.***

B. Energy law

Simplification of the total work. As we have seen, the work done by a conservative force can be simply found from its associated potential energy. How can this simplification be exploited in applying the kinetic-energy law?

The kinetic-energy law

$$\Delta K = W_{tot} \qquad \text{(B-1)}$$

relates the change of the kinetic energy K of a particle to the total work done on the particle. But this total work may be expressed as the sum of two parts so that one can write

$$W_{tot} = -\Delta U + W_{oth}\,. \qquad \text{(B-2)}$$

Here the first term on the right side represents work which is done by conservative forces and which can thus, by (A-6), be expressed in terms of the change ΔU of a corresponding potential energy. The second term W_{oth} on the right side represents all the remaining *other work*, i.e.,

W_{oth} = other work (due to all interactions *not* included in the potential energy). (B-3)

It is usually most convenient to use the change of potential energy ΔU to describe the work done by *all* conservative forces. The other work W_{oth} is then the work done by all other non-conservative froces (e.g., friction forces).

Energy and the energy law. By using (B-2), the kinetic energy law (B-1) can be written as

$$\Delta K = -\Delta U + W_{oth} .$$

Hence

$$\Delta K + \Delta U = \Delta(K + U) = W_{oth} . \quad \text{(B-4)}$$

This can be simplified by introducing the following definition of a quantity E, called the *energy* (or *total energy*) of the particle:

Def: ***Energy:*** $E = K + U .$ (B-5)

By using this definition, the relation (B-4) can be expressed in the following simple form called the *energy law*:

Energy law: $\Delta E = W_{oth} .$ (B-6)

Stated in words, this law says the following: *The change of a particle's energy is equal to other work due to interactions not included in the energy.* [If the potential energy accounts for all conservative interactions, the other work is then merely that due to all non-conservative interactions (such as friction)].

Properties of the energy. By its definition (B-5), the energy E is the sum of the particle's kinetic energy (which depends on the particle's speed) and of its potential energy (which depends on the particle's interactions with other objects). The energy is thus a hybrid quantity describing both motion and interaction.

The definition (B-5) implies that the energy of a particle is a number expressed in terms of the SI unit *joule*. This energy can be positive, negative, or zero (since the potential energy can have any sign and magnitude).

Importance of the energy law. The energy law is the central result of this chapter and the preceding one. It is extremely important and widely useful. Like Newton's law, it relates the motion of a particle to its interactions with other objects. But, unlike Newton's law, it describes the motion of a particle by its kinetic energy, and describes the interaction of the particle by its potential energy and other work done on it. In this way it relates directly a particle's speed and position.

Comparison with the kinetic-energy law. The energy law is a minor, although very useful, modification of the kinetic-energy law. It differs from the latter in the following respect: In the kinetic-energy law, conservative interactions are described by the work which they contribute to the total work W_{tot} displayed on the right side of (B-1). But in the energy law, the *same* conservative interactions are described by the potential energy included in the energy E displayed on the left side of (B-6).

The kinetic-energy law is merely a special case of the more general energy law. Indeed, if *no* interactions are included in the energy E, then U = 0 and E = K. Furthermore, *all* interactions must then be included in the other work, i.e., the other work W_{oth} is then the total work W_{tot}. Hence the energy law (B-6) becomes identical to the kinetic-energy law (B-1).

Conservation of energy. An important special case arises when all forces are conservative (e.g., when no work is done by friction forces). All the conservative interactions can then be described by their potential energy and all other work W_{oth} is zero. The energy law (B-6) then implies that $\Delta E = 0$. Thus the particle's energy remains unchanged, i.e.,

$$\text{if } W_{oth} = 0, \qquad E = K + U = \text{constant} . \qquad \text{(B-7)}$$

One then says that the particle's energy is *conserved.*

When a particle's energy is conserved, its kinetic and potential energies can each change, but their sum remains unchanged. In other words, the total energy remains constant while it is transformed from kinetic energy to potential energy, or vice versa.

Example: Projectile motion

Let us use the energy law to examine the familiar case of a projectile launched with some speed v_o from a point A on the ground. (See Fig. B-1.)

Implications of the energy law. The projectile's interaction with the earth can be described by its gravitational potential energy U = mgh (if m is the mass of the projectile and h is its height above the horizontal ground). The energy law (B-6) then asserts that

$$\Delta E = \Delta K + \Delta U = W_{oth}. \qquad \text{(B-8)}$$

If air resistance is negligible, $W_{oth} = 0$ since the gravitational interaction is the only one affecting the projectile. Thus (B-8) implies that the energy E remains constant, i.e., that

$$E = K + U = \frac{1}{2}mv^2 + mgh = \text{constant} \qquad \text{(B-9)}$$

throughout the entire motion of the projectile.

Changes of speed. While the projectile moves so that its height h above the ground increases, its gravitational potential energy U also increases. The projectile's kinetic energy K, and thus speed v, must then correspondingly decrease so as to keep the total energy E constant. (Conversely, while the projectile moves so that its height decreases, its speed must increase.)

Calculation of speed. According to (B-9), the energy of the projectile at any point P is the same as at the launch point A. Thus

$$\frac{1}{2}mv^2 + mgh = \frac{1}{2}mv_o^2 + 0$$

so that

$$v = \sqrt{v_o^2 - 2gh}.$$

This predicts the speed of the projectile at any point of its trajectory.

Impact speed. The projectile's potential energy of the point B is the same as at its launch point A (since both these points are at the same height). The constancy (B-9) of the energy then implies that the

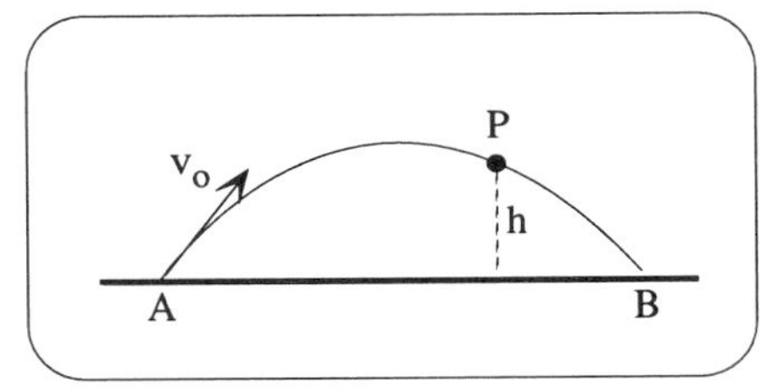

Fig. B-1. Motion of a projectile.

projectile's kinetic energy, and thus also its speed, must be the same at these points.

Effect of air resistance on impact speed. If air resistance is *not* negligible, the other work W_{oth} in (B-8) is that done by the friction force due to the air. This work is negative (since the friction force is opposite to the particle's velocity at any point). Hence (B-8) implies that the kinetic energy at B must be less than that at A (i.e., that the projectile strikes the ground with a speed smaller than its launch speed v_o).

➜ ***Go to Sec. 14B of the Workbook.***

C. Problem solving with the energy law

Problem-solving resources

Available mechanics laws. To predict motions or solve mechanics problems, one requires basic theoretical knowledge about the relation between motion and interactions.

We now have *two* basic mechanics laws which provide such theoretical knowledge, namely Newton's law and the energy law. The application of both of these laws, singly or in combination, allows us to solve many problems, including some which we could not solve previously.

Both of these mechanics laws relate motion and interaction, but focus on different aspects of these:

* **Newton's law ($m\vec{a} = \vec{F}_{tot}$).** This law describes a particle's motion by its acceleration, and describes its interactions by forces. *This law provides useful information relating a particle's motion and interactions at any instant.* (It thus allows one also to infer how the motion varies in the course of time.)
* **Energy law ($\Delta E = W_{oth}$).** This law describes a particle's motion by its kinetic energy, and describes its interactions by potential energy and work. *This law provides useful information relating a particle's speeds and positions at any two instants, without mention of the elapsed time.*

Application of the problem-solving method. The general problem-solving method discussed in Chapter 11 is still applicable. The only difference is that useful information about a problem can now be found by applying either Newton's law or the energy law (or both together). One may thus need to choose between these two laws. This choice is ordinarily easy (especially since the preceding paragraphs have indicated the conditions when these are likely to be useful). Furthermore, one can always try to foresee the likely consequences of applying these laws and choose the one which appears most promising.

Once one has chosen to apply one of these laws, the following subsidiary choices need to be made, as indicated in Fig. C-1.

In the case of Newton's law, one needs to specify to which *particle* this law is to be applied, at what *time*, and along what *direction*.

In the case of the energy law, one needs to specify to which *particle* this law is to be applied, and between which *times*. (No choice of direction is needed since this law relates numbers rather than vectors.)

In both cases, the particle needs to be adequately described by a system diagram before the law can be applied. (However, information about the particle's acceleration is not needed for application of the energy law.)

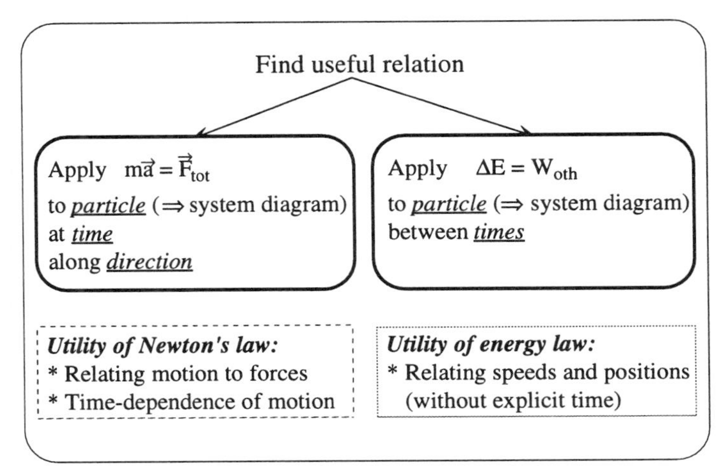

Fig. C-1. Options for finding a useful relation.

Prototype problem

The following problem illustrates how a typical problem can be solved by applying both of the preceding mechanics laws.

Problem statement. A pendulum consists of a bob, of mass m, suspended from a string. The pendulum is released from rest when its string is horizontal. What then is the magnitude of the tension force exerted by the vertical string when the pendulum bob passes its lowest position?

Solution. The following shows the solution of this problem.

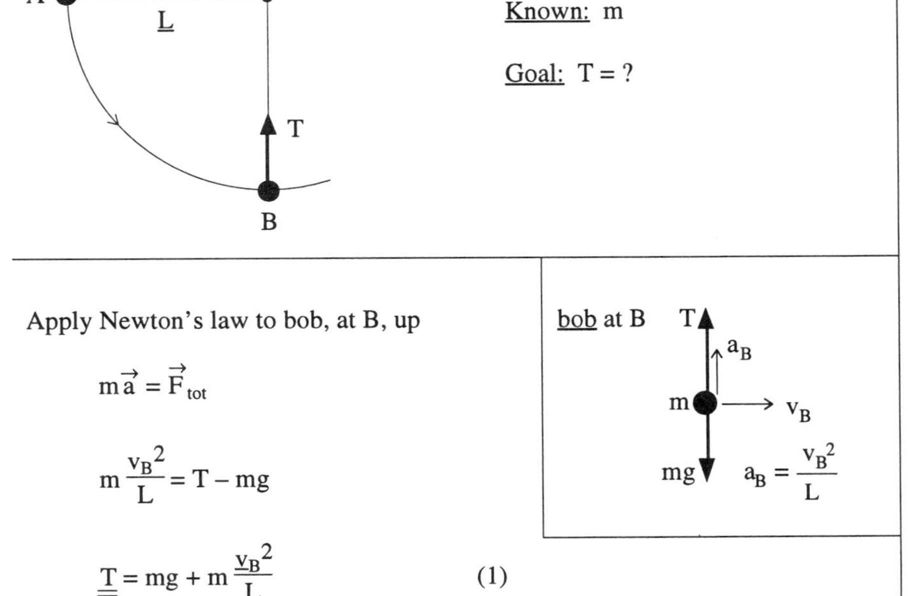

Analysis of problem

Construction of solution

Need info relating motion and interaction at the lowest point B. Hence apply Newton's law.

System diagram for bob at B. Here its speed is maximum and its acceleration is upward (toward center of the arc).

.

Wanted unknown T is underlined ttwice. Need to find unknown speed v_B (which is underlined once).

(Find v_B.) Apply energy, law, to bob, A→B

Want info relating speeds and positions at B and A. Hence apply energy law.

$$\Delta E = W_{oth}$$

$$\Delta K + \Delta U_g = W_T$$

$$[\frac{1}{2} m v_B^2 - 0] + [0 - mgL] = 0$$

$$\underline{v_B}^2 = 2gL \qquad (2)$$

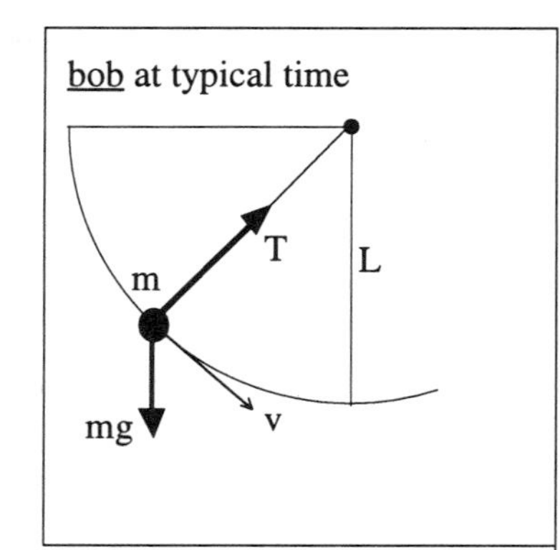

System diagram for bob at typical time during descent along its circular path.

U_g = gravitational potential energy of bob (measured from point B).

W_T = work done by tension force. (This work is zero since the tension force is always perpendicular to the bob's displacement.)

(Find T.) Eliminate v_B by (2) and (1):

$$T = mg + 2mg$$

$$\boxed{T = 3mg} \qquad (3)$$

Note that the tension force is 3 times larger than that exerted when the pendulum is hanging vertically at rest.

Checks (done in writing or mentally)

Checks

Units in (3) obviously OK.

Total force on the bob at B is up since acceleration is up. Hence upward tension force should be greater than downward gravity force on bob. Eq. (3) is OK.

Check that value of T is sensible.

→ ***Go to Sec. 14C of the Workbook.***

D. Conservative property of central forces

As we have seen, appreciable simplifications occur when forces are conservative (so that the work done by them is independent of the path). Some of these simplifications have been examined in the special case of a constant force, like the gravitational force near the earth. Do these simplifications also occur more generally in the case of forces which are *not* constant?

Fundamental forces, like gravitational and electric forces, are called *central* forces (because the force on one particle by another is directed along the line joining them and depends only on the distance between them). Are such central forces conservative?

Work done along an infinitesimal displacement. To examine this question, consider a particle fixed in position at some point O. As indicated in Fig. D-1, this particle exerts a force $\vec{F}$ on another particle P located at the point A which is at a distance R from O. If this force is central, it is directed along a radial direction (i.e., radially away from O if the force is repulsive, or radially toward O if it is attractive); furthermore, the magnitude F of this force depends only on the distance R from O to A. What then is the work d'W done by this force on the particle P when it moves through some infinitesimal

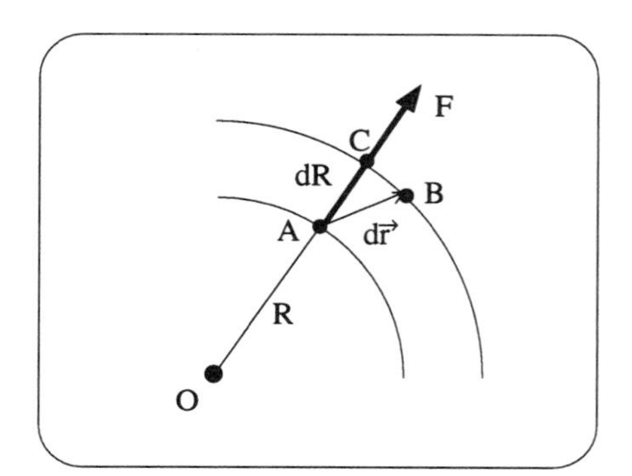

Fig. D-1. Infinitesimal displacement in the presence of a central force.

displacement $d\vec{r}$, from the point A to a neighboring point B (at a slightly different distance R + dR from O)?

According to the definition of work,

$$d'W = F\, dr_F \qquad \text{(D-1)}$$

where F is the magnitude of the force at the point A and dr_F is the component of the infinitesimal displacement along this force, i.e., along the radial direction. Since the displacement is very small, the short arc length BC in Fig. D-1 is very nearly a straight line perpendicular to AC. In this figure, the component dr_F of this displacement along the force is then just the small change dR of the particle's radial distance from O. Hence the work (D-1) done along the particle's small displacement is simply

$$d'W = F\, dR \qquad \text{(D-2)}$$

if the force $\vec{F}$ is repulsive as indicated in Fig. D-1, (or is $-F\,dR$ if the force is attractive). In other words, since the force is radial, it is only the radial part of the displacement which contributes to the work.

Work done along various infinitesimal displacements. Because the work along an infinitesimal displacement depends only on the radial component of this displacement, the *same* small work d'W is done along each of the infinitesimal displacements AC, AB, or AD in Fig. D-2. Furthermore, the magnitude of the force is the same at the points A and H which are both at the same distance from the particle at O. Hence the work done along the infinitesimal displacements HJ or HK is the same as the work done along any of the previous displacements AC, AB, or AD (since the radial part dR of all these displacements is the same).

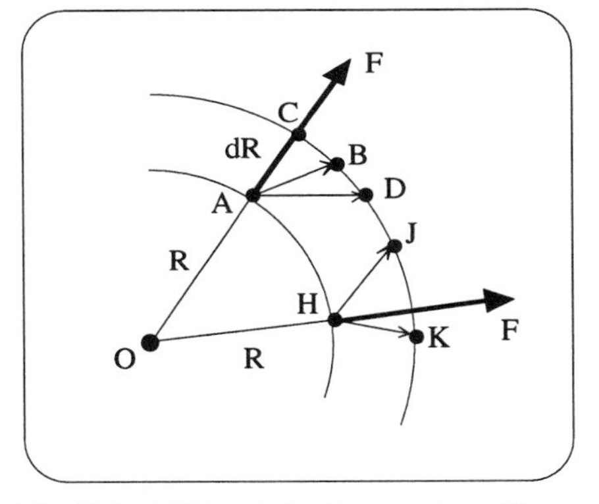

Fig. D-2. Different displacements yielding the same work.

Work done along different paths. Consider then the work done on the particle P along any two different long paths from A to B, such as the paths indicated in Fig. D-3. Each of these paths consists of a sequence of infinitesimal displacements labeled 1, 2, 3, ... for one of these paths, and 1', 2', 3', ... for the other path (where corresponding displacements start at the same distance R from O and have the same radial parts dR). Then the work done along displacement 1 is the same as that done along displacement 1'; the work done along displacement 2 is the same as that done along displacement 2'; and so forth. Hence the work done along the entire first path is the same as the work done along the entire second path. In other words, *the work done by a central force is independent of the path.*

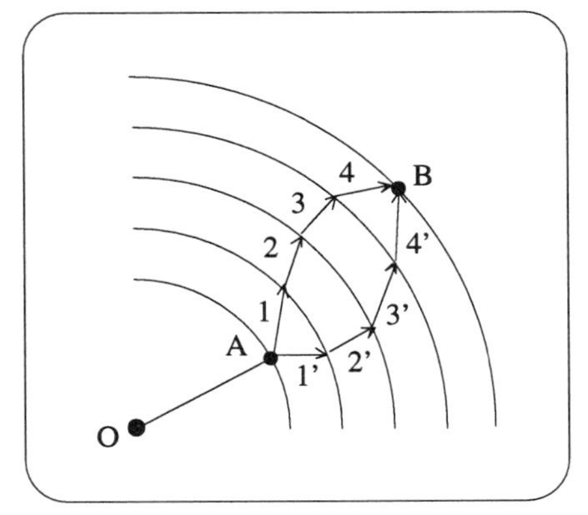

Fig. D-3. Two paths between the same two points.

Work done by several central forces. Lastly, suppose that a particle P is acted on by several central forces due to other particles located at various points O_1, O_2, ... When the particle P moves along any path, works W_1, W_2, ...are then done on it by the individual forces due to these other particles. The *total* work W done on P by the forces due to all these other particles is then simply

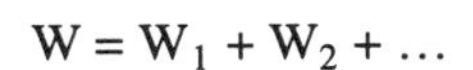

$$W = W_1 + W_2 + \ldots$$

But each of the individual works W_1, W_2, ... is independent of the path of P. Hence the work W due to the central forces exerted by *all* the other particles must also be independent of the path.

The considerations of this section can thus be summarized by the following conclusion:

Central forces, by one or more particles, are conservative.	(D-3)

In other words, the work is always independent of the path if the individual forces are directed along the lines joining the interacting particles and depend only on the distances between them.

→ ***Go to Sec. 14D of the Workbook.***

E. Gravitational and electric potential energies

Since central forces are conservative, the work done by them can be found from a knowledge of their potential energies. What then is the potential energy in the case of the important gravitational force (or in the case of the similar electric force)?

Consider two particles 1 and 2 separated by a distance R. The magnitude F of the force on particle 2 by particle 1 depends then on the distance R so that

$$F = \frac{C}{R^2}. \tag{E-1}$$

In the case of the gravitational force, we know from (12C-5) that this force is attractive and that the constant C is

$$C = G\, m_1 m_2 \tag{E-2}$$

where G is the gravitational constant, and m_1 and m_2 are the masses of the two particles.

Suppose that particle 1 is located at the point O and that particle 2 is located at any other point A, as illustrated in Fig. E-1. By the definition (A-3), the potential energy U_A of particle 2, due to its interaction with particle 1, is then given by

$$U_A = W_{AS} \tag{E-3}$$

where W_{AS} is the work done by the gravitational force on particle 2 when it moves from A to some standard position S. This standard position can conveniently be

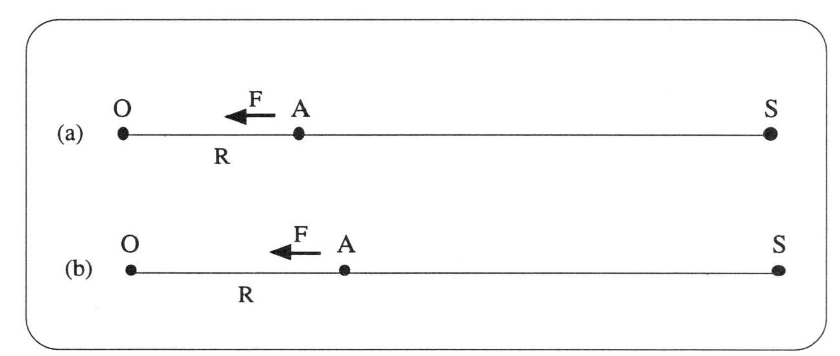

Fig. E-1. Gravitational force exerted on a particle at A by another particle at O.

chosen to be "infinitely" far away (i.e., so very far away from O that the interaction between the particles is negligibly small). Since the work is independent of the path, one can also simplify calculation of the work W_{AS} by choosing the straight-line path from A to S.

Qualitative properties of the potential energy

Sign of the potential energy. Suppose that the force is *attractive* as illustrated in Fig. E-1 (e.g., that this is a gravitational force). Then the work W_{AS} done on the particle is *negative* (since the direction of the force is always opposite to the successive small displacements of the particle moving from A to S). Hence the potential energy U_A of the particle, relative to the far-distant standard position S, must be *negative*.

On the other hand, suppose that the force is *repulsive* (e.g., that this is an electric force between particles having charges of the same sign). Then the work done by this force along any small displacement is positive. Hence the potential energy must also be *positive*.

Magnitude of the potential energy. If the distance R between the two particles is larger than in Fig. E-1a (i.e., if the point A is farther away from O, as indicated in Fig. E-1b), then the work W_{AS} is done along a shorter path from A to S. The *magnitude* of this work (irrespective of its sign) is thus correspondingly smaller. According to (E-3), the magnitude of the potential energy U_A is thus smaller if the interacting particles are farther apart.

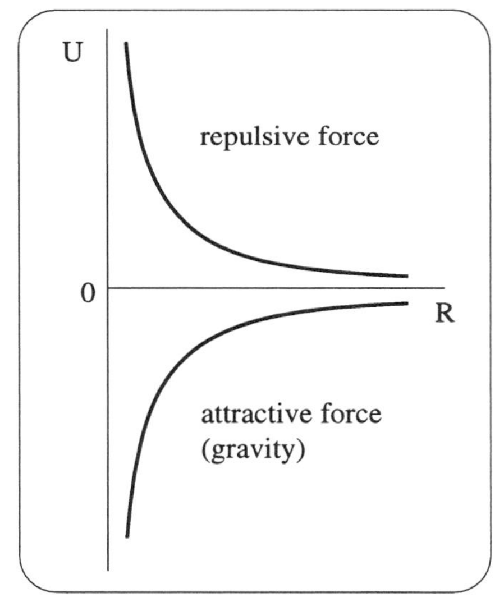

Fig. E-2. Qualitative dependence of the potential energy U on the distance between interacting particles. (U = 0 at the standard position where this distance is very large.)

As usual, the magnitude of the potential energy must be zero at the standard position S. Since the standard position has been chosen very far away, this implies that the potential energy is zero when the particles are very far apart.

Dependence of potential energy on distance. The preceding comments indicate how the potential energy U of a particle depends on its distance R from the other particle. This dependence is illustrated by the graphs in Fig. E-2.

Remarks on the sign of the potential energy.

The sign of the potential energy is also obvious from the following considerations: (a) According to (A-8), the potential energy always decreases along the direction of the force. (b) The potential energy is zero very far away, i.e., at the chosen standard position.

If the force is *attractive*, as indicated in Fig. E-1, the potential energy must thus decrease as the particle moves *closer* to the other particle at O. Starting from zero far away, it must then become increasingly negative.

Conversely, if the force is *repulsive*, the potential energy must decrease as the particle moves *farther* away from the particle at O. Thus the potential energy must be positive so as to approach zero far away.

Remarks on the choice of standard position

The gravitational potential energy is negative because we chose the standard position very far away. This choice is different from that which we made in the special case of the nearly constant gravitational force *near* the surface of the earth.

In that case the standard position must also be chosen near the surface of the earth. Depending on this choice, the potential energy can then be either positive or negative. For example, if the standard position is chosen at a mountain top high above sea level, the gravitational potential energy at all lower points is negative (just as it is in the general case where the standard position is chosen very far from the earth).

Example: Satellite speed in an elliptical orbit

Fig. E-3 illustrates the motion of a satellite orbiting the earth in an elliptical orbit under the sole influence of its gravitational interaction with the earth. The total energy E of the satellite must then be constant, i.e.,

$$E = K + U = \text{constant}$$

where U is the gravitational potential energy of the satellite due to the earth.

At a point B closer to the earth this gravitational potential energy is smaller (i.e., more negative) than at a point A farther away. Since the total energy E is the same at these points, the kinetic energy K of the satellite must be larger at B than at A. Hence the satellite speeds up as it gets closer to the earth.

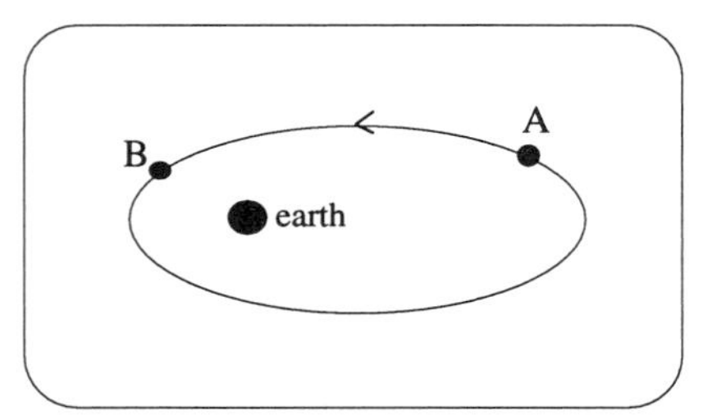

Fig. E-3. Elliptical orbit of a satellite around the earth.

Calculation of the potential energy

The potential energy due to the force (E-1) can be determined quantitatively by using the definition (E-3) to calculate the work W_{AS}.

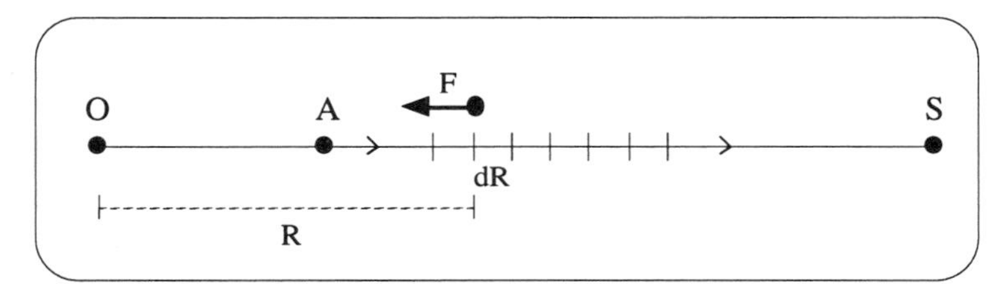

Fig. E-4. Successive small displacements of a particle from A to S.

Work done in a small displacement. Consider the case where the force on the particle is attractive, as illustrated in Fig. E-4. Suppose that the particle, located at a distance R from the other particle at O, moves a small distance dR to a slightly larger distance R + dR. If the force at this position has a magnitude F, the infinitesimal work done on the particle is then

$$d'W = -F\,dR \tag{E-4}$$

where the minus sign indicates that the particle's displacement is opposite to the force.

Work along entire path. The entire work W_{AS} done on the particle is then found by adding the infinitesimal works done in all its successive infinitesimal displacements from A to S. According to (E-3), the particle's potential energy at A is then

$$U_A = W_{AS} = \int_A^S (-F\,dR) = -\int_A^S F\,dR\,.$$

If the magnitude of the force is specified by (E-1), one then gets

$$U_A = -\int_A^S \frac{C}{R^2}\,dR = -C\int_A^S \frac{dR}{R^2}\,. \tag{E-5}$$

Calculation of the sum. A sum, like that in (E-5), could be easily calculated if each of its terms would be merely a small change. For then the sum of all the successive small changes would be simply equal to the total change. But each term in the sum (E-5) is more complex (equal to a small change dR divided by R^2). To calculate such a sum, one can try to express each term as a small change of some quantity. In (E-5) this can be done by expressing each term in terms of the infinitesimal change of the quantity (1/R), i.e.,

$$d\left(\frac{1}{R}\right) = -\frac{dR}{R^2}.$$

$$d\left(\frac{1}{R}\right) = \frac{1}{R'} - \frac{1}{R}$$
$$= \frac{1}{R+dR} - \frac{1}{R}$$
$$= \frac{R-(R+dR)}{R\,(R+dR)}$$
$$= -\frac{dR}{R\,(R+dR)}$$
$$= -\frac{dR}{R^2} \quad \text{since } dR << R.$$

Hence (E-5) can be written as

$$U_A = C \int_A^S d\left(\frac{1}{R}\right). \tag{E-6}$$

Here the sum of all the successive small changes of the quantity (1/R) is just equal to the total change of this quantity. Thus one gets

$$U_A = C\left(\frac{1}{R_S} - \frac{1}{R_A}\right).$$

Since the standard position S is very far away, R_S is extremely large so that $1/R_S = 0$. Hence one gets the simple result

$$U_A = -\frac{C}{R_A}. \tag{E-7}$$

Potential energy. Since A can be any point, the result (E-7) can be summarized as follows: The potential energy of the particle, at any distance R from the other particle, is

$$U = -\frac{C}{R}. \tag{E-8}$$

Gravitational force and potential energy. In the case of the gravitational force, the constant C is given by (E-2). When the particles are separated by a distance R, the magnitude of the gravitational force, and the value of the corresponding gravitational potential energy, are then

$$\boxed{F = G\frac{m_1 m_2}{R^2}} \tag{E-9}$$

and

$$\boxed{U = -G\frac{m_1 m_2}{R}.} \tag{E-10}$$

This gravitational potential energy agrees with our previous qualitative considerations, i.e., it is properly negative (since the force is attractive) and varies with the distance R in the manner indicated in the graph of Fig. E-2.

Note that the potential energy varies with the distance R like (1/R) rather than like ($1/R^2$). This makes sense since the potential energy has the units of work (i.e., the units of a force multiplied by a distance).

Potential energy due to a spherical body. As mentioned in (12C-6), the gravitational force on a particle located outside a spherical body is the same as if the entire mass of the body were concentrated at its center. Hence the gravitational potential energy due to such a body can be similarly found by imagining its entire mass to be concentrated at its center.

Electric force and potential energy. In the case of the electric force, comparison with (12D-3) shows that the constant C in (E-1) is equal to $k_e|q_1 q_2|$ where k_e is the electric force constant and q_1 and q_2 are the charges of the interacting particles. When the particles are separated by a distance R, the magnitude of the electric force, and the value of the corresponding electric potential energy, are then

$$\boxed{F = k_e \frac{|q_1 q_2|}{R^2}} \qquad \text{(E-11)}$$

and

$$\boxed{U = k_e \frac{q_1 q_2}{R}\,.} \qquad \text{(E-12)}$$

Here the expression (E-12) for the potential energy has the correct sign. Indeed, if both charges have the same sign, so that the electric force is repulsive, the potential energy (E-12) is properly positive. But if both charges have the opposite signs, so that the electric force is attractive, the potential energy (E-12) is properly negative.

Applications of gravitational potential energy

The general expressions (E-9) and (E-10) for the gravitational force and potential energy are useful whenever one deals with astronomical problems or with particles that do not remain close to the earth's surface. The following is an example.

Example: Escape speed from the earth

Suppose that some projectile is launched vertically upward from the surface of the earth (as illustrated in Fig. E-5). With what speed v_o must the projectile be launched so that it will escape from the earth into outer space? (Neglect the effects of air resistance due to the earth's atmosphere.)

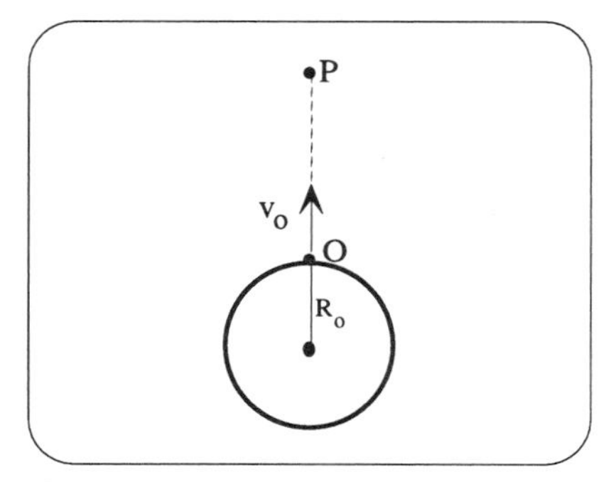

Fig. E-5. Projectile launched from the surface of the earth.

The only interaction of the projectile is its gravitational interaction with the earth. The energy law thus implies that the total energy of the projectile

$$E = K + U = \frac{1}{2}mv^2 - G\frac{Mm}{R} = \text{constant.} \qquad \text{(E-13)}$$

(Here m is the mass of the projectile, M is the mass of the earth, and R is the distance of the projectile from the center of the earth.) In particular, the energy E_o of the projectile must be the same as its energy E_P at its highest point (i.e., the point farthest from the earth). Hence

$$E_o = E_P\,. \qquad \text{(E-14)}$$

Ordinarily the projectile will rise with decreasing speed, reach its highest point, and then fall back down again. If it is to *escape* from the earth, it must still move with some appreciable speed even when it reaches a point very far from the earth. At such a large distance from the earth its speed v_P, and thus kinetic energy K_P, must then be positive. But the projectile's gravitational potential energy U_P at such a large distance is zero. Hence the energy $E_P = K_P + U_P$ of the projectile very far from the earth must be positive. Thus the conservation of energy (E-14) implies that the projectile escapes from the earth only if

$$\boxed{E_o > 0,} \qquad \text{(E-15)}$$

i.e., only if its initial total energy is positive.

If the projectile is to escape from the earth, its launch speed v_o must then be such that

$$\frac{1}{2} m{v_o}^2 - G\frac{Mm}{R_o} > 0$$

or

$$v_o > \sqrt{\frac{2GM}{R_o}} \qquad \text{(E-16)}$$

where R_o is the radius of the earth.

This result can be expressed more simply in terms of the magnitude g of the gravitational acceleration at the surface of the earth. There the magnitude of the gravitational force on the projectile is mg. Hence

$$mg = G\frac{Mm}{{R_o}^2}$$

or

$$GM = g{R_o}^2 . \qquad \text{(E-17)}$$

Thus (E-16) is equivalent to the condition that

$$\boxed{v_o > \sqrt{2gR_o}\,.} \qquad \text{(E-18)}$$

Numerically $\quad v_o > \sqrt{2\,(9.8\ \text{m/s}^2)\,(6.37 \times 10^6\ \text{m})}$

or

$$v_o > 11.2\ \text{km/s}\,. \qquad \text{(E-19)}$$

→ ***Go to Sec. 14E of the Workbook.***

F. Potential energy of spring forces

Properties of the force. The force on a particle by a spring opposes the deformation of the spring. As discussed in Sec. 10C and indicated in Fig. F-1, the force is thus always directed toward the equilibrium position O where the spring is undeformed and the force on the particle is zero. The magnitude of the force increases if the magnitude of the spring's elongation x increases (i.e., either if the spring is stretched so that x is positive, or if it is compressed so that x is negative). This is indicated in Fig. F-1 where x is the component of the particle's displacement from its equilibrium position.

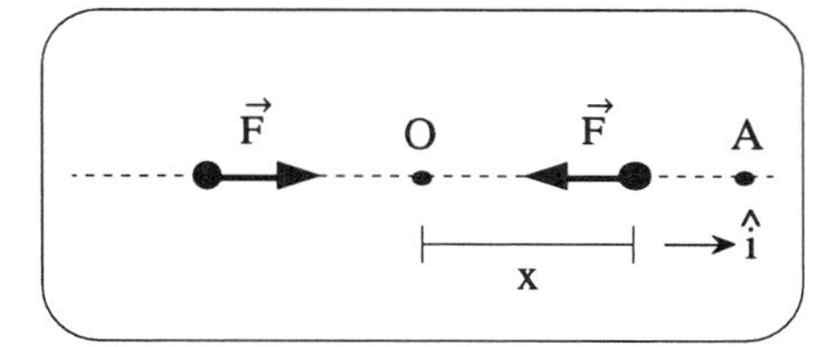

Fig. F-1. Forces exerted on a particle by a spring.

If the deformation of the spring is small, the magnitude F of the force is proportional to the magnitude |x| of the elongation. In accordance with (12A-1) one can then write

$$F = k\,|x| \qquad \text{(F-1)}$$

where k is the spring constant. The magnitude of the force is then the same if the magnitude of the elongation is the same, irrespective of whether the spring is stretched or compressed.

Potential energy. To find the potential energy of the particle due to this force, let us choose the standard position at the particle's equilibrium position O. (The potential energy is then zero at O where the force is zero.) According to its definition, the potential energy U_A at any other point A is then

$$U_A = W_{AO} \qquad \text{(F-2)}$$

where W_{AO} is the work done by the force on the particle moving from A to O.

Qualitative properties of the potential energy

In moving from a point A to the equilibrium position O, the particle always moves along the direction of the force on it. (This is true irrespective of whether the point A is located on the right or left side of the equilibrium position O in Fig. F-1, i.e., irrespective of whether the spring is stretched or compressed.) Hence the work W_{AO}, and thus also the potential energy U_A, is always positive.

The magnitude of the work is larger if the path from A to O is longer, i.e., if the magnitude of the elongation x is larger. Furthermore, the magnitude of this work depends only on the distance from A to O, irrespective of whether A is to the right or to the left of O. [This is so because the magnitude of the force in (F-1) depends only on the magnitude of the elongation and not on its sign.] Thus the potential energy U depends only on the magnitude |x| of the spring's elongation (i.e., on the magnitude of the particle's displacement from its equilibrium position O).

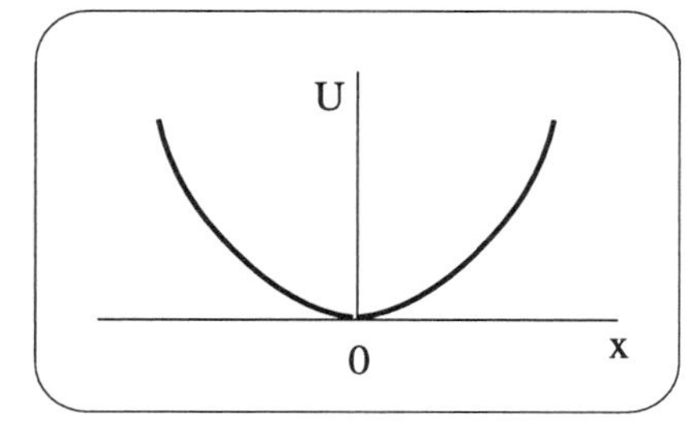

Fig. F-2. Qualitative graph of a spring's potential energy versus its elongation.

The preceding comments indicate that the potential energy U of the particle must depend on the elongation x of the spring in the symmetric fashion illustrated by the graph in Fig. F-2.

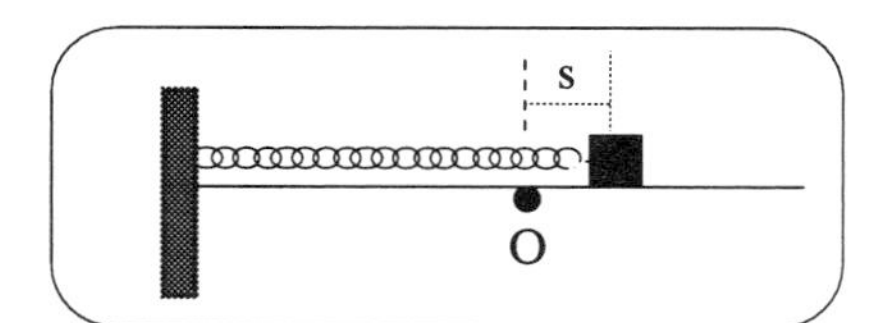

Fig. F-3. Particle connected to a horizontal spring.

Oscillations of a particle connected to a spring

A particle, free to move along a horizontal frictionless table, is connected to a horizontal spring, as indicated in Fig. F-3. The particle is initially at rest, while the spring is extended by an amount s, and is then released. What then is it subsequent motion?

The particle's energy is $E = K + U$ where K is its kinetic energy and U its potential energy due to its interaction with the spring. The energy law implies that this energy E remains constant since no other work is done by gravity or friction forces.

As indicated in Fig. F-4, the particle then oscillates back and forth so that its kinetic energy K and potential energy U change, but their sum remains constant. At the extreme points, where $x = s$ and $x = -s$, the particle's kinetic energy (and thus also speed) is momentarily zero and its potential energy is maximum. But when the particle passes through its equilibrium position O, its potential energy is minimum and its kinetic energy (and thus also speed) is correspondingly maximum.

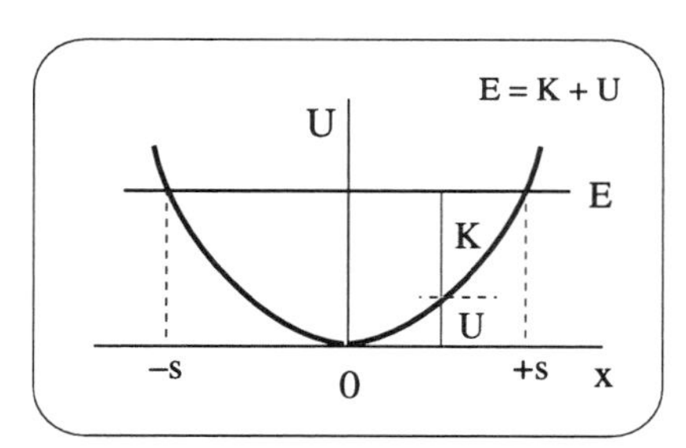

Fig. F-4. Energies of an oscillating particle.

Calculation of the potential energy

The potential energy due to the force (F-1) can be determined quantitatively by calculating the work specified in the definition (F-2).

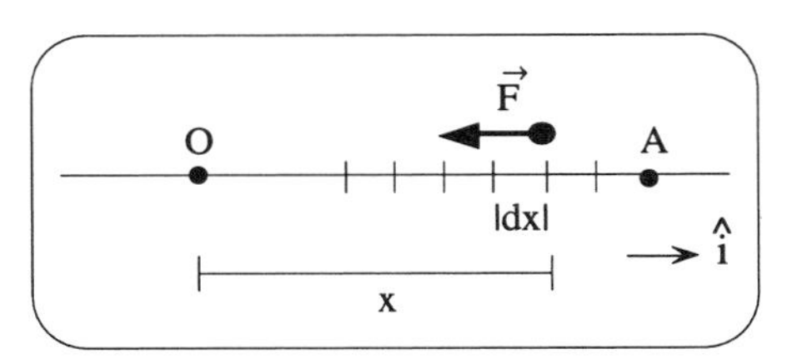

Fig. F-5. Successive small displacements of a particle from A to O.

If the particle in Fig. F-5 moves 1 mm to the left, the change dx = –1 mm and the magnitude |dx| of the particle's displacement is –dx = +1 mm.

Infinitesimal work. Consider the particle as it moves toward O through a small displacement from x to (x – |dx|), as illustrated in Fig. F-5. By (F-1), the magnitude of the force on the particle at this position is then kx. Also, the magnitude of the particle's displacement along this force is |dx| (i.e., –dx since the change dx of the particle's position coordinate is negative as the particle moves toward O). The infinitesimal work done in this small displacement is then positive and equal to

$$d'W = F\,|dx| = (kx)\,(-dx) = -\,kx\,dx\,. \qquad \text{(F-3)}$$

Work along entire path. The entire work W_{AO} done on the particle is found by adding the infinitesimal works done in all its successive small displacements from A to O. The particle's potential energy at A is then, by (F-2), equal to

$$U_A = W_{AO} = \int_A^O (-kx)\,dx = -\,k \int_A^O x\,dx\,. \qquad \text{(F-4)}$$

Here each term in the last sum can be expressed as an infinitesimal change of the quantity x^2. Indeed, by (5C-9),

$$d(x^2) = 2\,x\,dx\,.$$

Hence

$$U_A = -\,k \int_A^O (\tfrac{1}{2}\,dx^2) = -\tfrac{1}{2}\,k \int_A^O dx^2\,. \qquad \text{(F-5)}$$

But the sum of all the infinitesimal changes of x^2 is just equal to the total change of this quantity. Thus (F-5) is equal to

$$U_A = -\tfrac{1}{2}\,k\,(x_O{}^2 - x_A{}^2) = \tfrac{1}{2}\,k\,x_A{}^2 \qquad \text{(F-6)}$$

since $x_O = 0$.

Potential energy. The result (F-6) is true for any point A and can thus be summarized as follows: The potential energy of the particle, for any elongation x of the spring, is

$$\boxed{U = \tfrac{1}{2}\,kx^2\,.} \qquad \text{(F-7)}$$

Note that this quantitative result has the qualitative features portrayed in the graph of Fig. F-2.

➜ ***Go to Sec. 14F of the Workbook.***

G. Summary

Energy law: $\boxed{\Delta E = W_{oth}}$ (for any particle, relative to inertial frame)

Energy: $E = K + U$

Potential energy: $U_A = W_{AS}$ (for conservative force)

Work due to change of potential energy: $W = -\Delta U$

Potential energies due to various interactions: (See also Appendix D.)

Gravitational

Near the earth: $U = mgh$

General: $U = -G\dfrac{m_1 m_2}{R}$

Electric: $U = k_e \dfrac{q_1 q_2}{R}$

Due to spring: $U = \frac{1}{2}kx^2$ (for small elongations)

New abilities

You should now be able to do the following:

(1) Relate potential energy to the work done by conservative forces.

(2) Solve mechanics problems by using the problem-solving method together with Newton's law and the energy law. (The problem-solving method is summarized in Appendix G.)

Grand summary of mechanics laws

Two basic mechanics laws, relating the motion and interactions of a particle, have been discussed in the preceding chapters. The following chart summarizes these two mechanics laws and associated definitions.

Newton's law $\boxed{m\vec{a} = \vec{F}_{tot}}$

Validity: * For inertial frame.
Utility: * Relates motion and interactions at any instant.

Motion		Interactions	
$\vec{a} = d\vec{v}/dt$	{acceleration}	$\vec{F}_{tot} = \Sigma\,\vec{F}_s$,	{total force = sum of all forces}

Energy law $\boxed{\Delta E = W_{oth}}$

Validity: * For inertial frame.
Utility: * Relates speeds & positions at any two instants (without mention of time).

$E = K + U$

Motion		Interactions	
$K = \frac{1}{2}mv^2$	{kinetic energy of particle)	U	{potential energy}
		W_{oth}	{other work, by all forces not included in U}

➜ ***Go to Sec. 14G of the Workbook.***

Motion and Interaction of Systems

The preceding chapters dealt with the motion and interaction of single particles. However, one is usually interested in the motion of more complex systems (e.g., colliding cars, rolling wheels, gases or liquids) consisting of many particles moving relative to each other. How can one explain or predict the motion of such systems? The next few chapters will address this question. To do this, we shall use our previous knowledge about particles to derive more general mechanics laws about the motion and interactions of complex systems.

---

15 Momentum

Newton's law $m\vec{a} = \vec{F}_{tot}$, which specifies the relation between the motion of a particle and its interactions, is the most basic law of mechanics. Indeed, after being introduced in Chapter 9, it has been the basis of all our subsequent discussions of particle motions. Hence we shall begin our consideration of more complex systems by examining how this law can be extended to deal with systems consisting of several particles.

A. Momentum law

Systems. The next few chapters deal with the motions of various systems. The word "system" is here used to mean the following:

Def: **System:** Any set of particles selected for consideration. (A-1)

A system can thus be anything on which one wishes to focus one's attention (e.g., a single marble, a box filled with marbles, a wheel, a small part of a wheel, an automobile, a drop of liquid, etc.).

Whenever one selects a system for consideration, one divides the world into two distinct parts: the part consisting of all the particles *inside* the system, and the part consisting of all the particles *outside* the system. For example, in Fig. A-1 the particles 1, 2, and 3 have been chosen as the system of interest.

The particles 1, 2, and 3 (those indicated within the dashed boundary in Fig. A-1) are then the particles inside this system, i.e., the *internal* particles. The particles 1' and 2' (indicated by primes) are the particles outside this system, i.e., the *external* particles.

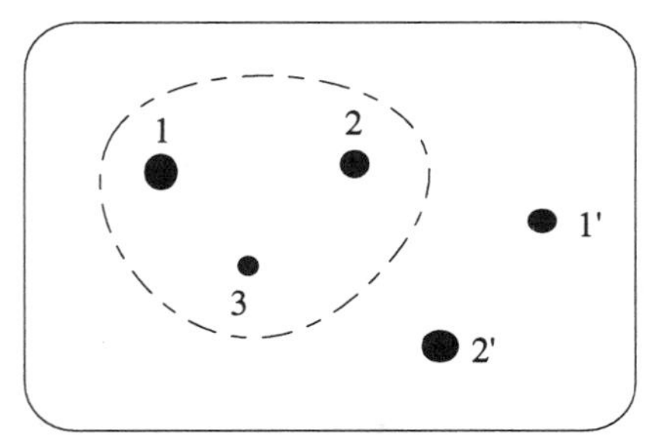

Fig. A-1. Particles inside and outside a system.

Derivation of the momentum law

Application of Newton's law. Consider any system of particles, such as the simple system in Fig. A-1. Newton's law

$$m\vec{a} = \vec{F}_{tot} \qquad \text{(A-2)}$$

can then be applied to every particle in the system to specify how its motion is influenced by its interactions with all other particles. For example, by applying Newton's law to particle 1, one gets

$$m_1\vec{a}_1 = [\vec{F}_{12} + \vec{F}_{13}] + [\vec{F}_{11'} + \vec{F}_{12'}]\,. \qquad \text{(A-3a)}$$

The total force on the right side of this equation is the vector sum of *all* the forces acting on particle 1. This sum consists of the sum of all the *internal* forces on this particle (i.e., the forces on this particle by all the other particles *inside* the system), and the sum of all the external forces on this particle (i.e., the forces on this particle by all the other particles *outside* the system). In (A-3a) the sum of all *internal* forces, indicated within the first pair of square brackets, is the sum of the forces exerted on particle 1 by particles 2 and 3. The sum of all *external* forces, indicated within the second pair of square brackets, is the sum of the forces exerted on particle 1 by particles 1' and 2'. (See Fig. A-2.)

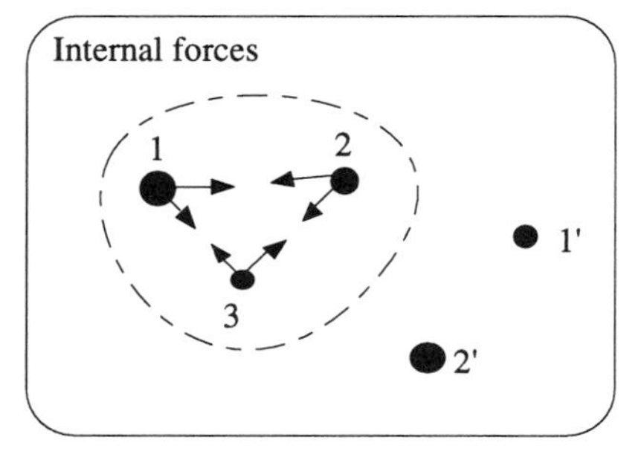

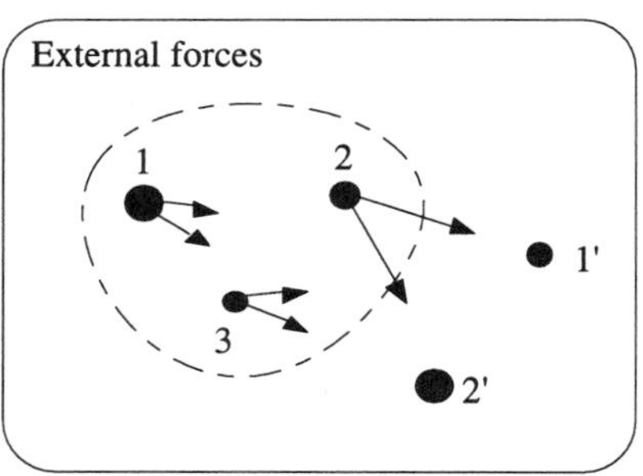

Fig. A-2. Internal and external forces on the system of Fig. A-1.

Similarly, applying Newton's law to particle 2 in the system, one gets

$$m_2\vec{a}_2 = [\vec{F}_{21} + \vec{F}_{23}] + [\vec{F}_{21'} + \vec{F}_{22'}] \qquad \text{(A-3b)}$$

where the internal forces on particle 2 are the forces on it by particles 1 and 2, and the external forces are the forces on it by particles 1' and 2'. Similarly, applying Newton's law to particle 3 in the system, one gets

$$m_3\vec{a}_3 = [\vec{F}_{31} + \vec{F}_{32}] + [\vec{F}_{31'} + \vec{F}_{32'}]\,. \qquad \text{(A-3c)}$$

Addition for all particles. To get a useful relation describing the entire system, one merely needs to add the equations (A-3a), (A-3b), and (A-3c) for all the particles in the system. Thus one obtains

$$m_1\vec{a}_1 + m_2\vec{a}_2 + m_3\vec{a}_3 = \vec{F}_{int} + \vec{F}_{ext}\,. \qquad \text{(A-4)}$$

Here we have introduced the following abbreviations:

Def: **Total internal force $\vec{F}_{int}$:** The sum of forces on all particles in the system by all other particles *inside* the system. (A-5)

Def: **Total external force $\vec{F}_{ext}$:** The sum of forces on all particles in the system by all particles *outside* the system. (A-6)

To avoid cumbersome writing, we can also use the conventional symbol Σ (the Greek letter *sigma*) to denote the sum on the left side of (A-4). Thus we can write (A-4) in the abbreviated form

$$\sum m_n \vec{a}_n = \vec{F}_{int} + \vec{F}_{ext} \qquad \text{(A-7)}$$

where the subscript n is supposed to assume the values n = 1, 2, and 3 (corresponding to the three successive terms in the sum for the three particles).

The relation (A-7) is similarly valid for a system consisting of *any* number of particles.

Cancellation of internal forces. The calculation of the sum of all internal forces is very simple because these forces always occur in pairs. For example, one has to add the force on particle 1 by particle 2, and then also the force on particle 2 by particle 1. Similarly, one has to add the force on particle 1 by particle 3, and then also the force on particle 3 by particle 1. But

$$\vec{F}_{12} + \vec{F}_{21} = 0$$

since $\vec{F}_{21} = -\vec{F}_{12}$ because of the relation between mutual forces. Similarly, the sum of mutual forces between *any* two particles in the system is zero.

Hence all the mutual forces, exerted on particles in the system by other particles in the same system, cancel each other. Thus the total internal force is zero and we arrive at the following important result

$$\boxed{\vec{F}_{int} = 0\,.} \qquad \text{(A-8)}$$

Correspondingly, (A-7) becomes simply

$$\boxed{\sum m_n \vec{a}_n = \vec{F}_{ext}\,,} \qquad \text{(A-9)}$$

a relation which involves merely the *external* forces on the system.

Momentum of a particle. The left side of (A-9) can be simplified by recalling that the acceleration $\vec{a}$ of a particle is defined as the rate of change of its velocity $\vec{v}$. Thus one can write for any particle

$$m\vec{a} = m\frac{d\vec{v}}{dt} = \frac{d(m\vec{v})}{dt}\,. \qquad \text{(A-10)}$$

This can be simplified by introducing the quantity *momentum*, conventionally denoted by the letter $\vec{p}$ and defined as follows:

Def: | ***Momentum of a particle:*** $\vec{p} = m\vec{v}\,.$ | (A-11)

(Stated in words, the momentum of a particle is merely the vector obtained by multiplying the velocity of the particle by its mass.) With this definition, (A-10) can be written as

$$m\vec{a} = \frac{d\vec{p}}{dt}. \tag{A-12}$$

Thus Newton's law for a particle can also be expressed in the form

$$\textit{Newton's law:} \quad \frac{d\vec{p}}{dt} = \vec{F}_{tot} \tag{A-13}$$

which states that *the rate of change of a particle's momentum is equal to the total force on it.*

Momentum of a system. Using the preceding definition of momentum, the left side of (A-9) can be written as

$$\sum m_n\vec{a}_n = \sum \frac{d\vec{p}_n}{dt} = \frac{d}{dt}\left(\sum \vec{p}_n\right) = \frac{d\vec{P}}{dt}. \tag{A-14}$$

Here we have used the fact (5C-6) that the sum of rates of change is equal to the rate of change of the sum. Furthermore, we have introduced the obvious definition that the *total momentum* $\vec{P}$ of a system is the vector sum of the momenta of all the particles in the system. In other words,

Def:

$$\textit{Momentum of a system:} \quad \vec{P} = \sum \vec{p}_n. \tag{A-15}$$

Momentum law. With the preceding definitions, the relation (A-9) can be written in the following simple form:

$$\textit{Momentum law:} \quad \frac{d\vec{P}}{dt} = \vec{F}_{ext}. \tag{A-16}$$

In words, this law states that *the rate of change of the momentum of any system is equal to the total external force on it.*

Discussion of the momentum law

The momentum law is an important mechanics law applicable to any system, no matter how complex. It is thus very useful for dealing with the motion of systems consisting of any number of particles.

Newton's law as a special case. If a system is merely a single particle, the momentum law is identical to Newton's law. Indeed, the total momentum of the system is then merely the momentum $\vec{p} = m\vec{v}$ of the particle, and its rate of change is just $m\vec{a}$. Furthermore, the total external force on the system is the total force $\vec{F}_{tot}$ on the particle since *all* forces on a single particle are external.

Application of the momentum law. Just like Newton's law or any other mechanics law, the momentum law relates motion and interactions. It describes the motion of a system by its total momentum, and describes the interactions of the system by the total external force on it.

The momentum and the total external force are both vectors. Hence the momentum law implies that the components of these vectors along any direction are also equal. When applying the momentum law to any component, one must thus make the choices indicated by the following underlined quantities:

Apply *momentum law* to <u>system</u> at <u>time</u> along <u>direction</u>. (A-17)

A very useful feature of the momentum law is that it involves only the *external* forces on a system and disregards the *internal* forces as irrelevant. The momentum law can, therefore, often yield simple conclusions about complex systems, even if the particles in these systems move and interact in very complicated ways.

Useful choice of system. The irrelevance of internal forces can be exploited by using the following guideline to choose the system to which the momentum law is applied:

Choice of system: Choose a system so that unknown or uninteresting forces are internal, and so that forces of interest are external.	(A-18)

In this way one can ignore forces which are unknown or of no particular interest, but can use the momentum law to find desired information about other forces.

The preceding comments indicate the importance of choosing carefully, and specifying clearly, the system to which the momentum law is applied.

Mechanics laws available for problem solving. Our work up to this point has provided us with the following mechanics laws available for solving problems: (a) The momentum law which is applicable to any system, no matter how complex. (This laws is the same as Newton's law if it is applied to a single particle.) (b) The energy law which can be applied to any single particle.

The following section will illustrate how these laws can be used, singly or in combination, to explain or predict various kinds of motions.

➜ ***Go to Sec. 15A of the Workbook.***

B. Conservation of momentum

If the total external force on a system is zero (e.g., if the system does not interact with any other objects), the momentum law (A-16) implies that the system's momentum does not change. Thus

$$\text{if } \vec{F}_{ext} = 0, \qquad \vec{P} = \text{constant}. \tag{B-1}$$

One then says that the momentum of the system is *conserved*.

Similarly, if a *component* of the total external force along some particular direction is zero, then the component of the system's momentum along this direction remains constant (even if components of the momentum along other directions do change).

If the total external force is zero, the momentum law then allows one to relate the motions of the system (i.e., the velocities of its particles) at any two instants. Furthermore, this is true irrespective of the complexity of the interactions among the particles in the system (since internal forces are always irrelevant in the momentum law). The preceding remarks allow some useful applications of the momentum law, as illustrated by the following examples.

Example 1: Recoil of a gun

The gun in Fig. B-1 has a mass M. It is mounted on wheels so that it can move with negligible friction along the horizontal ground. What happens to this gun when it fires a bullet, of mass m, which leaves the gun with a horizontal velocity $\vec{v}'$?

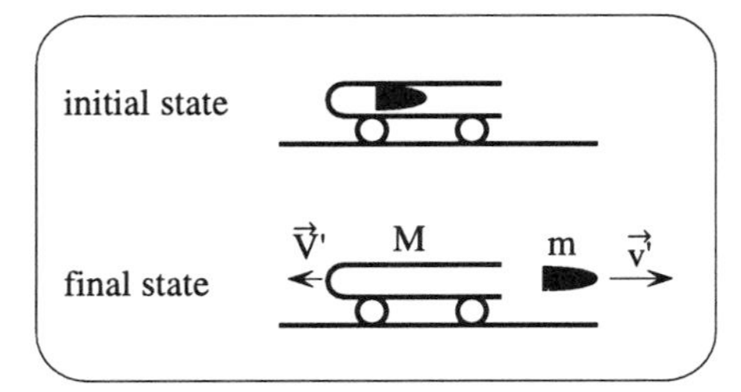

Fig. B-1. Firing of a gun.

Since friction is negligible, the total external force on the system, consisting of the gun and bullet, is zero along the horizontal $\hat{i}$ direction indicated in Fig. B-1. The momentum law $d\vec{P}/dt = \vec{F}_{ext}$ then implies that the component of the system's total momentum along this direction must remain unchanged. (This must be true regardless of the complexities of the explosion, and of the resulting internal forces, which cause the bullet to be propelled from the gun.) Thus

$$P_x' = P_x \qquad \text{(B-2)}$$

where P_x' is the system's final momentum component just after the bullet leaves the gun, and where P_x is its initial momentum component just before the firing of the gun. But $P_x = 0$ since both the gun and the bullet are at rest before the firing of the gun. Thus (B-2) implies that

$$MV_x' + mv_x' = 0 \qquad \text{(B-3)}$$

where V_x' is the component of the gun's velocity just after the bullet leaves the gun. Hence

$$\boxed{V_x' = -\left(\frac{m}{M}\right) v_x'.} \qquad \text{(B-4)}$$

The gun thus starts moving as a result of the explosion which expelled the bullet. The minus sign in (B-4) indicates that the velocity of the gun is directed opposite to that of the bullet, i.e., the gun "recoils". However, the magnitude of the gun's recoil velocity is appreciably less than that of the bullet's velocity if the mass m of the bullet is much smaller than the mass M of the gun.

As discussed in Sec. 9D, the accelerations of two interacting particles have at every instant opposite directions and magnitudes inversely proportional to their masses. The result (B-4) is merely an implication of this basic property of two-particle interactions.

Example 2: Collision between a car and a truck

A car, of mass m, travels east with a speed v, as indicated in Fig. B-2. At an icy intersection, it collides with a truck, having a mass 3m and traveling north with a speed $\frac{1}{2}v$. The two vehicles then lock together and slide jointly along the icy road. What is the velocity with which the resulting wreck slides?

Neglecting friction with the icy road, the total external force on the system consisting of the car and truck is zero. (This is also true along the vertical direction since the downward gravitational force by the earth is balanced by the upward force exerted by the ground.) The momentum law (A-16) then implies that the momentum of this system remains unchanged during the collision process (irrespective of the complexity of the internal forces that crunched the vehicles in the collision). Thus

$$\vec{P}' = \vec{P} \tag{B-5}$$

where $\vec{P}'$ is the momentum of the system after the collision and $\vec{P}$ is its momentum before the collision.

The momentum of the system *before* the collision is

$$\vec{P} = \vec{p}_c + \vec{p}_t = mv\,\hat{i} + (3m)\left(\frac{1}{2}v\right)\hat{j} \tag{B-6}$$

where $\vec{p}_c$ is the initial momentum of the car along the eastern $\hat{i}$ direction and $\vec{p}_t$ is the initial momentum of the truck along the northern $\hat{j}$ direction. The momentum of the system *after* the collision is

$$\vec{P}' = m\,\vec{V}' + 3m\,\vec{V}' = 4m\,\vec{V}' \tag{B-7}$$

since both the car and the truck then move jointly with the same final velocity $\vec{V}'$. By the conservation of momentum (B-5), the initial and final momenta of the system are then related as illustrated in Fig. B-3.

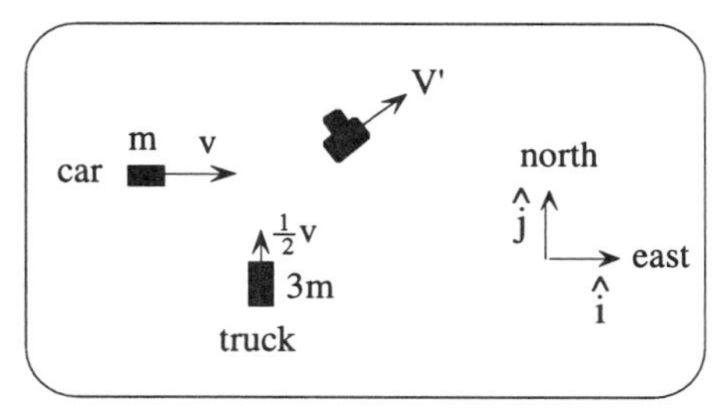

Fig. B-2. Collision of a car and truck.

Although the truck's speed is only half as large as that of the car, its mass is three times as large as that of the car. Hence the magnitude of the truck's momentum is *larger* than that of the car's momentum.

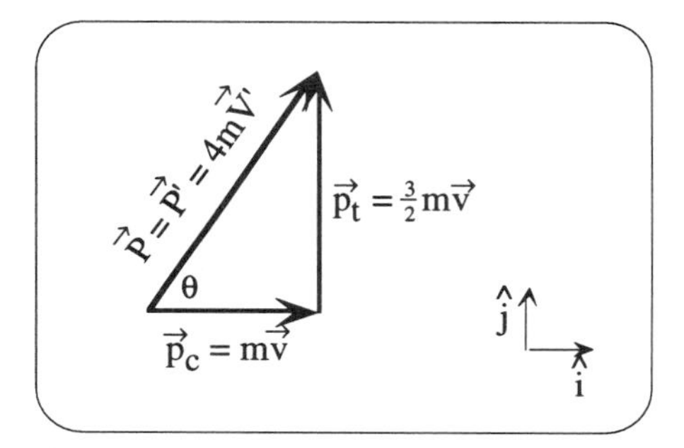

Fig. B-3. Initial and final momenta of the vehicles.

The conservation of momentum (B-5), applied along the $\hat{i}$ direction, implies that

$$P_x' = P_x \quad \text{or} \quad 4mV_x' = mv\,.$$

Thus
$$V_x' = \frac{1}{4}v\,. \tag{B-8}$$

Similarly, the conservation of momentum (B-5), applied along the $\hat{j}$ direction, implies that

$$P_y' = P_y \quad \text{or} \quad 4mV_y' = \frac{3}{2}mv\,.$$

Thus
$$V_y' = \frac{3}{8}v\,. \tag{B-9}$$

From the preceding information about the components of the wreck's velocity, one can readily find the direction and magnitude of this velocity. Thus the angle θ of this velocity, relative to the eastern $\hat{i}$ direction, is given by

$$\tan\theta = \frac{V_y'}{V_x'} = \frac{3}{2} = 1.5 \tag{B-10}$$

so that
$$\boxed{\theta = 56°\,.} \tag{B-11}$$

Similarly, the speed V' of the wreck is given by

$$V'^2 = V_x'^2 + V_y'^2 = \frac{1}{16}v^2 + \frac{9}{64}v^2 \tag{B-12}$$

The relations (B-10) and (B-12) are also immediately apparent from the vector diagram in Fig. B-3.

so that
$$\boxed{V' = \sqrt{\frac{13}{64}}\,v = 0.45\,v\,.} \tag{B-13}$$

➜ ***Go to Sec. 15B of the Workbook.***

C. Momentum change in short collisions

Change of momentum

The change of a system's momentum, during any infinitesimal time interval dt, can be obtained by multiplying both sides of the momentum law (A-16) by dt. Thus

$$d\vec{P} = \vec{F}_{ext}\, dt. \quad \text{(C-1)}$$

To find the system's change of momentum $\Delta\vec{P}$ during any long time interval Δt, one needs only to add all its successive infinitesimal momentum changes during this time. Thus

$$\Delta\vec{P} = \int d\vec{P} = \int \vec{F}_{ext}\, dt\,. \quad \text{(C-2)}$$

The change of momentum of the system can thus be calculated if one knows how the total external force varies during the time interval Δt.

Equivalently, one can write (C-2) in the simpler form

$$\boxed{\Delta\vec{P} = \vec{F}_{ext,av}\, \Delta t} \quad \text{(C-3)}$$

The product $\vec{F}_{av}\Delta t$ of an average force, multiplied by the short time during which it acts, is sometimes called the *impulse* exerted by that force.

where $\vec{F}_{ext,av}$ is the *average* external force during the time interval Δt. Comparison with (C-2) shows that this average force is defined so that

$$\vec{F}_{ext,av} = \frac{\int \vec{F}_{ext}\, dt}{\Delta t}\,. \quad \text{(C-4)}$$

For example, Fig. C-1 illustrates a time-varying external force and the corresponding average value of this force.

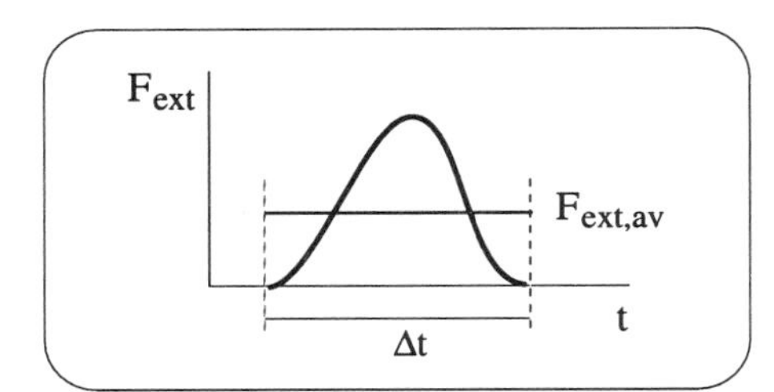

Fig. C-1. Magnitude of a total external force (along some fixed direction) varying during some time interval Δt.

The relation (C-3) indicates that the momentum change of a system depends on the average external force acting on the system and on the length of time during which this force acts.

Short collisions

Consider a collision between two particles like those illustrated in Fig. C-2 (or like the colliding car and truck considered in Sec. B). During the collision, the force exerted on each particle by the other may be very large. Hence the momentum of each particle can change appreciably, even during a very short time interval Δt [i.e., the right side of (C-3) can be large, despite the short time, because the average force on each particle is so large].

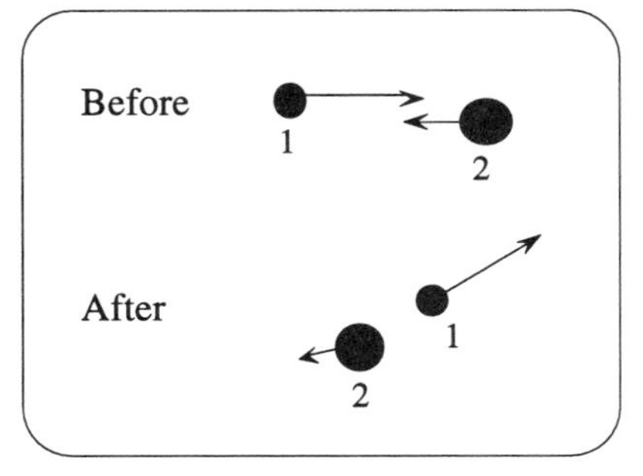

Fig. C-2. A collision between two particles 1 and 2.

Useful approximations. Although the velocities of the colliding particles change appreciably during a short collision time, this time is so short that the *positions* of the particles scarcely change at all.

The large mutual forces between the colliding particles are internal to the system consisting of both particles and thus don't affect the *total* momentum of this system. Furthermore, the *external* forces on this system are *not* larger than usual. (For example, the gravitational forces on the car and truck, or the forces exerted on these by the road, are not larger during the collision than at any other time.) Hence the right side of (C-2) or (C-3) is negligibly small if the time interval Δt is sufficiently short. Correspondingly, the momentum change $\Delta\vec{P}$ of the entire system of both particles is then negligibly small compared to the momentum change of either particle.

The preceding considerations lead to the following useful conclusion, valid to very good approximation if the collision time Δt is sufficiently short:

$$\text{During a short collision (if } \vec{F}_{ext,av}\,\Delta t \approx 0), \qquad \Delta\vec{P} \approx 0\,. \tag{C-5}$$

Hence the momentum of a system is, to good approximation, conserved in a short collision *even if the external forces on the system are non-zero* (as long as they are not excessively large).

For example, suppose that the friction forces exerted by the road would be appreciable in Example 2 of Sec. B. Then the conservation of momentum could still be applied during the short collision of the car and the truck. Thus one could relate their velocities just before their collision to their velocities just after their collision.

Problems involving collisions. The preceding comments suggest the following approach for solving mechanics problems involving a short collision: (a) Use the conservation of momentum (C-5) to relate the velocities of the particles immediately before their collision to their velocities immediately after their collision. (b) Use the familiar mechanics laws (e.g., Newton's law and the energy law) to find information about the particles' motion during the longer times before the collision and during the longer times after the collision.

The following example illustrates this approach.

Prototype example: Ballistic pendulum

Problem statement. A pendulum, consisting of a string attached to a wooden bob of mass M, hangs vertically at rest. A bullet, of mass m, is shot horizontally into the bob and becomes embedded in it. As a result, the bob starts moving along a circular arc, rising a vertical height h before coming momentarily to rest. Use this information to find the original speed of the bullet.

(Such a pendulum is called a *ballistic pendulum.* It is a practical device for determining the speeds of bullets by simple measurements, without any need for fancy technology to measure high speeds or very short time intervals.)

Solution. The following shows the solution of this problem.

Analysis of problem

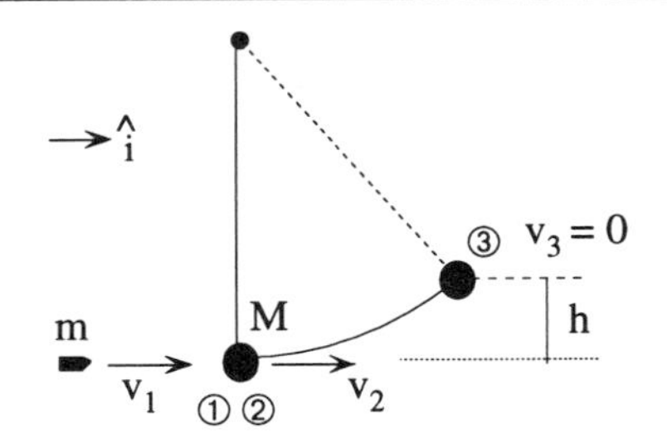

①: Just before collision
②: Just after collision (bob and bullet in it)
③: Bob at highest point

<u>Known:</u> m, M, h.

<u>Goal:</u> $v_1 = ?$

Construction of solution

Apply momentum law, to bob + bullet, ①→②

$$\Delta\vec{P} = \vec{F}_{ext,av}\,\Delta t = 0 \quad \text{(since } \Delta t \text{ very small)}$$

along $\hat{i}$

$$(M + m)\,v_2 - (0 + mv_1) = 0$$

$$\underline{\underline{v_1}} = \left(\frac{M + m}{m}\right)\underline{v_2} \qquad (1)$$

Apply momentum conservation, to the system consisting of the bullet and bob, during the short collision from the instant just before the bullet strikes the bob to the instant just after it becomes embedded in it. (External forces exerted by gravity and the string are irrelevant in this very short collision.)

Wanted unknown has been underlined twice, unwanted unknown underlined once.

(Find v_2.) Apply energy law, to bob + bullet, ②→③

Want to relate speeds and positions at the instants ② and ③. Hence apply energy law.

$$\Delta E = W_{oth}$$

$$\Delta K + \Delta U_g = W_T$$

Tension force by string does zero work.

$$[0 - \tfrac{1}{2}(M + m)v_2^2\,] + [(M + m)gh - 0] = 0$$

$$-\,v_2^2 + 2gh = 0$$

$$v_2^2 = 2gh$$

$$\underline{v_2} = \sqrt{2gh} \qquad (2)$$

(Find v_1.) Eliminate v_2. Put (2) into (1):

$$\boxed{v_1 = \left(\frac{M}{m} + 1\right)\sqrt{2gh}} \qquad (3)$$

Checks (done in writing or mentally)

<u>Checks</u>

Units of (3): $\frac{m}{s} = \sqrt{\frac{m}{s^2}(m)} = \frac{m}{s}$ OK.

Check extreme cases.

If h large: Expect v_1 large. Eq. (3) OK.

If $m << M$: Expect v_1 must be large to move heavier bob appreciably. Eq. (3) OK.

Summary. To solve this problem, we used the following two major steps: (1) We used the momentum law to relate the speeds (of the bullet and bob) immediately before and immediately after their short collision. (2) We then applied the energy law to the subsequent motion of the bob with its embedded bullet (and could thus relate its speed just after the collision to the ultimate height attained by it).

> ***Energy loss of the bullet***
>
> Note that the bullet loses most of its initial kinetic energy when it becomes embedded in the bob. (If the mass of the bob is very much larger than that of the bullet, the bob and embedded bullet barely move after their collision.) Indeed, while the bullet is slowed in penetrating the bob, negative work is done on the bullet by the forces exerted on it by the wooden material of the bob. This work is the *other* work in the energy law $\Delta E = W_{oth}$ applied to the bullet. Thus the energy $E = K + U$ of the bullet decreases, i.e., its kinetic energy K decreases. (Its gravitational potential energy U remains unchanged during the collision where the bullet remains at nearly the same position).

➜ ***Go to Sec. 15C of the Workbook.***

D. Motion of the center of mass

System-averaged acceleration and total external force. In (A-9) we obtained the following general relation between the accelerations of the particles in a system and the total external force on this system:

$$\sum m_n \vec{a}_n = \vec{F}_{ext} \,. \tag{D-1}$$

In Sec. B we then expressed the left side of this relation in terms of the rate of change of the system's momentum and thus obtained the momentum law (A-16).

However, the relation (D-1) can also be expressed in the following useful form

$$\boxed{M\vec{A} = \vec{F}_{ext}} \,, \tag{D-2}$$

similar to Newton's law for a particle, if we introduce the quantities M and $\vec{A}$. Here M is the *total* mass of the system (i.e., the sum of the masses of all particles in the system) so that

$$M = \sum m_n \,. \tag{D-3}$$

Furthermore, we have defined a *system-averaged acceleration* $\vec{A}$ so that

$$\vec{A} = \frac{\sum m_n \vec{a}_n}{M} \,. \tag{D-4}$$

This is the usual way of calculating an average over a system. For example, to find the average grade of students in a class, one multiplies every grade by the number of students receiving this grade, adds all these products, and then divides the result by the total number of students in the class.

In other words, this system-averaged acceleration is obtained by multiplying the acceleration of every particle in the system by its mass, adding all these products, and then dividing the result by the total mass of the system.

System-averaged velocity. In a fashion analogous to (D-4), the system-averaged velocity $\vec{V}$ of the system can be defined in terms of the velocities of all the particles in the system. Thus

$$\vec{V} = \frac{\sum m_n \vec{v}_n}{M} . \qquad \text{(D-5)}$$

The system-averaged acceleration of the system is then related to its system-averaged velocity in the following obvious way

$$\vec{A} = \frac{d\vec{V}}{dt} . \qquad \text{(D-6)}$$

Indeed, the rate of change of (D-5) is simply

$$\frac{d\vec{V}}{dt} = \frac{\sum m_n \frac{d\vec{v}_n}{dt}}{M} = \frac{\sum m_n \vec{a}_n}{M} = \vec{A} .$$

System-averaged position. The system-averaged position vector $\vec{R}$ of the system, also called the position of its *center of mass*, can be similarly defined in terms of the position vectors of all the particles in the system. Thus

Def: ***Position of center of mass:***

$$\vec{R} = \frac{\sum m_n \vec{r}_n}{M} . \qquad \text{(D-7)}$$

The system-averaged velocity of the system is then related to its system-averaged position vector in the following obvious way:

The symbol $\vec{R}_c$, with the subscript *c*, will sometimes also be used to denote the position of the center of mass.

$$\vec{V} = \frac{d\vec{R}}{dt} . \qquad \text{(D-8)}$$

Motion of the center of mass. The relation (D-2) specifies how the acceleration $\vec{A}$ of the center of mass is related to the total external force on the system. Since (D-2) is similar to Newton's law for a particle, one arrives at the following conclusion (illustrated in Fig. D-1):

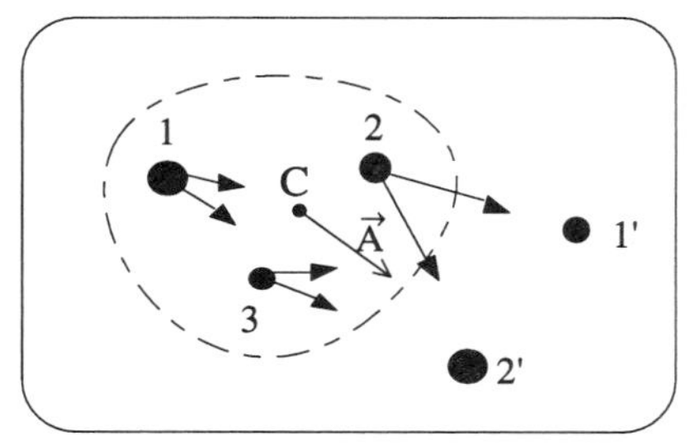

Fig. D-1. External forces on the system of Fig. A-1, and resulting acceleration $\vec{A}$ of its center of mass C.

The center of mass of a system moves like a particle having a mass equal to the total mass of the system, and acted on by a total force equal to the total external force on this system.	(D-9)

Note that it is irrelevant to which particles in the system these external forces are applied, or how far these particles are from the center of mass.

By (D-5) it also follows that $M\vec{V}$, the momentum of the center of mass, is just equal to the total momentum $\vec{P}$ of the system (i.e., to the sum of the momenta of all particles in the system). Thus

$$\vec{P} = M\vec{V} . \qquad \text{(D-10)}$$

Example 1: Throwing a twirling baton

A baton, thrown into the air, rotates about its center in the fashion indicated in Fig. D-2. But, during all this rotation, the center of mass of the baton (e.g., the center of the baton if it is a uniform rod) moves along the same trajectory as a single particle projected from the ground.

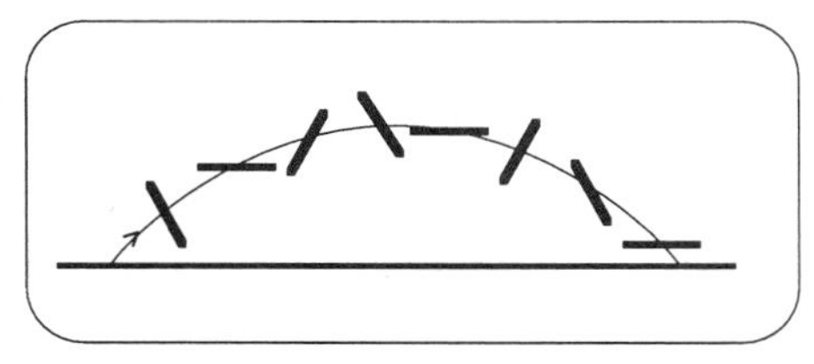

Fig. D-2. Trajectory of the center of mass of a rotating baton.

Example 2: Projectile exploding in mid-air

Fig. D-3 illustrates a projectile which explodes into three fragments at a certain instant during its flight. The resulting mutual forces between the fragments cause them to fly apart in different directions. But the external force on the system, consisting of the projectile or its fragments, is just the gravitational force — which remains the same during the entire explosion. Hence the center of mass of this system continues to move in the same simple way as if the explosion had never occurred.

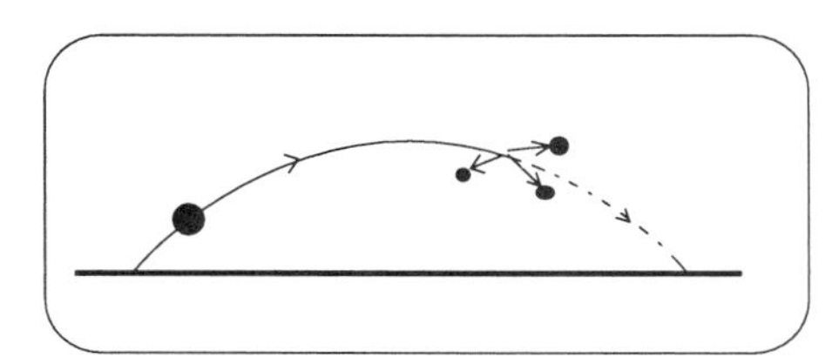

Fig. D-3. Trajectory of the center of mass of an exploding projectile.

Example 3: Collision of a car and truck

Example 2 of Sec. B discussed a car and truck which collide at an icy intersection and then jointly move off while sticking together. The total external force on the system consisting of the car and truck is zero if the friction force exerted on these vehicles by the road is negligible. During the entire motion (including the violent collision between the vehicles) their center of mass moves, therefore, simply with zero acceleration (i.e., it moves with constant velocity along a straight line).

➜ ***Go to Sec. 15D of the Workbook.***

E. Location of the center of mass

The position vector $\vec{R}$ of a system's center of mass can readily be found from its definition (D-7), which can also be written in the slightly simpler form

$$M\vec{R} = \sum m_n \vec{r}_n . \qquad \text{(E-1)}$$

The preceding relation between position vectors implies corresponding relations between the components of these position vectors along any direction. For example, if one introduces a coordinate system with coordinate directions $\hat{i}$ and $\hat{j}$, the position coordinates X and Y of the center of mass are related to the corresponding position coordinates x_n and y_n of the particles in the system so that

$$MX = \sum m_n x_n \quad \text{and} \quad MY = \sum m_n y_n . \qquad \text{(E-2)}$$

Simple properties of the center of mass

Positions measured from the center of mass. Suppose that the positions of all particles are measured from their center of mass C rather than from some other arbitrary origin O. (As indicated in Fig. E-1, the position vector of any particle, measured from C, can be indicated by the primed letter $\vec{r}'$.) The position vector $\vec{R}'$ of the center of mass, relative to its own position, is just zero. Hence (E-1) implies simply that

$$\text{relative to center of mass,} \quad \sum m_n \vec{r}_n' = 0. \tag{E-3}$$

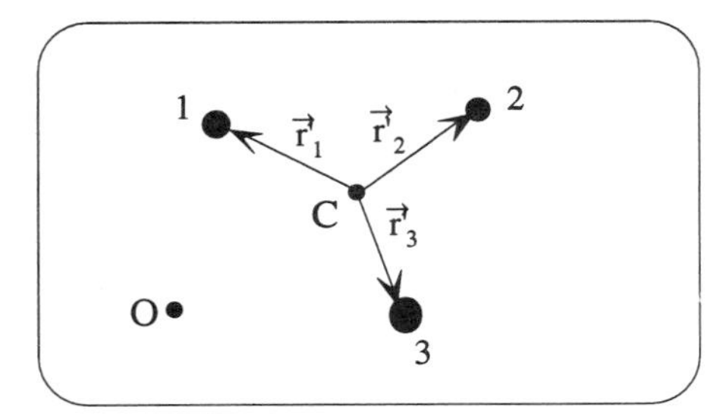

Fig. E-1. Position vectors of particles relative to their center of mass C.

This kind of a sum is thus particularly simple because the quantities are measured relative to the center of mass.

Center of mass of two particles. Where is the center of mass of two particles 1 and 2, having respective masses m_1 and m_2? Suppose that all position vectors are measured from the particles' center of mass C, as indicated in Fig. E-2. By (E-1) or (E-3), the position vectors $\vec{r}_1'$ and $\vec{r}_2'$ of the particles must then be related so that

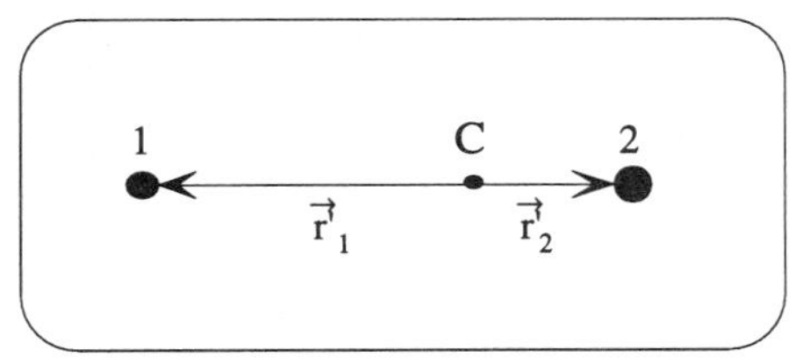

Fig. E-2. Center of mass of two particles (shown here for the case where $m_2 > m_1$).

$$M\vec{R}' = 0 = m_1\vec{r}_1' + m_2\vec{r}_2'$$

since the position vector $\vec{R}'$ of the center of mass, relative to itself, is zero. Hence

$$m_1\vec{r}_1' = -m_2\vec{r}_2'$$

or

$$\vec{r}_1' = -\left(\frac{m_2}{m_1}\right)\vec{r}_2'. \tag{E-4}$$

Here the minus sign indicates that the position vectors $\vec{r}_1'$ and $\vec{r}_2'$ have opposite directions. Thus the particles are located on opposite sides of the center of mass C (i.e., the center of mass is located between the particles). If the masses of the particles are the same, the center of mass is located halfway between the particles. More generally, the distances r_1' and r_2' of the particles from the center of mass are inversely proportional to the masses of the particles (i.e., the center of mass is closer to the particle of larger mass than to the particle of smaller mass).

Symmetric objects

The center of mass can always be calculated from its definition (E-1). However, in many important simple cases, the position of the center of mass can be found without any calculation at all.

Symmetry considerations. Suppose that an object has a plane of symmetry so that, for any particle on one side of this plane, there exists a particle of the same mas on the other side of the plane, at the same distance from it. (This is schematically illustrated in Fig. E-3.) Then the center of mass of the object cannot preferentially be located on one side of the plane rather than on the other. The symmetry of the situation then leads to the following conclusion:

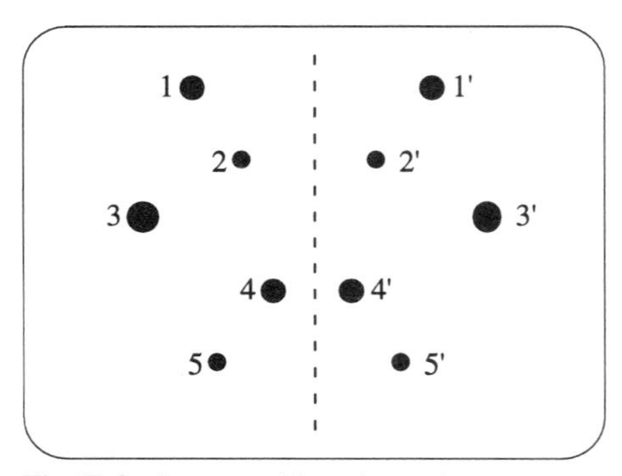

Fig. E-3. System with a plane of symmetry indicated by the dashed line. (Particles indicated by corresponding numbers have the same mass.)

If an object has a plane of symmetry, its center of mass must be located on this plane.	(E-5)

If an object is sufficiently symmetric, its center of mass can often be easily found without any calculation. The following are some examples.

Rectangular plate. A uniform rectangular plate has two planes of symmetry (passing through the center of the rectangle and parallel to its sides, as indicated in Fig. E-4a). The center of mass of the rectangular plate must then be located on both of these planes, i.e., it must be at the center of the rectangle.

Rectangular bar. Similar considerations apply to a uniform rectangular bar (i.e., to a three-dimensional body bounded by six rectangular faces, with opposite faces parallel to each other). There are then three planes of symmetry, each parallel to one pair of faces of the bar and all passing through the center of the bar. Hence the center of mass of the bar is at its center.

Disks or rings. A uniform disk, like that illustrated in Fig. E-4b, is very symmetric since *any* plane passing through the center of the disk is a plane of symmetry. (Only two such planes are shown in the figure.) Hence the center of mass of such a disk is located at the center of the disk. Similarly, the center of mass of a uniform ring, like that illustrated in Fig. E-4c, is at the center of the ring (i.e., at the center of the hole formed by the ring).

Spheres. Similar symmetry considerations imply that the center of mass of a uniform sphere (or spherical shell) is located at its geometric center.

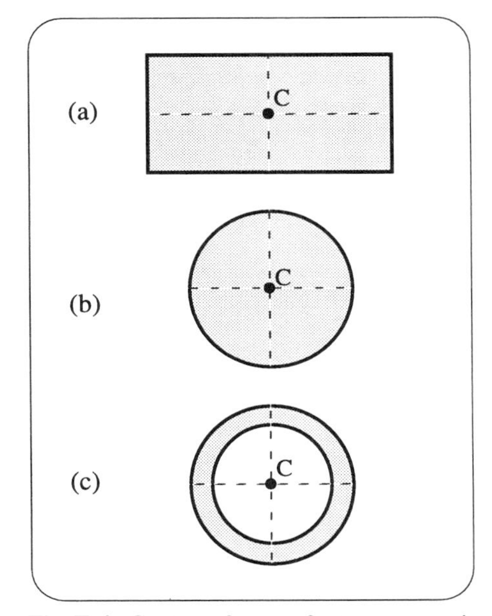

Fig. E-4. Centers of mass of some symmetric objects. (Planes of symmetry are incidated by dashed lines.)

Composite objects

Consider a system S which consists of two parts A and B, having masses M_A and M_B. The mass of system S is then $(M_A + M_B)$ and the position vector $\vec{R}$ of its center of mass is, by its definition (E-1), given by

$$(M_A + M_B)\,\vec{R} = \sum m_n \vec{r}_n$$

or

$$(M_A + M_B)\,\vec{R} = \sum_A m_n \vec{r}_n + \sum_B m_n \vec{r}_n\,. \qquad \text{(E-6)}$$

Here the sum over all particles in the system S has been written as the sum over all particles in part A of this system, plus the sum over all particles in part B of this system.

But the first sum on the right side of (E-6) is, by the definition (E-1), simply related to the position vector $\vec{R}_A$ of the center of mass of part A of the system. Similarly, the second sum on the right side is simply related to the position vector $\vec{R}_B$ of the center of mass of part B of the system. Thus (E-6) implies that

$$(M_A + M_B)\,\vec{R} = M_A\vec{R}_A + M_B\vec{R}_B$$

or

$$\vec{R} = \frac{M_A\vec{R}_A + M_B\vec{R}_B}{M_A + M_B}. \qquad \text{(E-7)}$$

This result can be stated in words as follows:

> The center of mass of a system consisting of several parts is the same as if each of these parts were concentrated in a single particle (having the mass of this part and located at the center of mass of this part). (E-8)

This result often makes it easy to find the location of the center of mass of a complex object. The following is an example.

Example: Center of mass of a clock pendulum

Fig. E-5 illustrates a pendulum of the kind used in old-fashioned pendulum clocks. The pendulum consists of a rod, of mass m and length L, attached to a brass disk of mass M and radius R. Where is the center of mass of this pendulum?

One can conveniently measure positions from the point of suspension O of the pendulum, and can specify coordinates along the downward $\hat{i}$ direction indicated in Fig. E-5. The center of mass of the rod is then located at its center, a distance L/2 from O. The center of mass of the bob is located at its center, a distance (L + R) from O. By (E-7) or (E-8), the position of the center of mass of this pendulum is then located at a distance X below O, where X is given by

$$X = \frac{m(L/2) + M(L + R)}{m + M}. \qquad \text{(E-9)}$$

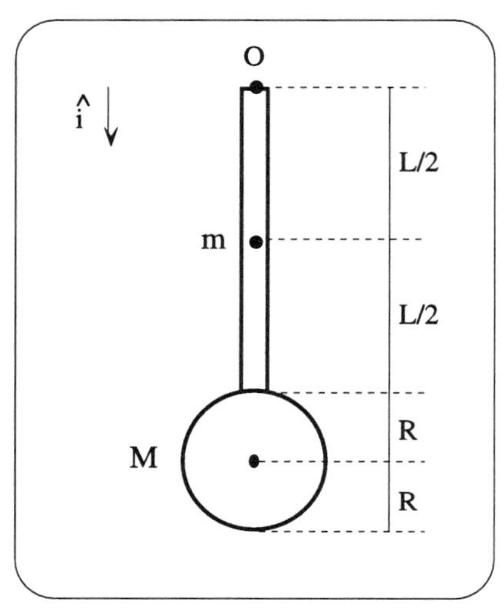

Fig. E-5. A clock pendulum.

➜ ***Go to Sec. 15E of the Workbook.***

F. Summary

Definitions

Momentum of a particle: $\vec{p} = m\vec{v}$

Momentum of a system: $\vec{P} = \sum \vec{p}_n = M\vec{V}$ (where $\vec{V}$ = velocity of CM)

Mass of system: $M = \sum m_n$ (sum of particle masses)

Center-of-mass (CM): $\vec{R} = \dfrac{\sum m_n \vec{r}_n}{M}$ (system-averaged position)

Momentum law: $\boxed{\frac{d\vec{P}}{dt} = \vec{F}_{ext}}$ (internal forces are irrelevant)

Equivalent forms:

Total change of momentum: $\Delta\vec{P} = \vec{F}_{ext,av}\,\Delta t$

Acceleration of center of mass: $M\vec{A} = \vec{F}_{ext}$

Special case: Conservation of momentum

$\Delta\vec{P} = 0$ if $\vec{F}_{ext,av}\,\Delta t = 0$

[i.e., either if $\vec{F}_{ext} = 0$

or if $\Delta t \approx 0$ (short collision)]

New abilities

You should now be able to do the following:

(1) Solve mechanics problems by applying the momentum law (sometimes together with the energy law), particularly in cases involving conservation of momentum in collisions.

(2) Find the location of the center of mass of simple systems.

(3) Predict or explain the motion of a system's center of mass.

→ ***Go to Sec. 15F of the Workbook.***

16 Energy of a System

The energy law, discussed in Chapter 14, allows one to relate directly the speeds and positions of a particle without the need to consider explicitly the elapsed time. By applying the energy law to every particle in a system, we shall now extend this law to systems consisting of any number of particles. In this way we shall obtain the energy law in its general form and thus arrive at one of the most widely important laws in all the sciences.

A. Energy law for a system

The energy law (14B-6) for a particle states that

$$\Delta E = W_{oth} \tag{A-1}$$

where the energy $E = K + U$ is the sum of the particle's kinetic and potential energies. Expressed in words, the energy law asserts that the change of a particle's energy is equal to the work done on the particle by all other interactions not included in its potential energy.

Derivation of the general energy law

The energy law can be applied to every particle in a system (e.g., to the system schematically indicated in Fig. A-1). Thus one can write, for each of the particles 1, 2, 3, ...

$$\Delta E_1 = (W_{oth})_1$$

$$\Delta E_2 = (W_{oth})_2$$

....

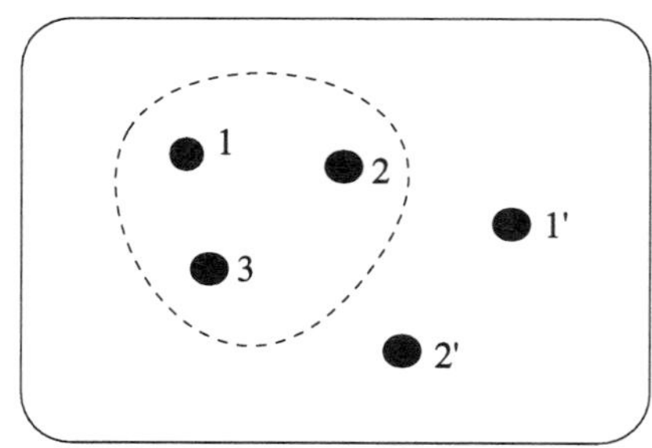

Fig. A-1. A simple system consisting of three particles 1, 2, and 3.

By adding these equations for all the particles in the system, one obtains

$$\Delta E_1 + \Delta E_2 + \ldots = (W_{oth})_1 + (W_{oth})_2 + \ \ldots .$$

By introducing the abbreviations E and W to denote the quantities on the left and right sides, one thus obtains the general energy law for any system:

General energy law:	$\Delta E = W_{oth}$.

(A-2)

This has exactly the same form as the energy law (A-1) for a single particle, except that each quantity denotes the *sum* of corresponding quantities for all particles in the system.

For example, the total energy $E = K + U$ of the system is just the sum of the energies of all the particles in the system. Thus the total kinetic energy K of the system is the sum of the kinetic energies of all the particles in the system, i.e.,

$$K = K_1 + K_2 + K_3 + \ldots . \qquad \text{(A-3)}$$

Similarly, the total potential energy U of the system is the sum of the potential energies of all the particles in the system. As usual, this total potential energy can be divided into internal and external parts by writing

$$U = U_{int} + U_{ext} \qquad \text{(A-4)}$$

where U_{int} is the total *internal* potential energy of the system (i.e., the sum of the potential energies of all particles in the system due to all other particles in the system) and U_{ext} is the total *external* potential energy of the system (i.e., the sum of the potential energies of all particles in the system due to all particles outside the system).

Similarly, the total other work can be written as

$$W_{oth} = (W_{oth})_{int} + (W_{oth})_{ext} \qquad \text{(A-5)}$$

where $(W_{oth})_{int}$ is the total *internal* other work done on the system and $(W_{oth})_{ext}$ is the total *external* other work done on the system.

Internal work and potential energy

When discussing the momentum law in Sec. 15A, we found that the total internal *force* on a system is simply zero because mutual internal forces cancel each other. However, the total internal *work* (done on particles in a system by other particles in the same system) is ordinarily *not* zero. Similarly, the total internal potential energy of a system is ordinarily also *not* zero.

Mutual work on a pair of particles. The relation between mutual forces does, however, lead to some simplifications. Indeed, consider any two particles 1 and 2 in the system. The mutual forces $\vec{F}_{12}$ and $\vec{F}_{21}$ on one particle by the other have equal magnitudes and opposite directions, as indicated in Fig. A-2. Consider then the mutual work $d'W_{mut}$ done by *both* particles (i.e., the *sum* of the works done on each particle by the other) when particle 1 is displaced by a small amount $d\vec{r}_1$ and particle 2 is displaced by a small amount $d\vec{r}_2$. This mutual work is

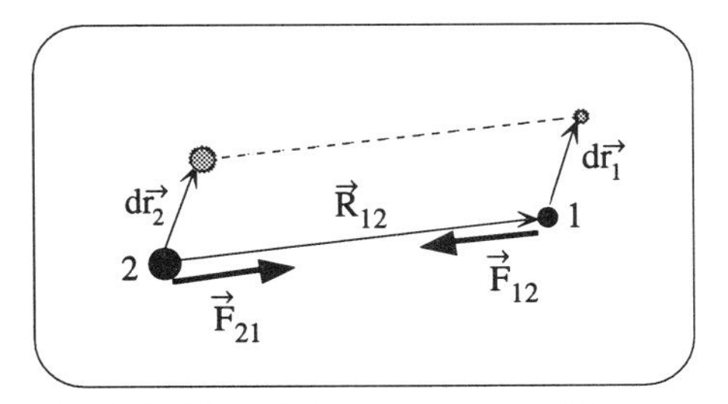

Fig. A-2. Mutual forces on two particles (shown here displaced by equal amounts so that their relative displacement is zero).

$$d'W_{mut} = d'W_{12} + d'W_{21} = \vec{F}_{12} \cdot d\vec{r}_1 + \vec{F}_{21} \cdot d\vec{r}_2$$

$$= \vec{F}_{12} \cdot d\vec{r}_1 - \vec{F}_{12} \cdot d\vec{r}_2 = \vec{F}_{12} \cdot (d\vec{r}_1 - d\vec{r}_2) . \qquad \text{(A-6)}$$

Suppose that both particles are displaced by the *same* amount, as indicated in Fig. A-2. The works done on the particles by their mutual forces have then equal magnitudes but opposite signs. In this special case, the mutual work done by the particles on each other is simply zero.

More generally, however, the displacements of the particles are *not* the same, i.e., their displacement relative to each other is *not* zero. Their mutual work (A-6) is then not zero, but equal to

$$d'W_{mut} = \vec{F}_{12} \cdot d(\vec{r}_1 - \vec{r}_2) = \vec{F}_{12} \cdot d\vec{R}_{12} \qquad \text{(A-7)}$$

where $\vec{R}_{12} = \vec{r}_1 - \vec{r}_2$ is the position vector of particle 1 *relative* to particle 2, and $d\vec{R}_{12}$ is the infinitesimal displacement of particle 1 *relative* to particle 2.

The mutual work, done on the particles by each other, depends thus only on their displacement *relative* to each other. As indicated by (A-7), this work is the same as if one of the particles (say, particle 2) remained at rest and the other particle moved relative to it.

Mutual potential energy of a pair of particles. The preceding result also implies that the mutual potential energy of a pair of interacting particles depends only on their positions relative to each other. This potential energy is, therefore, the same as the potential energy of one particle when the other one remains at rest (i.e., it is the same as any of the potential energies calculated in Chapter 14).

Internal potential energy of a system. Correspondingly, the internal potential energy of a system of particles (i.e., the sum of the potential energies of all the particles in the system due to their mutual interaction) is just the sum of the potential energies between all *pairs* of particles in the system.

Systems with constant interparticle distances. There are many systems whose constituent particles remain at constant distances from each other. The following are some examples where this is true to good approximation: (a) Rigid bodies (like rods, metal plates, steel blocks, etc.) are objects which don't change their shape or volume. The distance between any pair of atoms in such a rigid body thus remains constant. (b) A string usually has a fixed length, even if it changes its shape. The distance between neighboring particles in such a string thus remains constant. (c) A liquid (like water) is nearly incompressible so that its volume does not change. Hence the distance between neighboring atoms in such a liquid also remains constant.

If the distance R between two neighboring particles remains constant, a particle must move relative to the other one along a circular arc. As indicated in Fig. A-3, any infinitesimal displacement of this particle, relative to the other one, must then be perpendicular to the direction of the mutual force between the particles. Hence the mutual work (A-4) done on the particles by each other must be zero. Correspondingly, their mutual potential energy also remains unchanged.

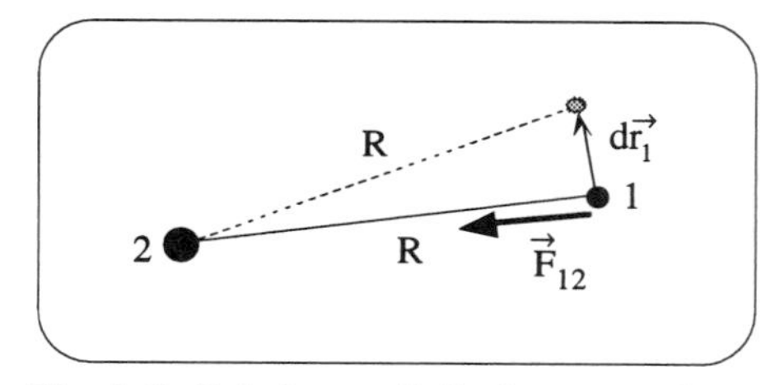

Fig. A-3. Relative small displacement of two particles separated by a constant distance R.

If interparticle distances are constant, $\qquad \Delta U_{int} = 0\,. \qquad$ (A-8)

Examples

Example 1. Blocks repelled by a compressed spring

Fig. A-4a shows two identical small blocks 1 and 2, each with mass m, lying on a horizontal table. The blocks are initially held at rest while touching a spring compressed a distance s. The spring has negligible mass and a spring constant k. When the blocks are released, they move apart. One wishes to find the final speeds of the blocks after they are no longer touch the spring. Friction forces are negligible.

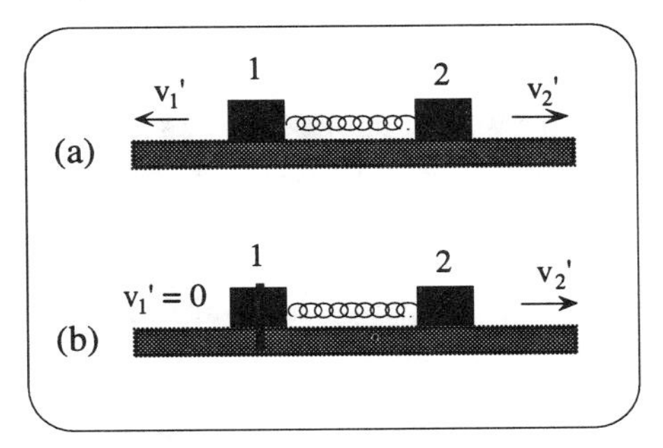

Fig. A-4. Blocks repelled by a compressed spring. (a) Both blocks free to move. (b) Block 1 fixed to the table.

Application of the energy law. The energy law (A-2), applied to the system consisting of the two blocks and the spring, yields

$$\Delta E = \Delta K + \Delta U = W_{oth}$$

or

$$(K_1' + K_2' - 0) + (0 - U_o) = 0\,. \tag{A-9}$$

Here ΔK, the change of the system's kinetic energy, is equal to the sum of the final kinetic energies K_1' and K_2' of the blocks since their initial kinetic energies are zero. The system's change of potential energy is its final zero potential energy (when the spring is undeformed) minus the initial potential energy $U_o = \frac{1}{2} ks^2$ of the compressed spring . The other work W_{oth} is zero since friction forces are negligible. Thus (A-9) implies that

$$K_1' + K_2' = U_o\,, \tag{A-10}$$

i.e., that the total final kinetic energy of both blocks is equal to the initial potential energy stored in the spring.

One block kept at rest. Consider first the special case, illustrated in Fig. A-4b, where the block 1 is nailed to the table. Hence it remains at rest and acquires no kinetic energy. Then block 2 alone moves after being released and (A-10) implies that

$$K_2' = U_o\,. \tag{A-11}$$

The potential energy of the spring is thus converted entirely into the kinetic energy of block 2. The final speed v_2' of block 2 must then be such that

$$\frac{1}{2} m v_2'^2 = \frac{1}{2} k\, s^2$$

or

$$v_2' = \sqrt{\frac{k}{m}}\,. \tag{A-12}$$

Both blocks free to move. Consider now the general case illustrated in Fig. A-4a where *both* blocks are free to move after they are released. Since they are identical, each of them acquires then the same kinetic energy $K_2' = K_1'$. Thus (A-9) implies that

$$2\, K_2' = U_o$$

or

$$K_2' = \frac{1}{2} U_o\,. \tag{A-13}$$

In other words, the potential energy of the spring is now converted into the equal kinetic energies of *both* blocks. Hence block 2 acquires only half as large a kinetic energy as in the previous case (A-11). As a result, the speed it acquires is *smaller* than in the previous case. (Its speed is smaller by a factor of $\sqrt{2}$ since the kinetic energy is proportional to the square of the speed.)

Example 2. Energy arguments applied to a lever.

Fig. A-5a shows a *lever* consisting of a rigid rod, of negligible mass, free to pivot around a fixed point O. A weight is used to apply a downward force of magnitude F_1 to the left end of the rod at a distance L_1 from O. What then is the magnitude F_2 of the downward force which must be applied to the right end of the rod, at a distance L_2 from O, so that the rod remains at rest?

Objects which remain at rest are commonly said to be *in equilibrium*.

An easy way of solving such problems is to exploit the property (A-8) that the internal potential energy of a rigid body remains unchanged. Thus, imagine that the rod is very slowly rotated about the pivot point O by some infinitesimal angle, as indicated in Fig. A-5b. Then the left side of the rod moves down a small vertical distance s_1, and the right side of the rod moves up a corresponding small vertical distance s_2. The kinetic energy K of the rod does not change in this very slow rotation. Its potential energy U also does not change since the internal potential energy of the rigid rod does not change. (It also has no gravitational potential energy since its mass is negligible.) Thus the energy change dE of the rod in this small rotation is zero. Correspondingly, the energy law (A-2) implies that dE = d'W = 0 where d'W is the other small work done by the two applied forces. Thus

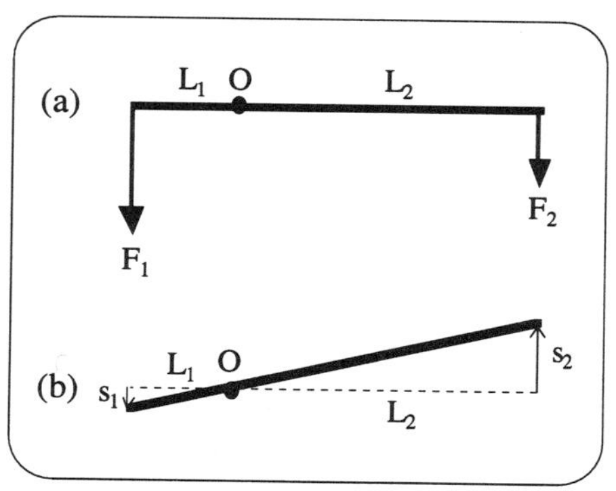

Fig. A-5. A lever free to pivot around the point O.

$$d'W = F_1 s_1 - F_2 s_2 = 0$$

or

$$\frac{F_2}{F_1} = \frac{s_1}{s_2}. \tag{A-14}$$

Because of the similar triangles in Fig. A-5b,

$$\frac{s_1}{s_2} = \frac{L_1}{L_2}. \tag{A-15}$$

Hence (A-14) implies that

$$\frac{F_2}{F_1} = \frac{L_1}{L_2} \quad \text{or} \quad F_2 = F_1 \frac{L_1}{L_2}. \tag{A-16}$$

This result indicates the utility of the lever. A small applied force of magnitude F_2 is sufficient to support a weight of much larger magnitude F_1, if the small force is applied at a much larger distance from the pivot point. (But, if the weight is to be moved, the smaller applied force must move through a larger distance. Thus one gains an advantage in requiring a smaller force, but must do the same amount of work.)

→ ***Go to Sec. 16A of the Workbook.***

B. System energies and center of mass

The energy of a system can often be easily found if one knows its center of mass. The following are some useful examples.

Kinetic energy

The velocity $\vec{v}$ of any particle in a system (relative to some inertial frame) may be expressed in terms of the velocity $\vec{V}_c$ of the system's center of mass (relative to the inertial frame) and the velocity $\vec{v}'$ of this particle's velocity relative to the center of mass. Indeed, according to (8E-4),

$$\vec{v} = \vec{V}_c + \vec{v}' . \tag{B-1}$$

The system's kinetic energy K, which involves the velocities of all the particles in the system, can then be expressed in terms of the velocities on the right side of (B-1). As shown below, one then obtains the following simple result

$$\boxed{K = K_c + K'} \tag{B-2}$$

where $K_c = \frac{1}{2}MV_c^2$ is the kinetic energy associated with the motion of the center of mass, and where K' is the system's total kinetic energy calculated with the particles' velocities *relative to the center of mass.*

Stated in words, the result (B-2) asserts that the kinetic energy of any system is equal to ⟨the kinetic energy which the system would have if its entire mass were concentrated at its center of mass⟩ plus ⟨the total kinetic energy of the particles moving *relative* to the center of mass⟩.

Argument leading to the result (B-2)

The kinetic energy K of a system is, by its definition, equal to the sum of the kinetic energies of all the particles in the system. Thus

$$K = \sum\left(\frac{1}{2}m_n\vec{v}_n^{\,2}\right) \tag{B-3}$$

where m_n is the mass of the n'th particle, $\vec{v}_n$ is its velocity, and the sum extends over all particles n = 1, 2, 3, ... in the system. But, by (B-1), the square of any such particle's velocity is

Remember that $\vec{v}^{\,2} = \vec{v}\cdot\vec{v} = v^2$.

$$\vec{v}_n^{\,2} = (\vec{V}_c + \vec{v}_n')^2 = \vec{V}_c^{\,2} + \vec{v}_n'^{\,2} + 2\vec{V}_c\cdot\vec{v}_n' .$$

When this expression is substituted into (B-3), the right side becomes decomposed into the following three separate sums

$$K = \frac{1}{2}\vec{V}_c^{\,2}\left(\sum m_n\right) + \sum\left(\frac{1}{2}m_n\vec{v}_n'^{\,2}\right) + \vec{V}_c\cdot\left(\sum m_n\vec{v}_n'\right)$$

or

$$K = \frac{1}{2}M\vec{V}_c^{\,2} + K' + \vec{V}_c\cdot\left(\sum m_n\vec{v}_n'\right). \tag{B-4}$$

Here we have taken constant factors outside each sum and used the fact that the sum of the masses of all particles is just the total mass M of the system.

But, by the definition (15D-5) of center of mass,

$$\sum m_n\vec{v}_n' = M\vec{V}_c' = 0 \tag{B-5}$$

since the velocity $\vec{V}_c'$ of the center of mass, relative to the center of mass, is just zero. Describing velocities relative to the center of mass is thus particularly simple since the last term in (B-4) is zero. Hence (B-4) becomes simply

$$K = K_c + K' \tag{B-6}$$

which is the result previously stated in (B-2).

Gravitational potential energy near the earth

One often wants to know the gravitational potential energy of some system near the surface of the earth. How can one find this potential energy for a complicated system (such as a table, a car, or a person)?

The gravitational potential energy, due to the earth, of any system is merely the sum of the gravitational potential energies of all the particles in the system. But the gravitational potential energy of any particle, having a mass m and located at a height h above some standard position, is equal to mgh. Hence the gravitational potential energy U_g of a system of such particles is

$$U_g = \sum m_n g h_n = g \sum m_n h_n \tag{B-7}$$

where m_n is the mass of the n'th particle, h_n is its height above the standard position, and the sum extends over all particles n = 1, 2, 3, ... of the system.

But, by the definition (14D-7) of center of mass,

$$\sum m_n h_n = MH_c \tag{B-8}$$

where H_c is the height of the system's center of mass above the standard position. Hence (B-7) becomes simply

$$\boxed{U_g = MgH_c\,.} \tag{B-9}$$

This result implies that *the gravitational potential energy of any system is the same as if its entire mass were concentrated at its center of mass.*

The preceding result makes it very easy to find the potential energy of any system if one knows the position of its center of mass.

➜ ***Go to Sec. 16B of the Workbook.***

C. Energy of atomic systems

Conservation of energy

The forces between all atomic particles (e.g., electrons, atoms, molecules, ...) are conservative. All the interactions between such atomic particles can, therefore, be described by their mutual potential energies. The energy law for any system of such particles can then be expressed in the form

$$\Delta E = \Delta K + \Delta U = W_{oth} \tag{C-1}$$

where K is the total kinetic energy of the system and where U is the total potential energy due to the mutual interactions between all the particles in the system. The other work W_{oth} is then due only to any remaining interactions with objects outside the system.

If a system of atomic particles is *isolated* (i.e., if it does not interact with any outside system), all such other work is zero. Hence $W_{oth} = 0$ so that the total energy of the system remains unchanged (i.e., is *conserved*). Thus

$$\boxed{\text{for an isolated system,} \qquad E = \text{constant.}} \qquad \text{(C-2)}$$

Energy conservation in collisions

Description of a collision process. A *collision* between two particles is a brief interaction occurring when they are sufficiently close together. For example, Fig. C-1 illustrates a typical collision between two particles 1 and 2 (e.g., two atoms or molecules). Such a collision process involves the following three successive time periods:

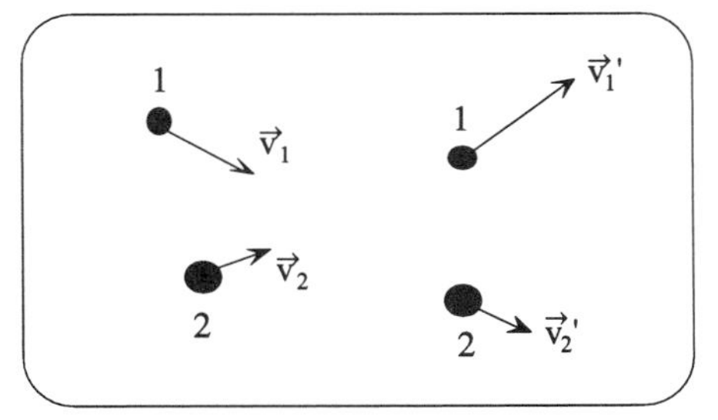

Fig. C-1. Collision between two particles.

(a) ***Initial time period*** (the time, before the collision, during which the particles are so far apart that they don't interact with each other). The particles then travel with constant initial velocities $\vec{v}_1$ and $\vec{v}_2$. Furthermore, their potential energy of interaction is zero (if the potential energy is measured from a standard state where the distance between the particles is very large).

(b) ***Collision time*** (the short time during which the particles are sufficiently close to each other that they interact appreciably). During this time the velocities of the particles change and the particles' potential energy of interaction is appreciable.

(c) ***Final time period*** (the time, after the collision, during which the particles are again so far apart that they don't interact with each other). The particles then travel with constant final velocities $\vec{v}_1'$ and $\vec{v}_2'$. Furthermore, their potential energy of interaction is again zero.

Energy conservation in a collision. By (C-2), the total energy of the system consisting of *both* particles must remain constant during all this time, irrespective of the complexity of the interactions involved in the actual collision. In particular, it must be true that

$$E = E' \qquad \text{(C-3)}$$

where E is the energy of the system *before* the collision and E' is its energy *after* the collision. Furthermore, these situations are particularly simple since they involve no interactions between the particles. Thus each of the energies in (C-3) is simply the sum of the energies of the two particles, i.e.,

$$\boxed{E_1 + E_2 = E_1' + E_2'.} \qquad \text{(C-4)}$$

Collision between elementary particles. An *elementary particle* is one which does not consist of other particles. (For example, such an elementary particle might be an electron or a proton.) Then the energy E of such a particle is just its kinetic energy K. In a collision between two such elementary particles, the energy conservation (C-4) then implies that

Although Newtonian mechanics is insufficient to deal with atomic particles (needing to be replaced by quantum mechanics), all these general energy arguments remain equally valid.

$$K_1 + K_2 = K_1' + K_2', \qquad \text{(C-5)}$$

Recent research has revealed that even a proton is not really elementary, but consists ultimately of still smaller particles.

i.e., that the total *kinetic* energy of the particles remains constant.

Collision between complex particles. Suppose, however, that a particle is complex so that it actually consists of other particles which can move relative to each other. (For example, the particle might be a nitrogen molecule which consists of two nitrogen atoms.) Then the energy E of such a particle does not merely consist of the kinetic energy K associated with the motion of its center of mass. Instead,

for a complex particle, $$E = K + E_{int} \qquad \text{(C-6)}$$

where E_{int} is the *internal* energy of the complex particle. For example, in the case of the nitrogen molecule, this internal energy would consist of the kinetic energy of the two nitrogen atoms moving relative to their center of mass (because they rotate about the center of mass and vibrate relative to each other) and of the potential energy of interaction between these atoms.

When one or both of the colliding particles are complex, the conservation of energy (C-4) does *not* necessarily imply the corresponding conservation (C-5) of the kinetic energies associated with the motions of their centers of mass. This is because the internal energies of complex particles may also change as a result of the collision. (For example, after a nitrogen molecule collides with another atom, its atoms might afterwards vibrate more rapidly than before.)

Elastic and inelastic collisions. One says that a collision is *elastic* if the total kinetic energy, associated with the motions of the centers of mass, remains constant. Conversely, one says that a collision is *inelastic* if this is not the case. (One says that it is *completely inelastic* if the particles no longer move relative to each other after the collision.)

The preceding comments imply that a collision between *elementary* atomic particles must be elastic. However, a collision involving complex particles is not necessarily elastic since some of their kinetic energy may be converted into the internal energy of these particles.

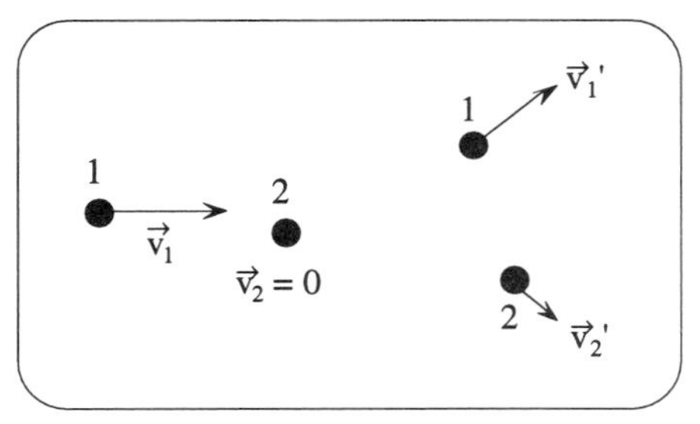

Fig. C-2. Collision of two identical particles.

Example: Elastic collision between identical particles

Fig. C-2 shows a particle 1, traveling with a velocity $\vec{v}_1$, which collides elastically with another identical particle 2 which is initially at rest. (For example, both of these particles might be protons.) What then can one say about the final velocities of these particles?

Conservation of momentum. Consider the isolated system consisting of both particles. Since the total external force on this system is zero, the momentum law implies that the system's total momentum remains unchanged. Hence the system's total initial momentum is equal to its total final momentum, i.e.,

$$m\vec{v}_1 + 0 = m\vec{v}_1' + m\vec{v}_2'$$

or $$\vec{v}_1 = \vec{v}_1' + \vec{v}_2' \qquad \text{(C-7)}$$

where we have used the fact that each particle has the same mass m, and where the final velocities of the particles have been denoted by $\vec{v}_1'$ and $\vec{v}_2'$. Fig. C-3 shows the vector diagram indicating the relationship (C-7) between these velocities.

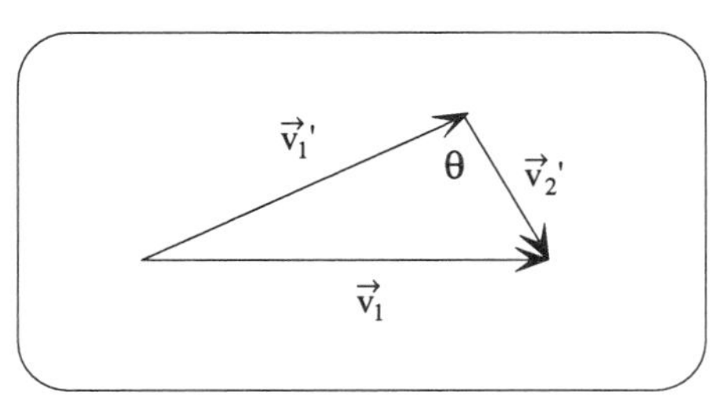

Fig. C-3. Relation between the velocities of two colliding particles.

Conservation of energy. Since the collision is elastic, there is no conversion of kinetic energy into internal energy. Hence the conservation of energy implies that the initial kinetic energy of the system of both particles is the same as its final kinetic energy. Thus

$$\frac{1}{2}mv_1^2 + 0 = \frac{1}{2}mv_1'^2 + \frac{1}{2}mv_2'^2$$

or

$$v_1^2 = v_1'^2 + v_2'^2 . \qquad \text{(C-8)}$$

This means that the Pythagorean theorem applies to the vector triangle in Fig. C-3. Hence the angle θ in this figure must be 90°.

Conclusion. The preceding considerations, based on the conservation of momentum and of energy, thus lead to the following general conclusion: In any elastic collision of two identical particles, where one of these is initially at rest, the particles must emerge with mutually perpendicular velocities.

This conclusion would not be true at very high energies where the particles move with speeds close to the speed of light. For then Newtonian mechanics needs to be modified by the theory of relativity.

➜ ***Go to Sec. 16C of the Workbook.***

D. Energy of macroscopic systems

Atomistic description of macroscopic systems

Macroscopic systems. All the systems encountered in everyday life (e.g., books, tables, automobiles, persons, ...) are *macroscopic*, i.e., they are very much larger than atoms or molecules. For example, the size of an atom is about 10^{-10} m. Thus even a fairly small everyday object, 1 cm in size, is very much larger than an atom or molecule, and contains about 10^{24} atoms. This is an enormously large number!

The word macroscopic, which means *large-scale*, must be carefully distinguished from the word microscopic which means *small-scale* (i.e., of atomic size).

Compare a macroscopic cube of edge length D = 1 cm = 10^{-2} m with an atomic-size cube of edge length d = 10^{-10} m. The macroscopic cube contains then $D^3/d^3 = 10^{24}$ atomic-size cubes.

Macroscopic versus atomistic descriptions. Any macroscopic system can be described from two points of view. (a) It can be described from a *macroscopic* point of view which considers only the system's large-scale properties (i.e., the properties that are readily observable in everyday life). (b) Alternatively, it can be described from an *atomistic* point of view which considers all the atomic particles in the system (i.e., all its atoms and molecules).

The atomistic point of view is seemingly more complex because it involves consideration of a very large number of atomic particles. However, it has some compensating simplicities because *all* systems, no matter how diverse, can be described from the same atomistic point of view. Knowledge about the motions of atomic particles can, therefore, be used to infer some general properties of all macroscopic systems. Furthermore, the interactions among atomic particles are rather simple since they can all be described by conservative forces and corresponding potential energies.

Energy law for a macroscopic system. If one adopts such an atomistic point of view, the energy law (C-1) can be applied to any macroscopic system (just as to an atomic system of the kind considered in preceding section). The energy change of such a system is thus given by

$$\Delta E = \Delta K + \Delta U = W_{oth} \, . \tag{D-1}$$

Here K is the total kinetic energy of all the atomic particles in the system and U is the total potential energy due to the mutual interactions between all these particles. The other work W_{oth} is then only due to any remaining *external* interactions not included in this potential energy.

If a macroscopic system is *isolated* so that it does not interact with any outside objects, $W_{oth} = 0$ and the total energy of the system must remain unchanged. Thus its total energy is *conserved*, i.e.,

$$\boxed{\text{for an isolated system,} \qquad E = \text{constant.}} \tag{D-2}$$

Atomistic description of the energy. The preceding statement of the energy law is correct only if the energy of the macroscopic system is described from an atomistic point of view which considers the motions and interactions of all the atomic particles in the system.

For example, consider a cup of water sitting on a table. From a *macroscopic* point of view, this cup has no energy (i.e., zero kinetic energy since it is at rest, and zero potential energy if its gravitational potential energy is measured from the table top). However, from an *atomistic* point of view, this cup of water has a very large amount of energy. Indeed, the molecules of water move randomly throughout the liquid. Hence they have a large amount of kinetic energy due to all their random motions, and a large amount of potential energy due to all their mutual interactions. Similarly, the molecules in the solid cup vibrate in random ways about their fixed equilibrium positions. Hence they too have a large amount of kinetic energy due to their random motions, and a large amount of potential energy due to their mutual interactions.

Macroscopic and non-macroscopic energies

A system's energy E, described atomistically as in the preceding paragraphs, can be decomposed into two distinguishable parts, a macroscopic energy E_{mac} and a non-macroscopic energy E_{nmac}. Thus one can write

$$E = E_{mac} + E_{nmac} \, . \tag{D-3}$$

Macroscopic energy. Here the macroscopic energy E_{mac} is that part of the energy which can be readily identified from a macroscopic point of view. For example, it might be the kinetic energy associated with the motion of the system's center of mass (e.g., the visible motion of a box sliding along the floor). Or it might be the sum of the kinetic energies of the visibly moving parts of a rotating piece of machinery. Or it might include the potential energy stored in a visibly compressed spring.

Non-macroscopic energy. By (D-3), the non-macroscopic energy is given by

$$E_{nmac} = E - E_{mac} , \quad \text{(D-4)}$$

i.e., it is that part of the total energy (of *all* the atomic particles in the system) which remains after the macroscopic part of the energy is subtracted.

Thermal energy. This non-macroscopic energy always includes some *thermal energy* E_{th}, i.e., energy associated with the random motion of the molecules in the system. For example, in a *gas* this thermal energy consists predominantly of the kinetic energy of the molecules moving randomly throughout the gas. In a *liquid* (where the molecules are much closer to each other) this thermal energy consists of the kinetic energy of the randomly moving molecules and also of the potential energy of interaction among them. In a *solid* (where the molecules always remain close to fixed positions) this thermal energy consists of the kinetic energy of the molecules vibrating randomly around these fixed positions and of the potential energy of interaction of these molecules.

These molecules may be single atoms (e.g., helium atoms) or more complex molecules consisting of several atoms (e.g., nitrogen molecules).

Chemical energy. In some cases, the non-macroscopic energy can also include *chemical energy* E_{ch} due to potential energy of interaction of atoms *within* molecules. For example, such chemical energy is stored in the molecules of gasoline. (This energy, released in the chemical reactions occurring in the cylinders of an engine, is ultimately used to drive a car.)

The non-macroscopic energy of a system consists thus typically of both thermal and chemical energies, i.e.,

$$E_{nmac} = E_{th} + E_{ch} . \quad \text{(D-5)}$$

→ ***Go to Sec. 16D of the Workbook.***

E. Energy transfers and transformations

As mentioned in (D-2), the energy of an isolated macroscopic system remains constant. However some of this energy can be transferred from one part of the system to another, or can be transformed from one form to another. Indeed, most processes in nature involve energy transfers or transformations. The following are some important examples.

Thermal energy transfer

Systems in thermal contact. Suppose that two macroscopic systems S' and S'' are at rest and in contact with each other, as schematically indicated in Fig. E-1. (For example, S' might be a porcelain cup and S'' might be tea in this cup.) Then the total energy E of the entire system consists of the thermal energies of these systems. (Chemical energies are irrelevant if no chemical reactions occur.) If this entire system is isolated, this total energy then remains constant, i.e.,

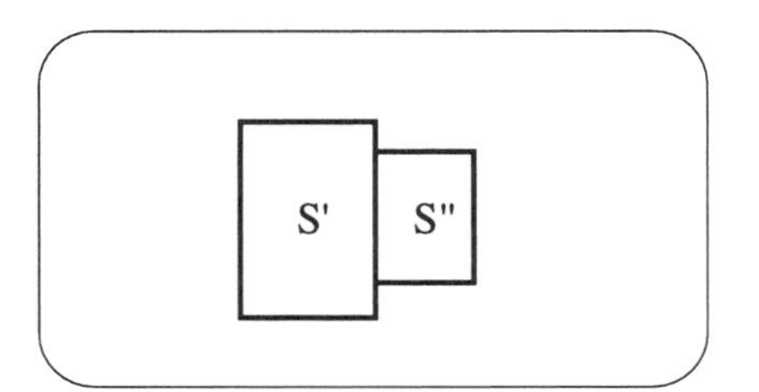

Fig. E-1. Two touching systems allowing transfer of thermal energy between them.

$$E = E_{th}' + E_{th}'' = \text{constant} \tag{E-1}$$

where E_{th}' and E_{th}'' are the thermal energies of the two systems.

Thermal energy changes of the systems. If molecules at the boundary between the two systems collide or otherwise interact with each other, energy can be transferred from one system to the other. The thermal energy of each system can then change, but the total energy of the entire system must remain constant. In other words, (E-1) implies that the thermal energies of the systems can change in such a way that

$$\Delta E_{th}' + \Delta E_{th}'' = 0$$

or

$$\Delta E_{th}' = -\Delta E_{th}'' . \tag{E-2}$$

If one system gains thermal energy, the other system must then lose a corresponding amount of thermal energy. (The system which gains thermal energy is called the "colder" of the two systems. The one which loses thermal energy is called the "warmer" of the two systems.)

Heat. The change of a system's thermal energy, resulting from a transfer of thermal energy, is called the *heat absorbed by the system* and is commonly denoted by the letter Q. Thus $Q = \Delta E_{th}$ and (E-2) can also be written in the form

$$Q' = -Q'' \tag{E-3}$$

In words, this merely says that the heat absorbed by one system is equal to the heat given off by the other system.

Approaching the most random distribution of thermal energy. If a transfer of thermal energy *can* occur between two systems, will it actually occur? The answer is provided by the following qualitative argument (which can be made more precise): Thermal energy is energy associated with the random motion of the molecules constantly interacting with each other. Under these conditions, it is extremely improbable that the energies of some molecules will remain large while the energies of some other molecules will remain small. Instead, the total thermal energy of an isolated system will, in the course of time, gradually become redistributed among all the molecules in the most random way (i.e., redistributed until the average thermal energy of each molecule becomes the same).

Occurrence of thermal energy transfer. A transfer of thermal energy between two systems occurs, therefore, only if the average thermal energy per molecule is *not* the same in the two systems. In that case, thermal energy is transferred from the system with the larger thermal energy per molecule to the system with the smaller thermal energy per molecule (i.e., from the warmer to the colder system).

Heat transfer and temperature. The average thermal energy per molecule in a system is closely related to the *temperature* of the system. The statements in the preceding paragraph can thus also be summarized in the following words: Heat flows between two systems if their temperatures are *not* the same. In this case heat flows from the system at the higher temperature to the system at the lower temperature, until the temperatures of the systems become the same.

Transformations of macroscopic energy into thermal energy

There are many cases where the macroscopic energy of a system is transformed into random thermal energy of the atomic particles in the system. (One sometimes describes this by saying that macroscopic energy is *dissipated.*) The following are some examples.

Pendulum oscillating in air. Fig. E-2 illustrates an oscillating pendulum suspended from the ceiling of a room. The pendulum bob has then appreciable macroscopic energy, both macroscopic kinetic energy (due to the motion of its center of mass, i.e., to the joint motion of all its molecules) and macroscopic gravitational potential energy due to its interaction with the earth.

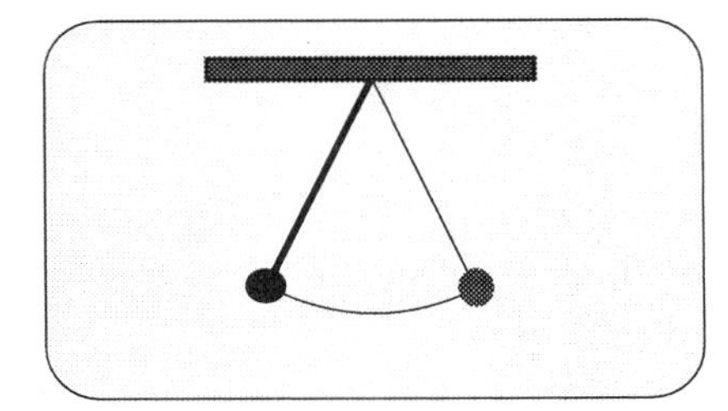

Fig. E-2. Pendulum oscillating in air.

However, the pendulum bob constantly collides with the molecules of the surrounding air in the room, and is gradually slowed down as a result of these collisions. Hence the macroscopic energy of the pendulum bob gradually decreases, while the random thermal energy of all the air molecules in the room increases. Finally the pendulum comes to rest, hanging vertically from the ceiling.

In this entire process, the total energy of the system consisting of the pendulum, the earth, and the air of the room has remained constant. But the initial macroscopic energy E_{mac} of the pendulum has disappeared, while the random thermal energy E_{th} of all the molecules in the air (and the molecules in the pendulum bob) has increased by a compensating amount. In other words, the conservation of energy for the entire system implies that these energy changes are related so that

Even when the pendulum hangs "at rest", it actually still jiggles erratically, by very small amounts, as it experiences random collisions with the molecules of the surrounding air.

$$(0 - E_{mac}) + \Delta E_{th} = 0$$

or

$$\Delta E_{th} = E_{mac}\,. \tag{E-4}$$

The initial macroscopic energy of the pendulum has thus been converted into increased random thermal energy of all the molecules in the system. (The temperature of the air and pendulum has, correspondingly, become slightly higher.)

Sliding crate coming to rest. After being given an initial push, a crate starts sliding along the horizontal floor. However, the crate quickly comes to rest as a result of interactions between the molecules near the bottom of the crate and the molecules near the surface of the floor. The total energy of all the molecules has remained constant in this process. However, the initial macroscopic energy of the moving crate has been converted into random thermal energy of all the molecules in the crate and in the floor.

Ball bouncing on the floor. As a ball bounces on the floor, the successive bounces become smaller until the ball comes to rest. Once again, the initial macroscopic energy of the ball has become converted into the random thermal energy of all the molecules in the ball, the floor, and the air.

Bullet hitting a wall. A bullet slams into a wall and stops when it becomes embedded in it. In this collision process, the initial macroscopic kinetic energy of the bullet has again been converted into the random thermal energy of the molecules in the bullet and in the wall. (Indeed, the resulting temperature rise may be large enough to be quite noticeable.)

Irreversibility of energy transformations

As pointed out in Sec. E1, the energy of an isolated system tends to become redistributed over its molecules in increasingly random form. Thus it is highly probable (as illustrated by the preceding examples) that the macroscopic energy of an isolated system will become transformed into random thermal energy of all the molecules in the system. However, it is exceedingly unlikely that the reverse will happen (i.e., that the random thermal energy of an isolated system will become transformed into macroscopic energy).

As a result, processes in nature tend to be *irreversible*. In other words, the macroscopic energy of an isolated system tends to get transformed into random thermal energy. But, once this has happened, it is practically impossible for the reverse to happen, i.e., for the random thermal energy to be transformed back into macroscopic energy. (This can only happen as a result of external intervention, i.e., if the system is *not* kept isolated, but allowed to interact in special ways with some other external system.)

The irreversibility becomes very apparent if one makes a movie of a naturally occurring process, and then plays the movie backward in a projector. It is then obvious that the events portrayed in the backward-played movie would never be observed in actuality.

The previous examples can be used to illustrate this irreversibility.

Pendulum oscillating in air. Suppose that a pendulum hangs vertically at rest suspended from the ceiling. Then it is exceedingly improbable that the pendulum will gradually start swinging from side to side by increasing amounts. Indeed, if we observed this occurring in a movie, we would immediately conclude that the movie is being played backwards (i.e., that it was actually made by filming a pendulum gradually slowing down as a result of its interactions with the surrounding air).

Strictly speaking, it is not really *impossible* that the pendulum might start swinging by increasing amounts. Indeed, imagine that, by happenstance, all the molecules moving randomly throughout the air would, at one time, all start moving in the same direction. Then their joint collisions with the pendulum bob could, indeed, get the bob moving by large amounts. But it is *exceedingly* improbable (i.e., practically speaking, impossible) that the air molecules would ever spontaneously all start moving in the same direction.

Crate sliding along the floor. Suppose that a movie shows a crate which initially lies at rest on a horizontal floor and then starts spontaneously sliding along this floor. Then we would again conclude that this is not a process observed in nature, but that the movie is being played backwards.

Ball bouncing on the floor. Similarly, suppose that a movie shows a ball which initially lies at rest on the floor and then starts spontaneously bouncing up and down by increasing amounts. Then we would again conclude that this is not a process observed in nature, but that the movie is being played backwards.

Example of complex energy transformations

Even seemingly simple processes may involve complex energy transformations. The following is an example.

Jumping. Consider the energy transformations occurring when a man, of mass M, jumps up from the ground. The energy of the combined system, consisting of the man and the earth, remains then constant. However, the following energy transformations occur.

(a) Initially, when the man is standing at rest on the ground, a certain amount E_{muscle} of chemical energy is stored in the man's leg muscles.

(b) When jumping, the man leaves the ground with some speed v. The chemical energy in his leg muscles has then become converted into his macroscopic kinetic energy $\frac{1}{2}Mv^2$.

(c) When the man reaches his maximum height h, he is momentarily at rest and has no macroscopic kinetic energy. His previous kinetic energy has then become converted into the macroscopic gravitational potential energy Mgh due to his interaction with the earth.

(d) Finally, when the man has landed on the floor and is again at rest, the preceding energy has become converted into the random thermal energy E_{th} of the molecules in the ground and the man's feet.

By the conservation of energy, all these energies are then related so that

$$E_{muscle} = \frac{1}{2}Mv^2 = Mgh = E_{th} \,. \qquad \text{(E-5)}$$

Jump height and animal size. The preceding argument can equally well be applied to any animal. Thus (E-5) implies that the height h to which an animal can jump is related to the energy stored in its leg muscles so that

$$h = \frac{E_{muscle}}{Mg} \,. \qquad \text{(E-6)}$$

This relation has the following interesting implication. Consider two animals A and B of similar construction, but of different sizes. For example, suppose that the volume of animal B is 100 times larger than that of animal A. Then the chemical energy E_{muscle} stored in the larger leg muscles of animal B is 100 times larger than that stored in the muscles of animal A. But the mass M of animal B is also 100 times larger than that of animal A. The ratio on the right side of (E-6) is, therefore, the *same* for both animals.

Thus we arrive at the following prediction: *All animals of the same construction, irrespective of their size, should be able to jump to the same height.*

For example, a kangaroo (which can be as high as 2 m) and kangaroo rat (which is about 0.1 m high) both have similarly constructed hind legs well adapted to jumping. Observations show that both animals can jump to the same height (about 2.5 m). In the case of the kangaroo rat, this jump height is actually much larger than the height of the animal itself. But this is not really a remarkable feat since the smaller kangaroo rat needs to lift less mass and can do this with less muscle energy.

➜ ***Go to Sec. 16E of the Workbook.***

F. Friction and energy dissipation

Friction forces and macroscopic objects

When a system is described in terms of its constituent atomic particles, all forces between these are conservative. Friction forces, which are *not* conservative, are only introduced as a convenient way of describing systems from a macroscopic point of view. They describe approximately the effects of molecular interactions that lead to the transformation of macroscopic energy into random thermal energy (i.e., that lead to the *dissipation* of macroscopic energy).

Center-of-mass motion vs. molecular motions. When we first introduced the energy law in Chapter 14, we dealt only with particles. Many of these were, in fact, macroscopic objects. But the motion of any such object could be adequately described by the motion of a single point, e.g., by the motion of its center of mass.

The energy which we then considered was only the energy E_c associated with the center of mass. Correspondingly, we expressed the energy law in the form

$$\Delta E = W_{oth,c}$$

where the other work $W_{oth,c}$ was other work done by external forces introduced to describe the motion of the *center of mass*. This other work included work done by various non-conservative forces (such as friction forces or forces involved in collisions). From an atomistic point of view, these forces actually describe interactions which lead to increases of the random thermal energy of the molecules in the object.

Analysis of frictional energy dissipation *(optional)*

Example of a box pulled along the floor. As a simple example, consider a box, of mass M, which is pulled along the horizontal floor with a constant horizontal force of magnitude F. A constant friction force of magnitude f is then exerted on the box by the floor (in a direction opposite to the applied force). Fig. F-1 indicates also all the other forces exerted on the box. What energy transformations occur when the box is pulled to the right through a distance D?

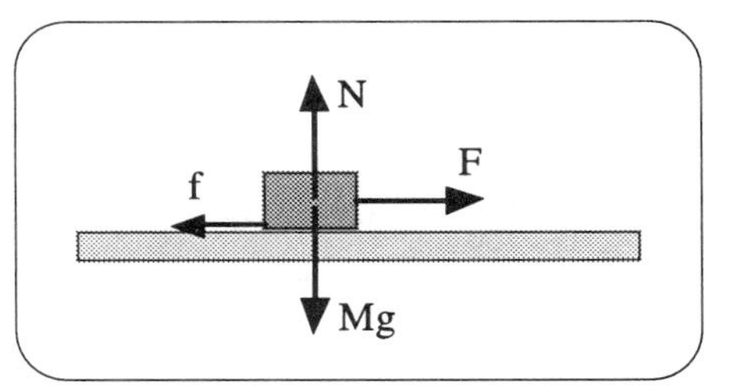

Fig. F-1. Forces on a box pulled along the floor.

Application of the energy law. Let us look at the situation from an atomistic point of view. By applying the energy law to the entire system consisting of the box and the floor, we get

$$\Delta K_c + \Delta E_{th} = FD . \qquad \text{(F-1)}$$

Here ΔK_c is the change of kinetic energy associated with the motion of the box's center of mass and ΔE_{th} is the change in the thermal energy of the system consisting of the box and the floor. The other work is then only the work FD done on the system by the applied force. (From an atomistic point of view, this is the entire work done on the system since interactions between the box and the floor are included in the internal thermal energy of the system.)

Thus (F-1) merely says that the work done by the applied force goes partly to increase the box's macroscopic kinetic energy, and partly to increase the thermal energy of molecules in the box and floor.

Energy change due to center-of-mass motion. The energy change associated with the motion of a system's center of mass can be found from the momentum law which implies that the acceleration $\vec{A}_c$ of the center of mass is related to the total external force on the system so that

$$M\vec{A}_c = \vec{F}_{ext} . \qquad \text{(F-2)}$$

This relation is similar to Newton's law for a single particle. Hence the same argument as that used in Sec. 13A leads to the following relation for the change of energy of the center-of-mass

$$\Delta K_c = W_{ext,c} . \qquad \text{(F-3)}$$

Here $K_c = \frac{1}{2} MV_c^2$ is the kinetic energy associated with the speed V_c of the center of mass, and $W_{ext,c}$ is the work done by the *total external force* on the system acting along the *path traversed by the center of mass.*

The result (F-3) can also easily be derived by using the dot product. By (F-2)

$$M\frac{d\vec{V}_c}{dt} = \vec{F}_{ext}$$

$$M\,d\vec{V}_c = \vec{F}_{ext}\,dt$$

$$M\,\vec{V}_c \cdot d\vec{V}_c = \vec{V}_c \cdot \vec{F}_{ext}\,dt$$

$$(M/2)\,d(\vec{V}_c^{\,2}) = \vec{F}_{ext} \cdot \vec{V}_c dt$$

$$d(\tfrac{1}{2}MV_c^2) = \vec{F}_{ext} . d\vec{R}_c$$

$$dK_c = d'W_{ext,c}$$

Energy change of the box's center of mass. The total external force on the box consists of the applied force and the friction force. (The vertical forces on the box cancel since the box has no vertical acceleration.) Hence the momentum law (F-2), applied to the box along the right direction in Fig. F-1, yields

$$MA_c = F - f .$$

By (F-3) the change of kinetic energy associated with the box's center of mass is thus equal to

$$\Delta K_c = (F - f)\, D\,. \tag{F-4}$$

Increase of thermal energy. By combining the relations (F-1) and (F-4) to eliminate the change of center-of-mass energy ΔK_c, one gets

$$\Delta E_{th} = f\, D\,. \tag{F-5}$$

This merely says that the change of the system's thermal energy is equal to the magnitude of the work done by the friction force on the box's center of mass. Since this magnitude is positive, (F-5) implies that the change in the system's thermal energy is also positive. As expected, the random thermal energy of the system thus *increases* as a result of the frictional interactions between the molecules near the surface of contact between the box and the floor.

Friction work done on the box. The work done by the friction force on the box's *center of mass* is simply $-fD$. Note, however, that this is *not* the same as the work W_{fric} done by friction on *all the particles in the box* (since every particle in it does not move through a distance D while acted on by the same friction force).

Calculation of this work W_{fric} is much more complex, but is required if one wants to apply the energy law to the box itself (rather than to the entire system consisting of the box and floor). The energy law, thus applied, asserts that

$$\Delta K_c + \Delta E_{th,box} = FD + W_{fric}\,. \tag{F-6}$$

Here $\Delta E_{th,box}$ is the change of thermal energy *of the box alone*. The right side of (F-6) is the work done on the box by the applied force and by friction. But calculation of the work W_{fric} done by friction on all the particles in the box requires a detailed knowledge of the interactions between the particles near the surface where the box touches the floor.

Simplified calculation of friction work done on the box

The friction work W_{fric} done on the box depends in detail on just how the particles in the box interact with the neighboring particles in the touching floor. Hence this friction force can only be calculated if one makes some specific assumptions about the nature of these interactions. For example, assume that the touching surfaces of the box and floor are actually rough, consisting of many mutually touching small bumps schematically illustrated in Fig. F-2a. Suppose also that a friction force of magnitude f_1 acts on each bump of the box surface where it touches a bump of the floor surface. (All these friction forces together produce the total friction force of magnitude f on the entire surface of the box.) As the box moves to the right by a small distance s, as indicated in Fig. F-2b, the bumps bend (and then snap back). Note, however, that in this motion the *contact points* of the bumps move to the right only by a distance s/2. Accordingly, the friction force f_1 on one of these bumps only does work of magnitude $f_1(s/2)$.

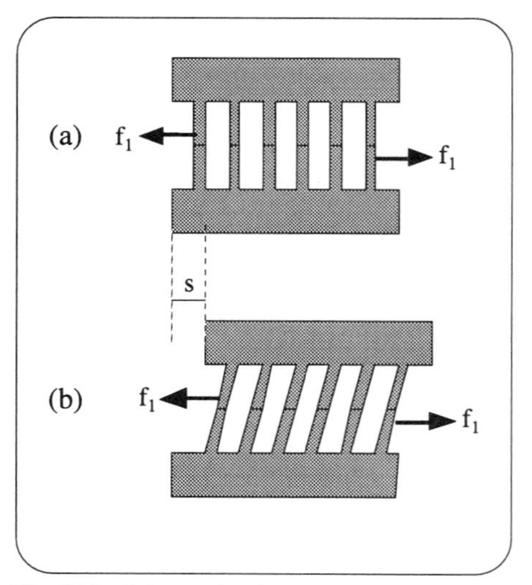

Fig. F-2. Schematic view, vastly magnified, of touching bumps near the surface where the box touches the floor.

When the box moves to the right by a total distance D, the friction force does then only work $-f(D/2)$ on all the molecules in the box. The energy relation (F-6), combined with (F-4), then implies that the change of thermal energy of the box is

$$\Delta E_{th,box} = \frac{1}{2} fD \,. \qquad \text{(F-7)}$$

By (F-5), this is half as large as the total change of thermal energy of the entire system consisting of the box and floor. (Thus the thermal energy of the floor increases by the same amount as the thermal energy of the box.)

➜ ***Go to Sec. 16F of the Workbook.***

G. Summary

Energy law for a system: $\boxed{\Delta E = W_{oth}}$

Energy of system: $E = K + U$

Kinetic energy: $K = \sum K_n$ (sum of all particle kinetic energies)

Potential energy: $U = U_{int} + U_{ext}$ (internal and external potential energies)

$U_{int} = \sum U_{ns}$ (sum over all pairs of internal particles)

Other work: $W_{oth} = W_{oth,int} + W_{oth,ext}$ (due to other internal or external interactions not included in U)

Energy and center of mass:

Kinetic energy: $K = K_c + K'$

$K_c = \frac{1}{2} M V_c^2$ (kinetic energy of center of mass)

K' = kinetic energy of motion relative to center of mass

Gravitational pot. energy: $U_g = MgH_c$ (near surface of the earth)

Energy of macroscopic system: $E = E_{mac} + E_{nmac}$

$(E_{nmac} = E_{thermal} + E_{chemical} \ldots)$

New abilities

You should now be able to do the following:

(1) Solve mechanics problems by applying the energy law (sometimes together with the momentum law) to systems consisting of a few particles.

(2) Apply the energy law to macroscopic systems to examine energy transfers or energy transformations.

Grand summary: Mechanics laws for any system

The following mechanics laws are now available to solve mechanics problems involving any system of particles: (a) Newton's law and the momentum law (which is the same as Newton's law when the system is a single particle). (b) The energy law.

These laws are summarized below. (See also Appendices E and F.)

Momentum law

* *Relates motion and external interactions*
* *Relates velocities at any two instants if total external force is zero*

$$\frac{d\vec{P}}{dt} = \vec{F}_{ext}$$

$m\vec{a} = \vec{F}_{tot}$ (for a particle) *{Newton's law}*

$M\vec{A}_c = \vec{F}_{ext}$ (for center of mass)

Energy law

* *Relates speeds and positions at any two instants (without mention of time)*

$$\Delta E = W_{oth}$$

→ ***Go to Sec. 16G of the Workbook.***

17 Rotational Motion

Rotating objects are very common and important. (For example, most vehicles have rotating wheels and most machines have rotating parts.) Any rotating object is a special kind of system consisting of many particles. Hence one should be able to predict its motion by applying the mechanics laws discussed in the preceding two chapters. But how can one describe the rotation in simple ways so that these laws can be readily applied? This chapter and the next will address this question and thus enable us to deal with many applications.

A. Angular description of motion

Angular versus linear descriptions of motion

Consider a wheel or some other object (like the rectangle in Fig. A-1) which rotates by some angle around some axis. The rotation of the object can then be described very simply by specifying the angle of rotation (e.g., by specifying that the rectangle in Fig. A-1 rotates counter-clockwise by 30° around the axis passing through the point O).

On the other hand, Newton's law and all other mechanics laws focus on *linear* motions, i.e., the motions of particles along straight or curved lines. But when an object rotates through some angle, the linear motions of its particles are quite complex because different particles in the object are displaced by different amounts and along different directions. (For example, in Fig. A-1 the particle B moves through a larger distance than particle A, and the particle C moves along a different direction than particle A.)

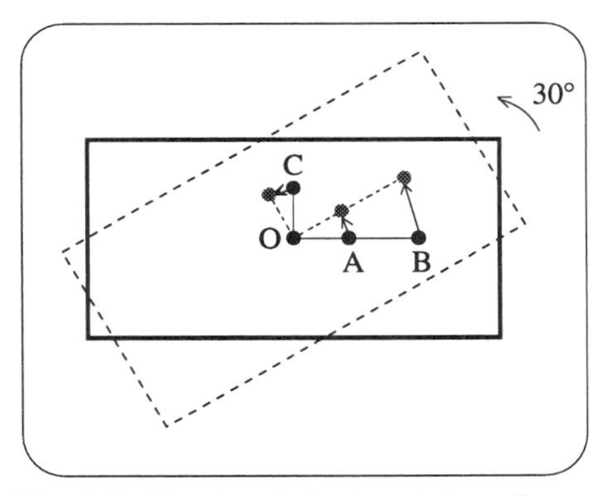

Fig. A-1. Counter-clockwise rotation of a rectangle by 30°.

The *linear* motion of particles is described in terms of distances, while their *rotational* motion is described in terms of angles.

To deal with rotational motion in a simple way, we must, therefore, describe rotational motion in terms of angles — and then also express our mechanics laws in terms of angles.

Angular displacement and position

Angular displacement. One can specify a rotation by an *angular displacement* defined as follows:

Def:

Angular displacement: A quantity specified by a rotation *axis*, an *angle* of rotation, and a *sense* of rotation.	(A-1)

The *rotation axis* is the line about which the object rotates. (The particles on this line are thus unaffected by the rotation.)

For the sake of simplicity, we shall restrict our discussion of rotation to the case where the rotation axis has a fixed direction (which we can then always assume to be perpendicular to the plane of the paper).

The *sense* of rotation (clockwise or counter-clockwise) needs to be specified to avoid ambiguity. For example, in Fig. A-2, both rotations of the rod are by 30° about the axis passing through O. However, the rotation in Fig. A-2a is counter-clockwise and that in Fig. A-2b is clockwise.

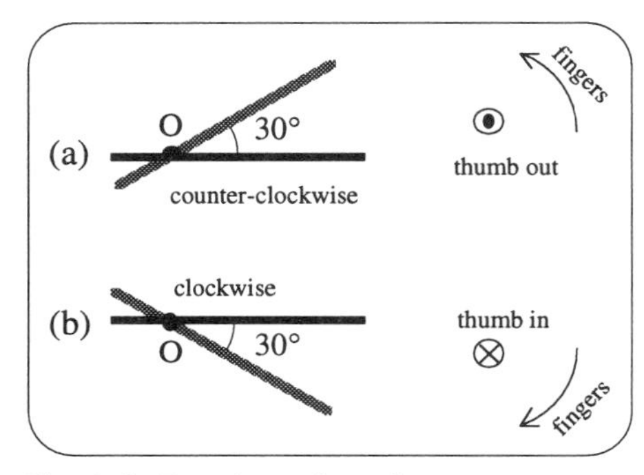

Fig. A-2. Rotations of a rod.

One can specify a particular sense (e.g., the counter-clockwise sense) as the standard sense. Then any angular displacement in this sense can be considered positive, and any angular displacement in the opposite sense can be considered negative. (For example, one can then say that the angular displacement of the rod in Fig. A-2a is 30°, and that the angular displacement of the rod in Fig. A-2b is –30°.)

Directional specification of angular displacement

An alternative specification of the angular displacement is useful for more complex rotations and for people who (in our age of digital watches) can no longer remember which way the hands of a clock move. This specification is based on the fact that the curled fingers of a hand specify a sense of rotation which is clearly associated with the direction specified by its thumb. The *right* hand is conventionally used. Thus one can say that the angular displacement in Fig. A-2a points *out* of the paper (i.e., the rotation sense is such that the right thumb points out of the paper). Similarly, one can say that the angular displacement in Fig. A-2b points *into* the paper.

One can then also specify the angular displacement by a *vector* whose magnitude is equal to the angle of rotation and whose direction is specified by the direction of the right thumb. (The direction of this thumb thus specifies both the direction of the rotation axis and the sense of rotation.)

Angular position. By measuring the angular displacement of an object from some standard orientation, one can specify the object's angular displacement ϕ (denoted by the Greek letter *phi*).

Def:

Angular position: An object's angular displacement relative to some standard orientation.	(A-2)

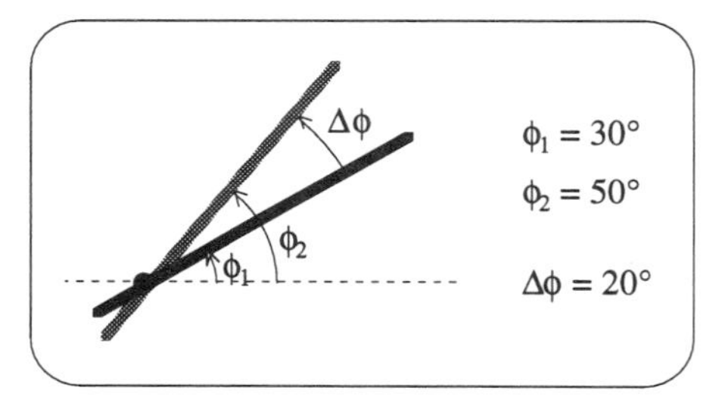

Fig. A-3. Angular positions and angular displacement of a rod.

For example, the two successive angular positions of the rod in Fig. A-3 are $\phi_1 = 30°$ counter-clockwise and $\phi_2 = 50°$ counter-clockwise (relative to the orientation specified by the dashed line).

An angular displacement between two angular positions can be expressed as the difference between these positions, i.e., the angular displacement is

$$\Delta\phi = \phi_2 - \phi_1 \,. \qquad \text{(A-3)}$$

For example, in Fig. A-3, the angular displacement $\Delta\phi$ of the rod, rotating from its angular position of 30° to its angular position of 50°, is 20° counter-clockwise.

Units of angle measurement. An angle can be measured in various units. For example, a *revolution* is that angle corresponding to one complete rotation which returns the object to its previous angular position. The common unit *degree* (abbreviated by °) is then defined so that

$$360 \text{ degree} = 1 \text{ revolution}\,. \tag{A-4}$$

However, for many purposes it is most convenient to specify an angle as a *pure number* (i.e., one *not* involving units referring to some standard of comparison). This can be done, as indicated in Fig. A-4, by determining the length s of the arc subtended by the angle on any circle of radius r. The angle ϕ can then be specified by the ratio s/r.

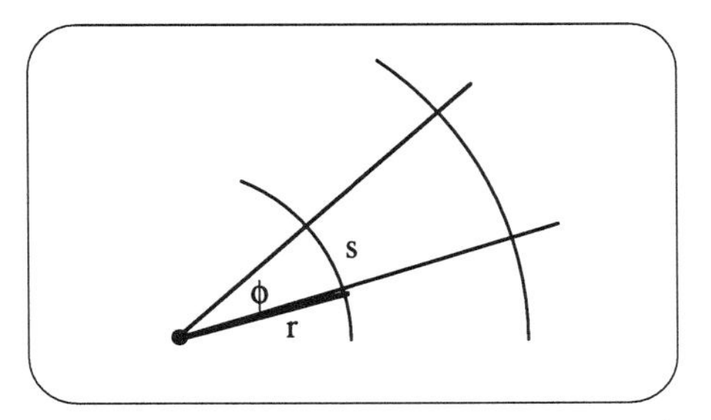

Fig. A-4. Angle specification by the ratio of arc length to radius.

Def: **Angle** *(pure number):* $\phi = \dfrac{s}{r}$ (A-5)

Note that the size of the chosen circle is irrelevant. (For example, if one chooses a circle of two times larger radius, the length of the subtended arc is two times larger. Hence the ratio s/r is the same.)

Sometimes the unit *radian* is used to indicate an angle specified by the ratio s/r. But this unit is merely a fancy way of writing the number 1. Thus it makes no difference whether one says that $\phi = 1.5$ or says that $\phi = 1.5$ radian.

Another familiar case, where a unit is merely introduced to denote a simple number, is the unit *dozen* which is defined so that dozen = 12.

Since one revolution corresponds to an arc length equal to a circle's circumference $2\pi r$, the corresponding angle is $s/r = (2\pi r)/r = 2\pi$. Thus

$$1 \text{ revolution} = 360 \text{ degree} = 2\pi = 2\pi \text{ radian}\,. \tag{A-6}$$

This relation makes it easy to express the same angle in terms of different units.

Rates of change of rotational quantities

Angular velocity. To specify how *rapidly* the angular position of an object changes, we proceed in a fashion totally analogous to that used in Sec. 4C to specify how rapidly the position of a particle changes. Suppose that, during an infinitesimal time interval dt, the object experiences a corresponding small angular displacement $d\phi$. Then we can define the *angular velocity* of the object as the rate of change of its angular position with time. In other words,

Def: ***Angular velocity:*** $\omega = \dfrac{d\phi}{dt}$ (A-7)

where we have used the conventional symbol ω (the Greek letter *omega*) to denote the angular velocity.

The unit of angular velocity is thus radian/second or simply (1/second), which can be abbreviated as s^{-1}.

Angular acceleration. Similarly, we can specify how rapidly the angular velocity of an object changes. This can be done by defining the *angular acceleration* of an object as the rate of change of its angular velocity with time. In other words,

Def: **_Angular acceleration:_** $\alpha = \dfrac{d\omega}{dt}$ (A-8)

where we have used the conventional symbol α (the Greek letter *alpha*) to denote the angular acceleration.

The unit of angular acceleration is thus radian/second^2 or simply (1/second^2), which can be abbreviated as s^{-2}.

Relating angular displacements and their rates of change

Obtaining information about rates of change. Chapter 5, which dealt with the motion of a particle along a straight line, discussed how information about a particle's position can be used to find corresponding rates of change (e.g., to find information about the particle's velocity or acceleration).

That previous discussion is equally applicable to rotational motion. For example, information about how an object's angular position changes with time can similarly be used to find corresponding rates of change (e.g., to find information about the object's angular velocity or angular acceleration).

Obtaining information from rates of change. Chapter 5 also showed how information about rates of change could be used to predict the motion of a particle along a straight line. For example, it discussed how knowledge about motion with constant acceleration can be used to predict the particle's velocity and position at any time. The arguments used in that chapter are again equally applicable to the rotational motion of an object.

Example of motion with constant angular acceleration. For instance, suppose that an object rotates with a constant angular acceleration α, and that its angular velocity ω_o and angular acceleration α_o are known at some initial time t = 0. The definition of angular acceleration then implies that

$$\frac{d\omega}{dt} = \alpha = \text{constant}\,. \qquad \text{(A-9)}$$

From this we can immediately infer, analogously to (5F-4), that at any time

$$\omega - \omega_o = \alpha t\,. \qquad \text{(A-10)}$$

[Indeed, we can readily check that this implies that $d\omega/dt$ has the known value α specified by (A-9). It also satisfies the requirement that $\omega = \omega_o$ when t = 0.]

The definition of angular velocity then implies, by (A-10), that

$$\frac{d\phi}{dt} = \omega = \omega_o + \alpha t\,.$$

From this we can infer, analogously to (5F-7), that the angular displacement at any time is

$$\Delta\phi = \phi - \phi_o = \omega_o t + \frac{1}{2}\alpha t^2\,. \qquad \text{(A-11)}$$

[Again one can check that this implies that $d\phi/dt$ has the known value specified by (A-10), and that it satisfies the requirement that $\phi = \phi_o$ when $t = 0$.]

By eliminating the time t between the relations (A-10) and (A-11), we can then also obtain the following relation, analogous to (5F-9), which connects the angular velocity and angular displacement

$$\omega^2 - \omega_o{}^2 = 2\alpha\,\Delta\phi\,. \tag{A-12}$$

→ ***Go to Sec. 17A of the Workbook.***

B. Relations between angular and linear motions

If an object rotates about a fixed axis, every particle in the object moves around a circle centered on that axis. If a particle is located at a distance r from this axis, this particle then moves around a circle of radius r, as indicated in Fig. B-1. What is the relation between the angular motion of the object and the linear motion of any such particle?

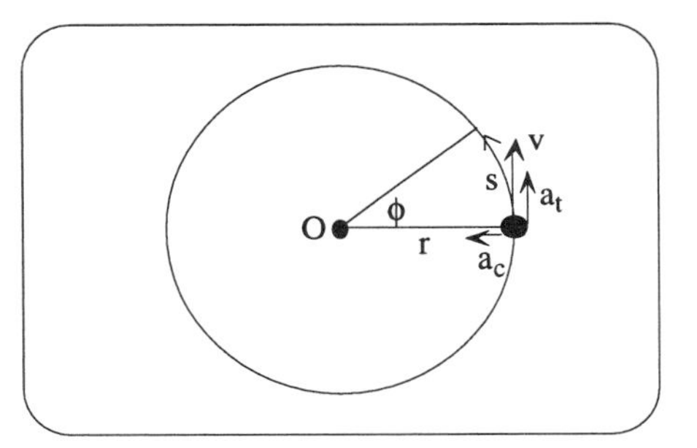

Fig. B-1. Motion of a rotating particle.

Arc distance and angle. Suppose that the particle rotates about the axis through an angular displacement ϕ (specified as a pure number, i.e., in radians). By (A-5), the particle then traverses an arc of length s such that $\phi = s/r$. Thus

$$\boxed{s = r\phi\,.} \tag{B–1}$$

As an object rotates through some angle, particles at a larger radius r (i.e., farther from the rotation axis) thus move through a proportionately longer distance of arc than particles closer to the axis.

Linear and angular velocity. During an infinitesimal time interval dt, the particle moves through an infinitesimal displacement $d\vec{r}$ tangent to its circular path (i.e., perpendicular to the particle's position vector relative to the axis). The magnitude of this displacement is equal to the infinitesimal distance ds which this particle moves along its circular path.

Hence the particle's velocity $\vec{v}$ is tangent to the particle's circular path. Furthermore, the magnitude v of this velocity (i.e., the particle's speed) is equal to

$$v = \frac{ds}{dt} = \frac{r\,d\phi}{dt} = r\frac{d\phi}{dt}$$

so that

$$\boxed{v = r\omega} \tag{B-2}$$

where we have used the definition (A-7) of angular velocity.

A particle at a larger distance r from the rotation axis moves thus with a proportionately larger speed than a particle closer to the axis.

Linear and angular acceleration. A particle moving around a circular path has an acceleration with components parallel and perpendicular to its velocity.

The tangential component of the acceleration, along the velocity, describes how rapidly the magnitude v of the particle's velocity changes. According to (8C-4), this component a_v along the velocity is equal to

$$a_v = \frac{dv}{dt} = \frac{d(r\omega)}{dt} = r\frac{d\omega}{dt}$$

or

$$\boxed{a_v = r\alpha\,.} \qquad \text{(B-3)}$$

Here we have used (B-2) and the definition (A-8) of the angular acceleration.

The component of the acceleration perpendicular to the particle's velocity is directed toward the center of the circle and describes how rapidly the direction of the particle's velocity changes. According to (8C-4), this component a_c toward the center is equal to

$$a_c = \frac{v^2}{r} = \frac{(r\omega)^2}{r}$$

or

$$\boxed{a_c = r\omega^2\,.} \qquad \text{(B-4)}$$

A particle farther from the rotation axis thus has acceleration components (and thus also a resultant acceleration) which are proportionately larger those of a particle closer to the axis.

➜ ***Go to Sec. 17B of the Workbook.***

C. Rotational kinetic energy

The preceding section shows how the motions of particles in a rotating object can be related to the angular quantities describing the rotation. Hence we are now in a good position to apply our familiar mechanics laws to such rotating objects. In particular, we shall spend the rest of this chapter discussing applications of the energy law.

Rotational kinetic energy and moment of inertia

Kinetic energy of a rotating object. Consider a solid object rotating about a fixed axis with some angular velocity ω. To apply the energy law to such a rotating object, we need to find its kinetic energy K. This kinetic energy is merely the sum of the kinetic energies of all rotating particles in the object, as indicated in Fig C-1. Thus

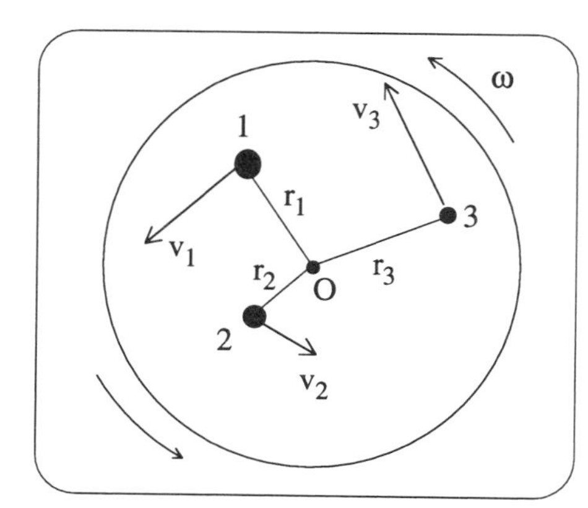

Fig. C-1. A few particles in an object rotating about an axis through O.

$$K = \sum \left(\tfrac{1}{2} m_n v_n^2\right) \qquad \text{(C-1)}$$

where m_n is the mass of the n'th particle, v_n is its speed, and the sum extends over all particles n = 1, 2, 3, ... in the object.

The speed v_n of any particle depends on its distance r_n from the rotation axis in the fashion specified by (B-2). Thus (C-1) becomes

$$K = \sum \left[\tfrac{1}{2} m_n (r_n \omega)^2\right] = \tfrac{1}{2} \sum (m_n r_n^2 \omega^2) = \tfrac{1}{2} \left(\sum m_n r_n^2\right) \omega^2$$

where the constant factors have been taken outside the sum. This last result can then be written in the form

$$\boxed{K = \tfrac{1}{2} I \omega^2} \qquad \text{(C-2)}$$

where we have introduced the abbreviation I to denote the following quantity called the *moment of inertia of the object about the axis of rotation*:

Def: **Moment of inertia:** $I = \sum m_n r_n^2$. (C-3)

Moment of inertia. The relation (C-2) indicates that the moment of inertia is a *quantity which describes the relationship between an object's rotational kinetic energy and its angular velocity.* This quantity depends on the masses and positions of all the particles in the object. In the case of a rigid object, the relative positions of the particles in the object remain always the same. Hence the moment of inertia of a rigid object is merely some constant characterizing that object.

According to the definition (C-3), each particle of mass m, at a distance r from the rotation axis, has a moment of inertia mr^2. The moment of inertia of the entire object is then merely the sum of the moments of inertia of all the particles in it.

Note that a particle located at a larger distance r from the rotation axis contributes more to the moment of inertia than a particle of the same mass located closer to the axis. This is because particles farther from the rotation axis move faster than those closer to the axis, and thus contribute more to the kinetic energy. [For example, according to (B-2), a particle twice as far from the axis moves with twice the speed, and thus contributes a kinetic energy which is four times as large.]

Moment of inertia of a dumbbell. The moments of inertia of some simple objects can readily be calculated. For example, Fig. C-2 illustrates a dumbbell consisting of two small metal balls connected by a thin rod of length L. The mass of the rod is negligible compared to the mass M of each ball. According to the definition (C-3), the moment of inertia of this dumbbell about its center O is

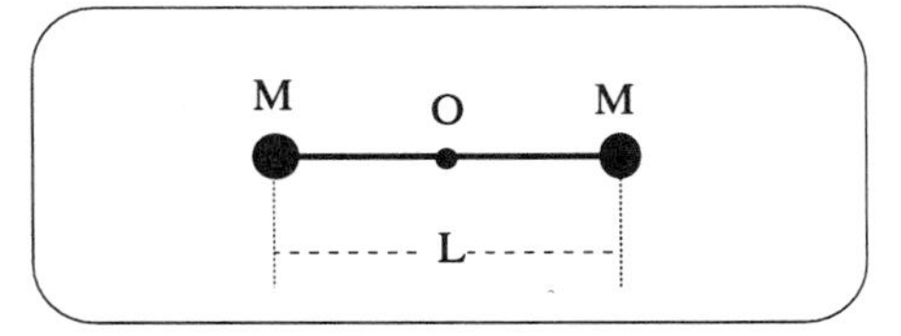

Fig. C-2. A dumbbell.

$$I = M\left(\frac{L}{2}\right)^2 + M\left(\frac{L}{2}\right)^2 = \frac{1}{2} M L^2 \; . \qquad \text{(C-4)}$$

Moment of inertia of a ring. Fig. C-3 illustrates a bicycle wheel, of mass M and radius R. The spokes of this wheel have negligible masses so that the entire mass M of the wheel is concentrated in its tire (i.e., in a thin ring of radius R). According to the definition (C-3), the moment of inertia of this ring about its center O is then merely the sum of the moments of inertia of all the particles in the ring, i.e.,

$$I = m_1R^2 + m_2R^2 + m_3R^2\ldots = (m_1 + m_2 + m_3\ldots)R^2$$

or $$\boxed{I = MR^2} \qquad \text{(C-5)}$$

since all particles in the ring are at the same distance R from its center O.

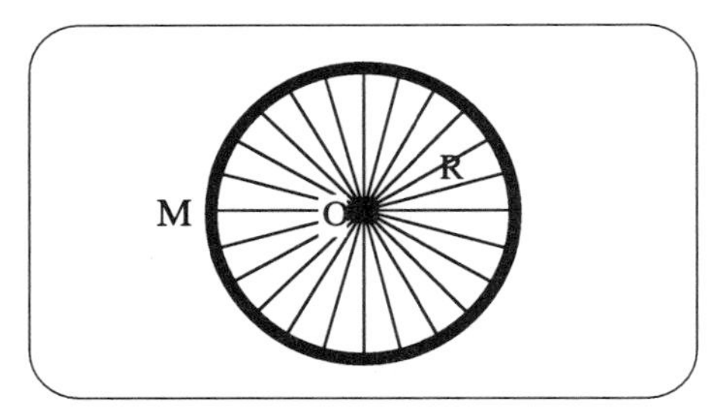

Fig. C-3. A bicycle wheel.

A similar argument shows that the moment of inertia about the central axis of a thin cylindrical shell, of mass M and radius R, is also given by (C-5).

Example: Weight descending from a pulley

The following prototypical problem illustrates the application of the energy law to a system involving rotational motion.

Problem statement. A block, of mass m, is attached to a thin string which is wrapped around a pulley free to rotate around a fixed horizontal axle. This pulley has a radius R and a moment of inertia I about its axle. All friction forces are negligible.

(a) What then is the speed attained by the block after it is released from rest and has descended a vertical distance h?

(b) What is this speed if the pulley is a spoked wheel whose mass M is concentrated at its outer rim?

Solution. The following illustrates the solution of this problem.

Analysis of problem

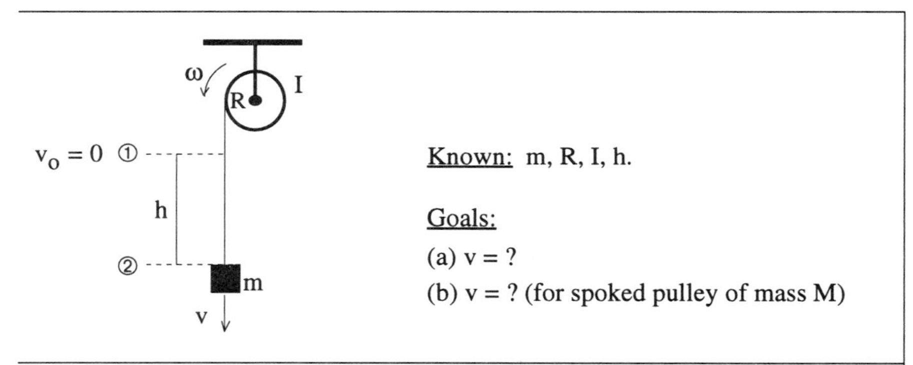

Construction of solution

Apply energy law, to system (pulley + string + block), ①→②

$$\Delta E = W_{oth}$$

$$\Delta K_p + \Delta K_b + \Delta U_g = 0$$

The change of kinetic energy is that of the rotating pulley plus that of the descending block.

$$(\tfrac{1}{2}I\omega^2 - 0) + (\tfrac{1}{2}mv^2 - 0) - mgh = 0$$

$$I\omega^2 + mv^2 = 2mgh \qquad (1)$$

But $$v = R\omega \qquad (2)$$

The speed of a rim point on the pulley is the same as the speed v of the block.

By (1): $I\left(\frac{v}{R}\right)^2 + mv^2 = 2mgh$

$$\left(\frac{I}{mR^2} + 1\right)v^2 = 2gh$$

$$\boxed{v = \sqrt{\frac{2gh}{1 + \frac{I}{mR^2}}}} \qquad (3)$$

For spoked pulley of mass M

$$I = MR^2 \qquad (4)$$

$$\therefore \qquad \boxed{v = \sqrt{\frac{2gh}{1 + \frac{M}{m}}}} \qquad (5)$$

Checks

Checks (done in writing or mentally)

Units of (3): $\frac{m}{s} = \sqrt{\frac{m}{s^2}(m)} = \frac{m}{s}$ OK.

Check extreme cases.

If h large: Expect large v. Eq. (3) OK.

If I or M large: Expect small acceleration of pulley particles, hence small v. Eq. (3) OK.

➜ ***Go to Sec. 17C of the Workbook.***

D. Moments of inertia of some common objects

Many problems involving rotating objects can be solved by applying energy arguments, provided that the moments of inertia of these objects are known. As we shall see, the moments of inertia of many common simple objects can be readily calculated from the definition (C-3).

Moment of inertia of a rod

Moment of inertia of a rod about its end. Fig. D-1 shows a uniform rod, of mass M and length L, which is free to rotate about a perpendicular axis passing through one of its end points A. To find the moment of inertia I_A of the rod about this axis, one needs only to apply the definition (C-3). Thus one can imagine that the rod is subdivided into many small particles, and can then calculate the sum of the moments of inertia of all these particles. This calculation is shown below and yields the simple result

$$I_A = \frac{1}{3}ML^2. \qquad \text{(D-1)}$$

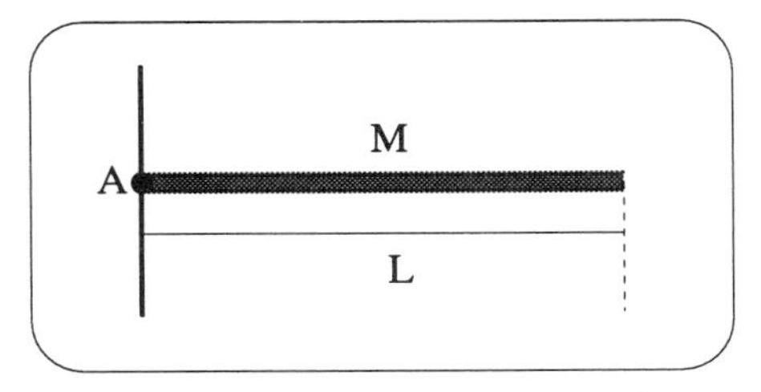

Fig. D-1. A rod of mass M and length L.

Calculation of the moment of inertia of a rod

As indicated in Fig. D-2, a typical small particle might be an infinitesimal segment of the rod, one which starts at a distance x from the axis and has an infinitesimal length dx. Since the mass of the entire rod is M, and the small rod segment constitutes only a fraction (dx/L) of the entire length L of the rod, the mass of the small segment is M(dx/L). The moment of inertia I_{seg} of this small segment, at a distance x from the axis, is then

$$I_{seg} = \left[M\left(\frac{dx}{L}\right)\right] x^2 = \left(\frac{M}{L}\right) x^2\, dx\,.$$

The moment of inertia I_A of the *entire* rod is found by finding the sum of the contributions of all such segments, beginning with the segment at x = 0 and ending with the segment at x = L. Thus

$$I_A = \left(\frac{M}{L}\right) \int_0^L x^2\, dx\,. \tag{D-2}$$

Each term in this sum can be expressed as an infinitesimal change of the quantity x^3. Indeed, by (5C-10), $d(x^3) = 3x^2\, dx$. Hence (D-2) is equal to

$$I_A = \left(\frac{M}{3L}\right) \int_0^L d(x^3) = \left(\frac{M}{3L}\right) [L^3 - 0] \tag{D-3}$$

since the sum of all the infinitesimal changes is just equal to the total change. Thus (D-3) yields the previously stated result (D-1).

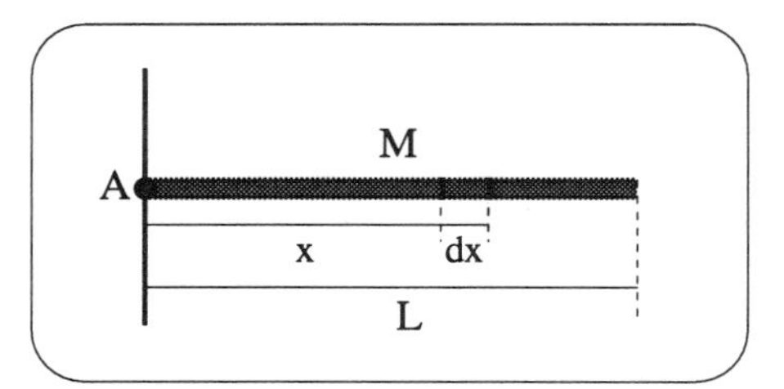

Fig. D-2. A small segment of a rod.

Moment of inertia of a rod about its center. What is the moment of inertia I of a uniform rod, of mass M and length L, about an axis passing through the *center* C of the rod? As indicated in Fig. D-3, such a rod can be viewed as consisting of two rods, joined at a common end point C, each having a length L/2 and a mass M/2. The moment of inertia I is then simply the sum of the moments of inertia of these two half rods, i.e., by (D-1)

$$I = \frac{1}{3}\left(\frac{M}{2}\right)\left(\frac{L}{2}\right)^2 + \frac{1}{3}\left(\frac{M}{2}\right)\left(\frac{L}{2}\right)^2$$

or

$$\boxed{I = \frac{1}{12} ML^2\,.} \tag{D-4}$$

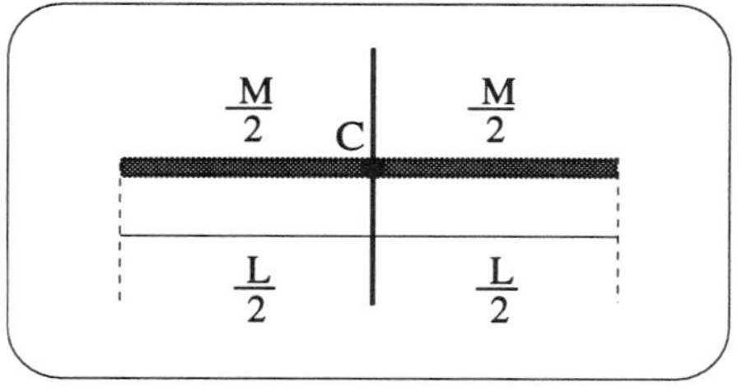

Fig. D-3. A rod, of mass M and length L, with its center at C.

Moment of inertia of a rectangular plate. The preceding result can be used to find the moment of inertia of other common objects. For example, consider a uniform rectangular plate of mass M and length L. What is the moment of inertia of this plate about an axis through its center C (when this axis lies in the plane of the plate and is perpendicular to the side of length L, as indicated in Fig. D-4)? As indicated in this figure, the plate can be viewed as consisting of a set of adjacent rods, each of length L, having masses M_1, M_2, M_3,.... The moment of inertia I of the rectangular plate is then equal to the sum of the moments of inertia of all these rods. Thus (D-4) implies that

$$I = \frac{1}{12} M_1 L^2 + \frac{1}{12} M_2 L^2 \ldots = \frac{1}{12}(M_1 + M_2 + \ldots)L^2$$

or

$$I = \frac{1}{12} ML^2\,. \tag{D-5}$$

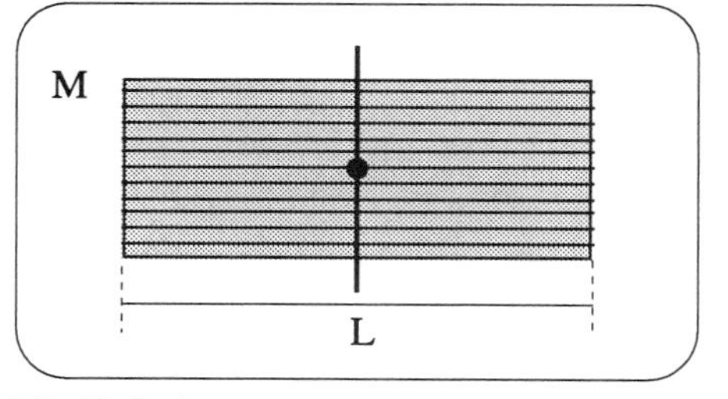

Fig. D-4. A rectangular plate viewed as a set of adjacent rods.

This result is exactly the same as the result (D-4) for a rod, except that the mass M is here the mass of the entire rectangular plate.

Example D-1: Angular velocity of a hinged rod

A rod, of mass M and length L, is hinged at one of its ends A so that it can rotate about it. The rod is initially horizontal and at rest (as indicated in Fig. D-5) and is then released. Friction forces are negligible. What is the angular velocity of the rod at the instant when it is vertical?

By applying the energy law to the rod, between the initial instant when the rod is horizontal and the instant when it is vertical, one gets

$$\Delta K + \Delta U_g = 0$$

or

$$[\tfrac{1}{2} I_A \omega^2 - 0] + [0 - Mg\,(\tfrac{1}{2}L)] = 0$$

where I_A is the moment of inertia of the rod about its end point A. By using the result (D-1) for this moment of inertia, the desired angular velocity ω of the rod is then found to be

$$\omega^2 = \frac{MgL}{I_A} = \frac{MgL}{\frac{1}{3}ML^2} = \frac{3g}{L}$$

or

$$\omega = \sqrt{\frac{3g}{L}}\,. \qquad \text{(D-6)}$$

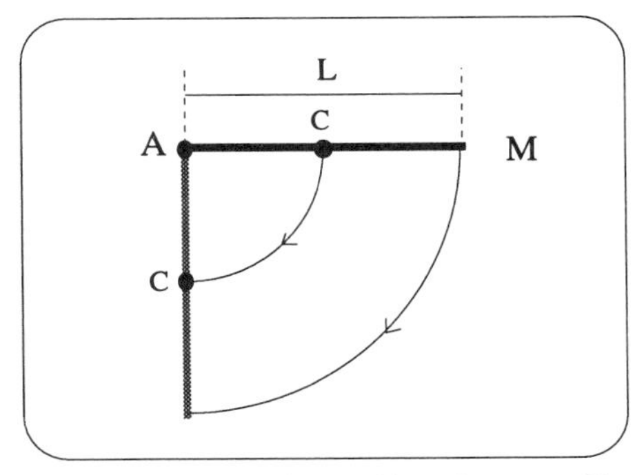

Fig. D-5. Motion of a rod hinged at one of its ends.

As the rod descends to the vertical position, its center of mass descends by a vertical distance L/2. Its gravitational potential energy decreases accordingly.

Moment of inertia of a disk

What is the moment of inertia of a uniform disk, of mass M and radius R, about an axis passing through its center and perpendicular to the disk? If all particles in the disk were located at the outside rim, so as to form a ring, the moment of inertia would be simply MR^2. But many particles in the disk are located much closer to the rotation axis and thus contribute less to the moment of inertia. Hence one expects that the moment of inertia of the disk should be *smaller* than MR^2.

As illustrated in Fig. D-6, a disk can be viewed as a set of concentric rings of varying radii, ranging from a radius zero to a radius R. By (C-5), the moment of inertia of any such ring, of mass m and radius r, is simply mr^2. The moment of inertia I of the entire disk is thus the sum of the moments of inertia of all these rings. This sum can be readily calculated, as shown below, and yields the result

$$\boxed{I = \tfrac{1}{2} MR^2}\,. \qquad \text{(D-7)}$$

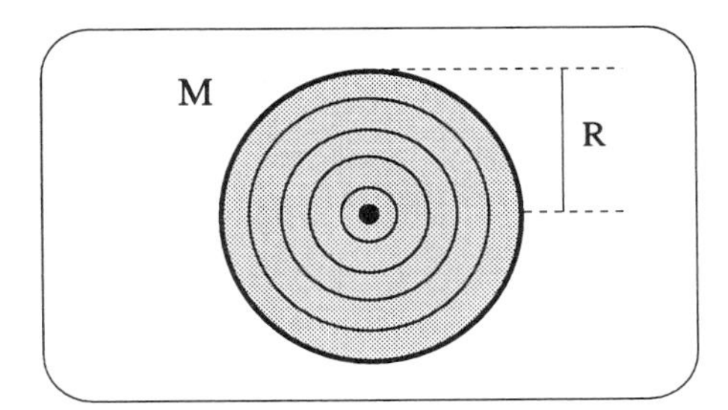

Fig. D-6. A disk viewed as a set of adjacent thin rings.

Calculation of the moment of inertia of a disk

A typical thin ring has a radius r and an infinitesimal thickness dr (shown much exaggerated in Fig. D-7). The small area of this ring is $2\pi r\,dr$ (since it is essentially just a rectangle, of height dr, bent around into a circle having the length $2\pi r$ equal to the circumference of the ring). The area of the thin ring thus constitutes a fraction $(2\pi r\,dr/\pi R^2)$ of the entire area πR^2 of the disk. Hence the mass of the ring is $M(2\pi r\,dr/\pi R^2)$. The moment of inertia I_{ring} of this ring is then

$$I_{ring} = \left[M\left(\frac{2\pi r\,dr}{\pi R^2}\right)\right] r^2 = \left(\frac{2M}{R^2}\right) r^3\,dr\,.$$

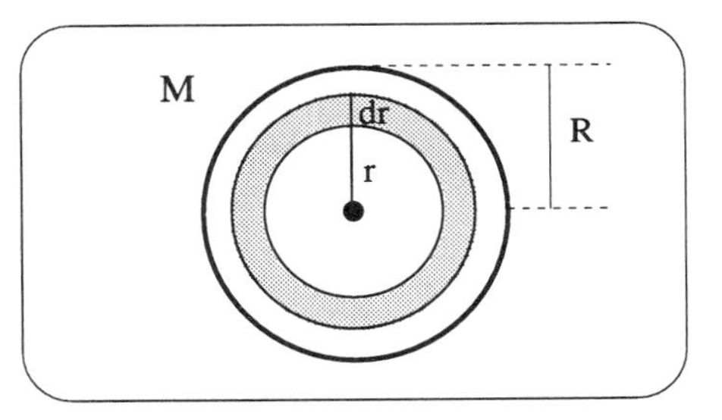

Fig. D-7. A thin ring within a disk.

The moment of inertia I of the *entire* disk is found by finding the sum of the contributions of all such rings, beginning with the ring of radius r = 0 and ending with the ring of radius r = R. Thus

$$I = \left(\frac{2M}{R^2}\right) \int_0^R r^3 \, dr \, . \qquad \text{(D-8)}$$

Each term in this sum can be expressed as an infinitesimal change of the quantity r^4. Indeed, by (5C-10), $d(r^4) = 4r^3\,dr$. Hence (D-7) is equal to

$$I = \left(\frac{2M}{R^2}\right)\left(\frac{1}{4}\right) \int_0^R d(r^4) = \left(\frac{M}{2R^2}\right)[R^4 - 0] \qquad \text{(D-9)}$$

since the sum of all the infinitesimal changes is just equal to the total change. Thus (D-9) yields the previously stated result (D-7).

A similar argument shows that the moment of inertia about the central axis of a uniform circular cylinder, of mass M and radius R, is also given by (D-7).

Example D-2: Weight descending from a pulley

The example in Sec. C discussed the problem of a weight descending from a pulley. Suppose that this pulley is a solid disk of mass M so that its moment of inertia is given by (D-7). According to the answer found for that problem, the speed of the descending weight would then be

$$v = \sqrt{\frac{2gh}{1 + \frac{I}{mR^2}}} = \sqrt{\frac{2gh}{1 + \frac{M}{2m}}} \, . \qquad \text{(D-10)}$$

→ ***Go to Sec. 17D of the Workbook.***

E. Moments of inertia about different axes

Moments of inertia about parallel axes

Parallel-axes relation. By its definition (C-3), the moment of inertia I of an object, about an axis passing through a point O, is equal to

$$I = \sum m_n r_n^2 \, . \qquad \text{(E-1)}$$

The moment of inertia can thus be found by calculating the indicated sum.

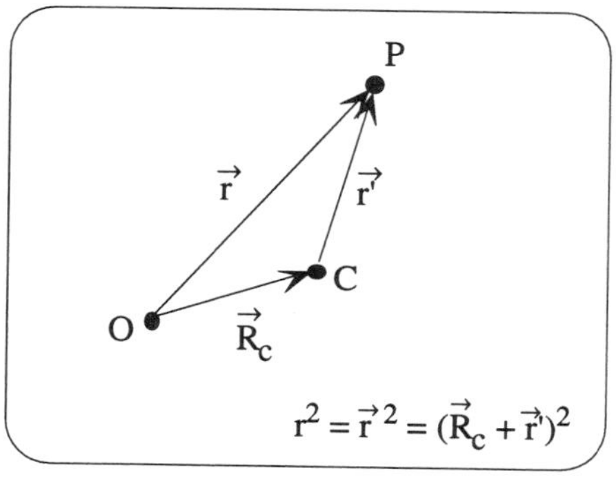

But, as indicated in Fig. E-1, the position vector $\vec{r}$ of a particle relative to the point O is related to the position vector $\vec{r}\,'$ of the particle relative to some other point O'. Hence there is also a relation between the moments of inertia about parallel axes passing through these points. As shown below, this relationship is particularly simple if the point O' is the center of mass C and then yields the following s simple result.

Fig. E-1. Position of a particle P specified relative to a point O or relative to the center of mass C.

Parallel-axes relation:	$I = MR_c^2 + I'$.

(E-2)

Here M is the mass of the object, R_c is the distance of its center of mass from the axis through O, and I' is the moment of inertia of the object about a parallel axis passing through its center of mass.

In words, (E-2) says that the moment of inertia of an object about any axis is equal to ⟨the moment of inertia which it would have if all its mass were concentrated at its center of mass⟩ plus ⟨the moment of inertia about its center of mass⟩.

Argument leading to the parallel-axes relation

All the vectors here are component vectors perpendicular to the axis.

As indicated in Fig. E-1, the position $\vec{r}$ of every particle can be specified in terms of its position $\vec{r}\,'$ relative to the center of mass. In the definition (E-1) of moment of inertia, the square of a particle's distance from the axis through O is then equal to

$$r_n{}^2 = \vec{r}_n{}^2 = (\vec{R}_c + \vec{r}_n{}')^2 = \vec{R}_c{}^2 + \vec{r}_n{}'^2 + 2\vec{R}_c \cdot \vec{r}_n{}' \,.$$

When this expression is substituted into (E-1), the right side becomes decomposed into three separate sums. Thus

$$I = \left(\sum m_n\right)\vec{R}_c{}^2 + \sum\left(m_n \vec{r}_n{}'^2\right) + 2\,\vec{R}_c \cdot \left(\sum m_n \vec{r}_n{}'\right)$$

or

$$I = \frac{1}{2} M\vec{R}_c{}^2 + I' + 2\,\vec{R}_c \cdot \left(\sum m_n \vec{r}_n{}'\right). \tag{E-3}$$

Here we have taken constant factors outside each sum and used the fact that the sum of all particle masses is the total mass M of the system.

But, by the definition (15D-5) of center of mass,

$$\sum m_n \vec{r}_n{}' = M\vec{R}_c{}' = 0 \tag{E-4}$$

since the position vector $\vec{R}_c{}'$ of the center of mass, relative to itself, is zero. Describing positions relative to the center of mass is thus particularly simple. Hence (E-3) becomes

$$I = M\vec{R}_c{}^2 + I' \tag{E-5}$$

which is the result previously stated in (E-2).

Utility of the parallel-axes relation. We already calculated the moments of inertia about the centers of mass of a rod or a disk. (One can similarly calculate the moments of inertia about the centers of some other objects.) The moment of inertia about a parallel axis through any other point can then immediately be found from the relation (E-2), *without any further complex calculation.* This is why the parallel-axes relation (E-2) is very useful. The following are some examples.

Examples

By (D-5), the moment of inertia I' about an axis through the center of mass of a rod, of mass M and length L, is $ML^2/12$. According to the parallel-axes relation (E-2), the moment of inertia of this rod, about an axis through its end point A in Fig. E-2a, is then

$$I_A = M\left(\tfrac{1}{2}L\right)^2 + \tfrac{1}{12}ML^2 = \tfrac{1}{3}ML^2 \,. \tag{E-6}$$

This result agrees properly with (D-1).

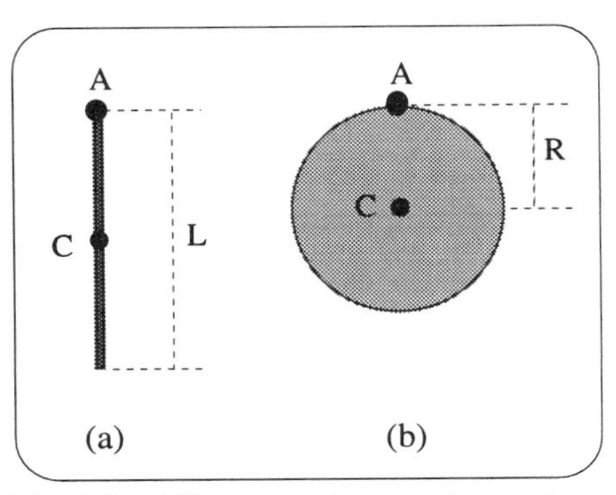

Fig. E-2. Different rotation axes for a rod and for a disk. (The axes are perpendicular to the rod and to the plane of the disk.)

By (D-7), the moment of inertia I' about an axis through the center of mass of a uniform disk, of mass M and radius R, is $MR^2/2$. According to the parallel-axes relation (E-2), the moment of inertia of this disk, about an axis through a point A on its rim in Fig. E-2b, is then

$$I_A = MR^2 + \frac{1}{2}MR^2 = \frac{3}{2}MR^2 . \qquad \text{(E-7)}$$

Moments of inertia about perpendicular axes

What is the moment of inertia I of a thin two-dimensional plate, about an axis which is perpendicular to this plate and passes through some point O? As is apparent from Fig. E-3, the definition (C-3) of moment of inertia implies that

$$I = \sum m_n r_n^2 = \sum m_n (x_n^2 + y_n^2) = \sum m_n x_n^2 + \sum m_n y_n^2$$

or

$$\boxed{I = I_x + I_y} \qquad \text{(E-8)}$$

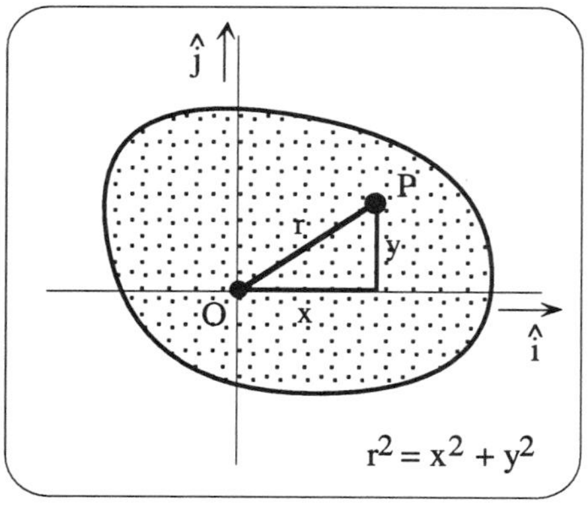

where I_x and I_y are the moments of inertia of the plate through two mutually perpendicular axes which lie in the plane of the plate and pass through the point O.

Fig. E-3. A particle P, in a flat plate, located at a distance r from a perpendicular axis passing through O.

Example

Fig. E-4 shows a uniform rectangular plate, of mass M, which has sides of lengths L_1 and L_2. What is the moment of inertia of this plate, about an axis passing through its center C and perpendicular to the plate? This moment of inertia can immediately be found from our knowledge (D-3) of the moments of inertia about axes parallel to the rectangle's sides. Indeed, by (E-8),

$$I = \frac{1}{12}ML_1^2 + \frac{1}{12}ML_2^2 = \frac{1}{12}M(L_1^2 + L_2^2) . \qquad \text{(E-9)}$$

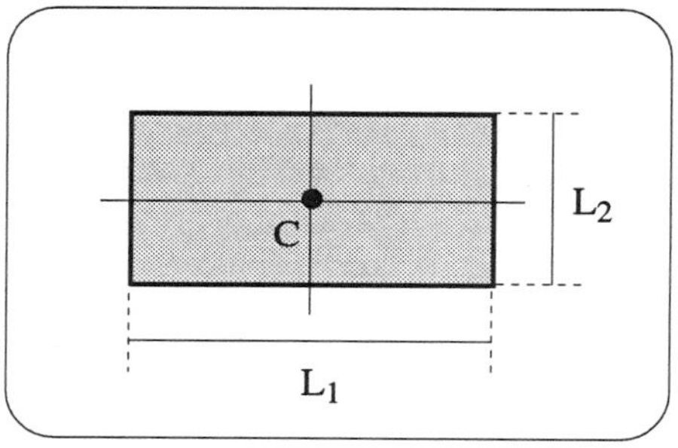

Fig. E-4. A rectangular plate.

➜ ***Go to Sec. 17E of the Workbook.***

F. Rotational work and torque

To apply the energy law $\Delta E = W_{oth}$ to a rotating object, one may need to find the work done on the object by various forces. We now discuss how this work can be calculated.

Infinitesimal work on a rotating particle

Fig. F-1 indicates a particle (e.g., a particle inside a rotating object) which is located at a distance r from a rotation axis. Suppose that this particle rotates by an infinitesimal angle $d\phi$ about this axis while it is acted on by a force $\vec{F}$. What then is the work d'W done on the particle by this force?

This work is, by its definition, equal to

$$d'W = \vec{F}\cdot d\vec{r} = F_{dr}\,|d\vec{r}| \qquad \text{(F-1)}$$

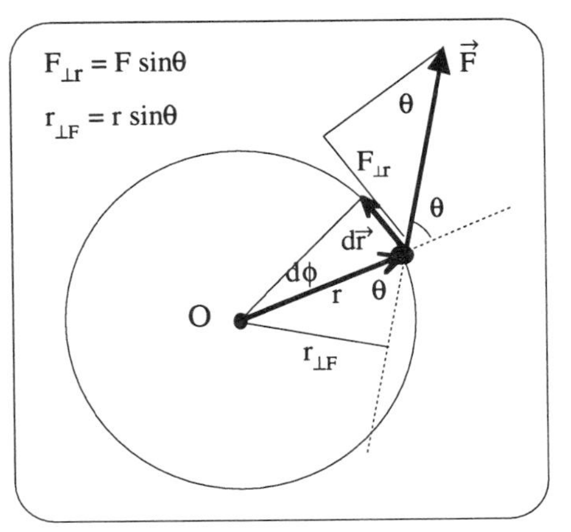

Fig. F-1. Infinitesimal rotation of a particle.

where $|d\vec{r}|$ is the magnitude of the particle's infinitesimal displacement and F_{dr} is the component of the force along this displacement. Here the magnitude of the particle's infinitesimal displacement is just the small distance $ds = r\,d\phi$ which the particle moves along its circular path. Furthermore, the direction of this displacement is tangent to the particle's path, i.e., perpendicular to its radial position vector $\vec{r}$. To make this explicit, the component F_{dr} of the force along this displacement can be denoted by $F_{\perp r}$. Hence (F-1) can be written as

$$d'W = F_{\perp r}\, r\, d\phi = rF_{\perp r}\, d\phi$$

or

$$\boxed{d'W = \tau\, d\phi\,.} \qquad \text{(F-2)}$$

Here we have introduced the quantity τ (denoted by the Greek letter *tau*) which is called torque and defined as follows:

Def: **Torque:** $\tau = rF_{\perp r}\,.$ (F-3)

Power. The relation (F-2) implies that the power $\mathcal{P}$ delivered to the rotating particle by the force is equal to

$$\mathcal{P} = \frac{d'W}{dt} = \tau\,\frac{d\omega}{dt} = \tau\,\omega\,. \qquad \text{(F-4)}$$

Torque

The quantity *torque*, defined in (F-3), involves the force on a particle as well as the distance of this particle from the axis of rotation. Torque, like force, is thus also a quantity describing interaction. However, it describes more specifically how the interaction affects *rotation*.

Properties of the torque. The definition (F-3) implies that the torque has the following properties: (a) The torque is zero if the force is zero, and is larger if the force is larger. (b) The torque depends only on the *component* of the force perpendicular to the position vector of the particle relative to the rotation axis. Thus the torque is zero if the force is directed along the radial direction from the axis, and (for a given magnitude of the force) is maximum if the force is perpendicular to this radial direction. (c) The torque on a particle depends on the distance of this particle from the rotation axis. Thus the torque is zero if the particle is located at the rotation axis, and (for a given force) is larger if the distance of the particle from the axis is larger.

> ***Example***
>
> To show how the torque describes the effectiveness of a force in producing rotation, consider the familiar example of a force used to push open a door.
>
> If this force is applied parallel to the door, as indicated by the force $\vec{F}_1$ in Fig. F-2, the door cannot be opened, no matter how large this force might be. (Indeed, the torque produced by this force about the hinge would then be zero.)

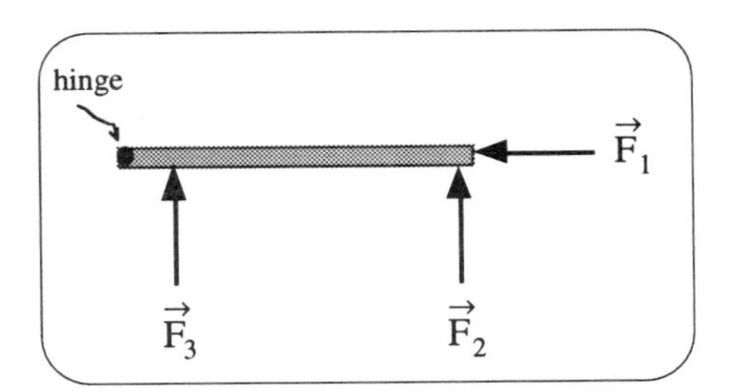

Fig. F-2. Top view of a door free to rotate about a hinge.

Suppose that a force of this magnitude is applied at the same point, but along different directions. It would then be most effective in opening the door if it is applied perpendicular to the door, as indicated by the force $\vec{F}_2$ in Fig. F-2. (Indeed, the torque produced by this force would then be maximum.)

However, if this force is applied at a distance closer to the hinge, the force would be less effective in opening the door, and could not open the door at all if applied directly to the hinge. (Indeed, the torque produced by this force would then be smaller or zero.)

Specification of torque. A complete specification of torque requires the phrase

torque on <u>particle</u> by <u>force</u> about <u>axis</u> (F-5)

where one needs to specify each of the underlined entities.

Sign of the torque. The torque defined by (F-3) is a *number* describing how effectivea force is in producing rotation about a fixed axis. This number can be positive or negative depending on the sense of the rotation produced by the force. For example, suppose that one chooses to call counter-clockwise rotations positive. Then a torque (like that in Fig. F-1), which tends to produce a counter-clockwise rotation, is considered positive; conversely, a torque which tends to produce a clockwise rotation is considered negative.

In more complex situations, where the rotation axis can change its direction, the torque is defined as a vector having a direction along the axis of rotation. The torque defined in (F-3) is then only one *component* of this vector.

Units of torque. The definition (F-3) implies that the SI unit of torque is Nm (i.e., newton multiplied by meter).

Note that the unit of torque is the same as that of work, and could thus also be expressed in terms of the unit *joule*. However, this is ordinarily not done since the meaning of torque is very different from that of work.

Alternative definitions of torque. The definition (F-3) can be expressed in various equivalent ways, all of which may be useful. For example, as is apparent from Fig. F-1, the component $F_{\perp r}$ of the force, perpendicular to the particle's position vector, is equal to $F \sin\theta$ (where θ is the angle between the particle's position vector $\vec{r}$ and the force $\vec{F}$). Thus the definition (F-3) of torque can be expressed in the following equivalent forms:

$$\boxed{\tau = rF_{\perp r} = rF\sin\theta = Fr_{\perp F}\,.} \qquad \text{(F-6)}$$

In the last expression we have used the fact, apparent from Fig. F-1, that $r\sin\theta = r_{\perp F}$, the component of the particle's position vector perpendicular to the force $\vec{F}$. (This component $r_{\perp F}$ is sometimes called the *lever arm* of the force about the axis.)

Calculation of work

A knowledge of the work (F-2) done in an infinitesimal rotation allows one readily to calculate the work done in more complex situations.

Work in a large rotation. If a particle rotates through a large angle, the work W done it is just the sum of the successive infinitesimal works. Thus

$$W = \int d'W = \int \tau\, d\phi. \qquad \text{(F-7)}$$

The calculation of work is particularly simple if the torque is constant. Thus,

if τ is constant, $$W = \tau \int d\phi = \tau\, \Delta\phi \qquad \text{(F-8)}$$

where $\Delta\phi$ is the total angular displacement of the particle.

Work on a system. When an entire system rotates, the *total* work done on it is the sum of the works done it by all forces, i.e., the work done on it by the torques on all the particles. When this system is a rigid solid object (so that the distances between its particles remain unchanged), the internal forces do no work. The total work is then merely that done by the torques due to external forces. (Furthermore, some of this work may be expressed in terms of a potential energy and thus not be included in the *other* work specified by the energy law $\Delta E = W_{oth}$.)

Example: Wheel pulled by a string

The following prototypical problem illustrates the application of work and the energy law to a system involving rotational motion.

Problem statement. A uniform wheel, of mass M and radius R, is free to rotate around a horizontal axle. The wheel is initially at rest and a constant force of magnitude F_o is applied to a string wrapped around the wheel. What is the angular velocity attained by the wheel after it has rotated by n revolutions?

Solution. The following illustrates the solution of this problem.

Analysis of problem

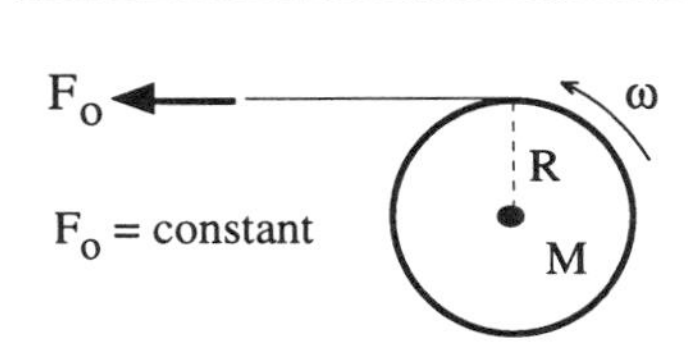

Known: M, R, F_o, n revolutions.
$\omega = 0$ when $t = 0$.

Goals: $\omega = ?$ (F_o, R, M, n)

Construction of solution

We want to relate angular speeds and positions at two instants, without an explicit interest in the elapsed time. Hence the energy law should be useful.

Apply energy law, to wheel, initial → final

$$\Delta E = W_{oth}$$

$$\Delta K + 0 = \tau\, \Delta\phi = (RF_o)\, \Delta\phi$$

Torque is constant since force is constant.

$$\left(\tfrac{1}{2} I\, \omega^2 - 0\right) = (RF_o)\,(2\pi n)$$

Each revolution corresponds to a rotation angle of 2π.

$$\omega^2 = \frac{4\pi R F_o n}{I} \qquad (1)$$

For solid disk, $I = \frac{1}{2} MR^2$

By (1) $$\boxed{\omega = \sqrt{\frac{8\pi F_o n}{MR}}} \qquad (2)$$

Checks (done in writing or mentally)

Checks

Units of (2): $\frac{1}{s} = \sqrt{\frac{kg\ m/s^2}{kg\ m}} = \frac{1}{s^2} = \frac{1}{s}$ OK

If n or F_o large: Expect large ω. Eq. (2) OK.

If M large: Expect slow angular acceleration, small ω. Eq. (2) OK.

Check extreme cases.

➜ ***Go to Sec. 17F of the Workbook.***

G. Summary

Angular description of motion:

Angular position: ϕ

Angular velocity: $\omega = \frac{d\phi}{dt}$

Angular acceleration: $\alpha = \frac{d\omega}{dt}$

Relations between linear and angular quantities:

Arc distance: $s = r\phi$

Speed: $v = r\omega$

Acceleration components: $a_v = r\alpha$, $a_c = r\omega^2$

Rotational energy and work:

Rotational kinetic energy: $K = \frac{1}{2} I\omega^2$ (I = moment of inertia)

Moment of inertia: $I = \sum m_n r_n^2$.

Parallel-axes relation: $I = MR_c^2 + I'$ (I' = moment about CM)

Rotational work: $d'W = \tau\, d\phi$

Torque: $\tau = rF_{\perp r} = rF \sin\theta = Fr_{\perp F}$ (See Fig. G-1.)

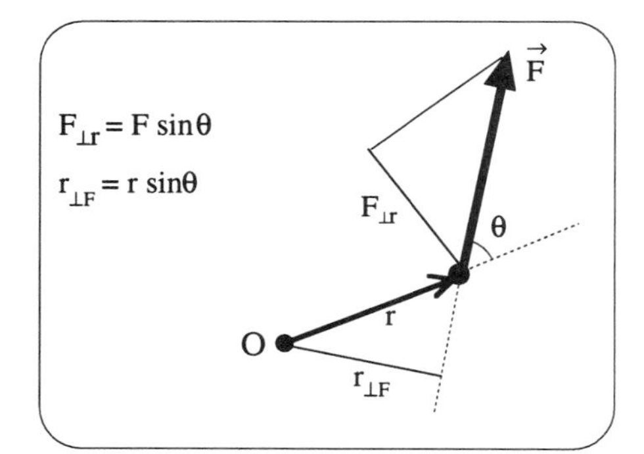

Fig. G-1. Force $\vec{F}$ on a particle located at a distance r from an axis at O.

New abilities

You should now be able to do the following:

(1) Describe the rotation of an object, about a fixed axis, in terms of its angular position, angular velocity, and angular acceleration — and relate these descriptions in the case of constant angular acceleration.

(2) Find the moment of inertia of an object when one knows its moment of inertia about its center of mass, or in other simple situations that do not require the calculation of complex sums of over many particles.

(3) Apply the energy law, together with knowledge about rotational kinetic energy and rotational work, to find information about the motion of rotating objects.

➜ ***Go to Sec. 17G of the Workbook.***

18 Angular-Momentum Law

The preceding chapter discussed how the energy law can be applied to rotating systems. This law provides useful knowledge relating speeds and positions, but does not provide detailed information about a system's motion in the course of time. To obtain such information, the present chapter investigates the implications of Newton's law applied to the particles of a rotating system. This investigation yields the "angular-momentum law" which is analogous to the momentum law previously obtained in Chapter 15. We shall then have obtained all the basic mechanics laws needed to deal with the motions of rotating objects and complex systems.

A. Angular-momentum law for a particle

Derivation of the angular-momentum law

Application of Newton's law. Fig. A-1 illustrates a particle, of mass m, rotating about a fixed axis and located at a fixed distance r from this axis. According to Newton's law, the acceleration $\vec{a}$ of this particle is then related to the total force $\vec{F}_{tot}$ on it so that

$$m\vec{a} = \vec{F}_{tot} . \qquad \text{(A-1)}$$

Perpendicular component of Newton's law. The particle moves in a direction tangent to its circular path, i.e., *perpendicular* to its radial position vector. Newton's law (A-1), applied along this perpendicular direction, implies that

$$m\, a_{\perp r} = (F_{tot})_{\perp r} . \qquad \text{(A-2)}$$

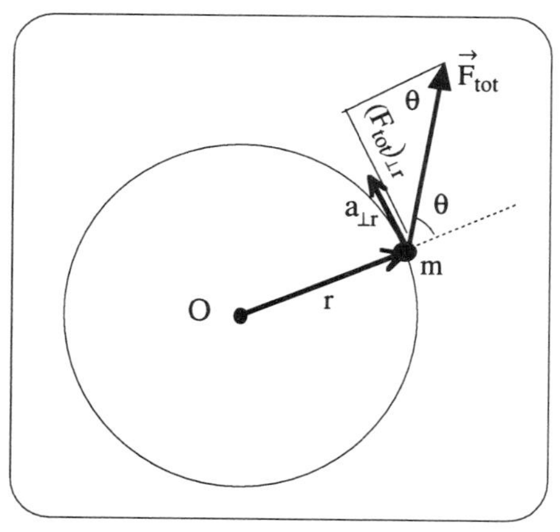

Fig. A-1. Particle rotating about an axis passing through O.

Here the component $a_{\perp r}$ of the acceleration perpendicular to the radius (i.e., its tangential component) is just $r\alpha$ if α is the particle's angular acceleration.

Expression in terms of torque. In (A-2) the component of the total force, perpendicular to the radius, is related to the torque produced by this force. To make this relation explicit, one needs only to multiply both sides of (A-2) by the radius r. Thus one gets

$$r\, m\, (r\alpha) = r\, (F_{tot})_{\perp r}$$

or $$mr^2\alpha = \tau_{tot} \,. \tag{A-3}$$

Here τ_{tot} is the torque produced on the particle by the total force. (This is also the total torque on the particle, i.e., the sum of the torques produced by all forces on it.)

The component of the total force, perpendicular to the radius, is equal to the sum of the components of all the individual forces. Hence the torque produced by this total force is equal to the sum of the torques produced by the individual forces.

Angular-momentum law. The left side of (A-3) can be expressed in the form

$$mr^2\alpha = mr^2\frac{d\omega}{dt} = \frac{d(mr^2\omega)}{dt} = \frac{d\ell}{dt} \tag{A-4}$$

where the quantity ℓ is called the angular momentum of the particle in accordance with the following definition:

Def: **Angular momentum of a particle:** $\ell = mr^2\omega$ (A-5)

With this definition, the relation (A-3) can be stated in the following form:

Angular-momentum law *(for a particle):* $\dfrac{d\ell}{dt} = \tau_{tot} \,.$ (A-6)

This angular-momentum law describes the motion of a particle by its angular momentum, and its interaction by the total torque on it. Thus it is a mechanics law which specifies the relation between motion and interaction in a form particularly useful for rotational motion.

It can be shown that the angular-momentum law (A-6) is valid *even if the particle's distance r from the rotation axis changes.*

Angular momentum

Alternative expressions for angular momentum. The angular momentum of a particle can be expressed in several equivalent forms. For example, the definition (A-5) implies that

$$\ell = I\omega \tag{A–7}$$

where $I = mr^2$ is the particle's moment of inertia about the rotation axis. Note that the angular velocity ω, and hence the angular momentum ℓ, can be positive or negative, depending on the sense of rotation. (For example, suppose that one chooses to call the counter-clockwise sense positive, as indicated in Fig. A-2. Then ω and ℓ would be positive if the particle moves counter-clockwise, and would be negative if the particle moves clockwise.)

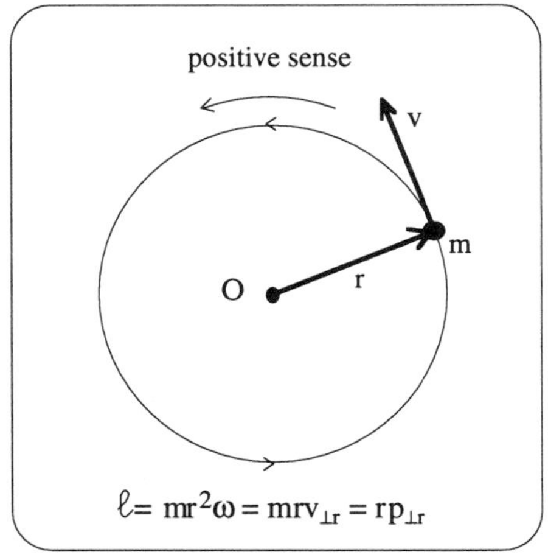

Fig. A-2. Particle rotating around a circular path.

A particle moving in a circular path with a speed v, as indicated in Fig. A-2, has a component of velocity $v_{\perp r}$ along the sense of rotation perpendicular to the radius, i.e., tangent to the path. (This component is equal to + v or −v, depending on the sense of rotation.) Since $v_{\perp r} = r\omega$, the angular momentum in (A-5) can also be written as

$$\boxed{\ell = mrv_{\perp r} = rp_{\perp r}} \qquad \text{(A-8)}$$

where $p_{\perp r} = mv_{\perp r}$ is the component of the particle's momentum along the sense of rotation, perpendicular to the particle's position vector $\vec{r}$.

The relation (A-8) is the general definition of angular momentum, true even if the distance r changes so that the particle does not travel around a circle.

The last expression in (A-8) shows that a particle's angular momentum ℓ is closely related to its linear momentum. (This is why the quantity ℓ is called *angular momentum.*)

Properties of angular momentum. According to its definition, the angular momentum is a positive or negative number. By (A-8), its SI units are kg m^2/s.

The magnitude of the angular momentum, like that of a torque, depends crucially on a particle's distance r from the rotation axis. It is zero if the particle lies on this axis, and is larger if the distance from this axis is larger.

In more complex situations, where the rotation axis can change its direction, the angular momentum is defined as a vector having a direction along the axis of rotation. The angular momentum defined in (A-5) is then only the *component* of this vector along the direction of the rotation axis.

→ ***Go to Sec. 18A of the Workbook.***

B. Angular-momentum law for a system

Derivation of the angular-momentum law

General form of the angular-momentum law. To extend the angular-momentum law (A-6) to a system consisting of any number of particles (like the system schematically indicated in Fig. B-1), one only needs to apply this law to every particle in the system and then to add the resulting equations. As shown in greater detail below, one obtains in this way the angular-momentum law in its general form:

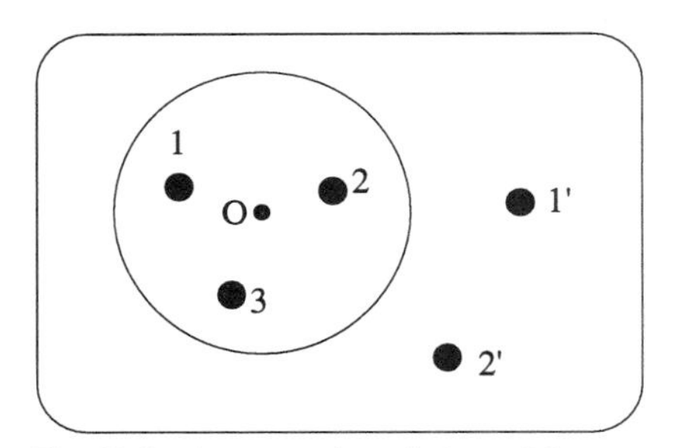

Fig. B-1. A system (consisting of three particles 1, 2, and 3) interacting with external particles 1' and 2'. The system's angular momentum can be considered about any point, such as O.

Angular-momentum law: $$\frac{dL}{dt} = \tau_{ext}\,. \qquad \text{(B-1)}$$

Angular momentum of a system. Here L is the total angular momentum of the system, defined as the sum of the angular momenta of all of its particles. Thus

Def: ***Angular momentum of a system:*** $$L = \sum \ell_n\,. \qquad \text{(B-2)}$$

By (A-5), this angular momentum is equal to

$$L = \Sigma m_n r_n^2\ \omega = (\Sigma m_n r_n^2)\ \omega$$

or

$$\boxed{L = I\omega} \qquad \text{(B-3)}$$

where we have used the definition (17C-3) of the moment of inertia I.

Total external torque. On the right side of (B-1), the symbol τ_{ext} denotes the total *external* torque on the system, i.e., the sum of the torques on all particles in the system by all particles *outside* the system. (The sum of all *internal* torques, exerted on particles in the system by other particles in the same system, is zero because of the relation between mutual forces.)

Argument leading to the general angular-momentum law

By applying the angular-momentum law (A-6) to particle 1 in a system, like that indicated in Fig. A-1, one gets

$$\frac{d\ell_1}{dt} = [\tau_{12} + \tau_{13}] + [\tau_{11'} + \tau_{12'}] \qquad \text{(B-4a)}$$

where the right side indicates the total torque on particle 1. This torque consists of the total *internal* torque on this particle (i.e., of the sum of torques on this particle by all other particles *in* the system) and of the total *external* torque on this particle (i.e., of the sum of the torques on this particle by all the particles *outside* the system). Similarly, the angular-momentum law applied to particle 2 implies that

$$\frac{d\ell_2}{dt} = [\tau_{21} + \tau_{23}] + [\tau_{21'} + \tau_{22'}] . \qquad \text{(B-4b)}$$

Applying the angular-momentum law to every particle in the system, and then adding all the equations (B-4a), (B-4b), ... then yields the result

$$\frac{d\ell_1}{dt} + \frac{d\ell_2}{dt} + \ldots = \tau_{int} + \tau_{ext}$$

or
$$\frac{d}{dt}(\ell_1 + \ell_2 + \ldots) = \frac{dL}{dt} = \tau_{int} + \tau_{ext} \qquad \text{(B-5)}$$

where τ_{int} is the total *internal* torque on the system. This is the angular-momentum law (B-1) if one can show that this total internal torque is zero.

To show this, we need merely convince ourselves that the sum of the mutual torques, exerted on any pair of particles in the system, is zero. Fig. B-2 shows such a pair of particles, e.g., particles 1 and 2. The mutual forces on these particles have equal magnitudes and opposite directions. Furthermore, if these forces are directed along the line joining the particles, the distance $r_{\perp F}$ from the axis, perpendicular to either of these forces, is the same. As indicated in Fig. B-2, the torque τ_{12} exerted on particle 1 by particle 2 has then the same magnitude, but opposite direction, as the torque τ_{21} exerted on particle 2 by particle 1. Hence the sum of the mutual torques

$$\tau_{12} + \tau_{21} = 0.$$

Correspondingly, all internal torques cancel each other in pairs.

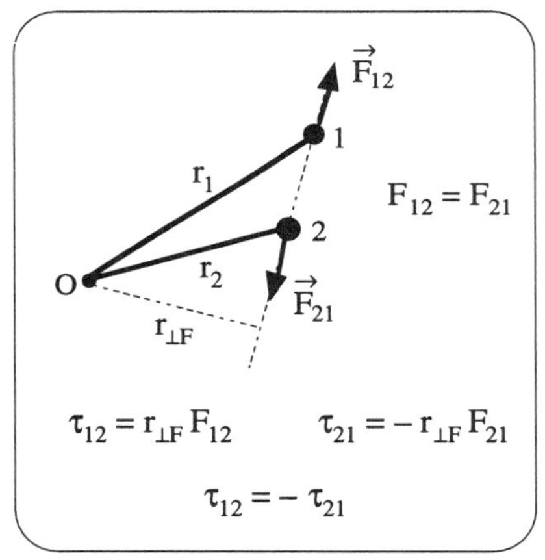

Fig. B-2. Mutual torques on a pair of particles.

Discussion

Characteristics of the angular-momentum law. Stated in words, the angular-momentum law (B-1) asserts that *the rate of change of the angular momentum of a system is equal to the total external torque on this system.*

In this law, the angular momenta and torques must all be calculated about the *same* axis.

The angular-momentum law, like all mechanics laws, provides a relation between the motion of a system and its interactions with other systems. To do this, it describes the motion of a system by its angular momentum, and the interactions of this system by the torques exerted on it. The angular-momentum law is thus especially useful for dealing with rotational motion.

Comparison with the linear momentum law. The *angular*-momentum law relates a system's change of *angular momentum* and the total external *torque* on it. This law is analogous to the previously studied *linear*-momentum law $d\vec{P}/dt = \vec{F}_{ext}$ which relates the rate of change of *linear momentum* and the total external *force* .

The angular-momentum law, like the linear-momentum law, involves only the *external* interactions of the system, but not any internal ones. As we shall see, the kinds of applications of the angular-momentum law are thus quite similar to those of the linear-momentum law.

Angular-momentum law for a rigid object. If a system is a rigid object, the distance of every particle from the axis of rotation remains unchanged. Hence the moment of inertia of the object is merely a constant. By (B-3), the rate of change of the angular momentum is then

$$\frac{dL}{dt} = \frac{d(I\omega)}{dt} = I\frac{d\omega}{dt} .$$

The angular-momentum law can then be expressed in the following simple form:

$$\boxed{\text{if } I = \text{constant}, \qquad I\alpha = \tau_{ext} .} \tag{B-6}$$

This establishes a very simple relation between the angular acceleration α of an object and the total external torque on it.

The relation (B-6) indicates that the moment of inertia characterizes an object by specifying how its angular acceleration is related to an external torque applied to it. For example, an object with a large moment of inertia is more difficult to set rotating than an object with a small moment of inertia.

Gravitational torque. In applying the angular-momentum law, one often needs to know the total external torque exerted on a system by the gravitational force near the surface of the earth. This gravitational torque τ_g is just the sum of the gravitational torques on all particles in the system, i.e.,

$$\tau_g = \sum \tau_n = \sum [r_n (m_n g) \sin\theta_n] = g \sum m_n x_n . \tag{B-7}$$

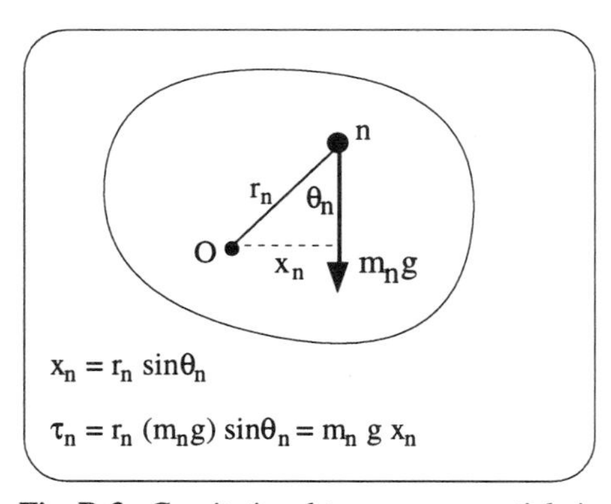

Fig. B-3. Gravitational torque on a particle in a system. (The particle is labeled by n.)

As indicated in Fig. B-3, here $x_n = r_n \sin\theta_n$ is the horizontal lever arm of the force on the n'th particle. The sum in (B-7) extends over all particles n = 1, 2, 3, ... in the system. But the last sum in (B-7) is related to the horizontal position X_c of the center of mass of the system of total mass M. Thus (B-7) is equal to

$$\tau_g = g\,(MX_c) = Mg\,X_c . \tag{B-8}$$

Thus we arrive at the following useful result:

The torque exerted on a system by the gravitational force near the earth is the same as if the entire mass of the system were concentrated at its center of mass.	(B-9)

➜ ***Go to Sec. 18B of the Workbook.***

C. Analogies between linear and rotational motions

There are many analogies between the motion of a particle (or of a system's center of mass) along a straight line, and the rotational motion of an object about a fixed axis. These analogies are summarized in the following table. (They make it easy to remember relations about rotational motion because these are so similar to familiar corresponding relations about particle motion.)

The relations for linear motion are expressed in terms of components of the quantities along a direction parallel to this line.

Linear motion		**Rotational motion**	
Position	x	ϕ	Angular position
Velocity	$v_x = dx/dt$	$\omega = d\phi/dt$	Angular velocity
Acceleration	$a_x = dv_x/dt$	$\alpha = d\omega/dt$	Angular acceleration
Mass	m	I	Moment of inertia
Momentum	$p_x = mv_x$	$L = I\omega$	Angular momentum
Momentum law	$dp_x/dt = F_{ext,x}$	$dL/dt = \tau_{ext}$	Angular-momentum law
(The external force on a single particle is the total force on it.)	$ma_x = F_{ext,x}$	$I\alpha = \tau_{ext}$	*(For constant moment of inertia.)*
Kinetic energy	$K = \frac{1}{2} mv_x^2$	$K = \frac{1}{2} I\omega^2$	Kinetic energy
Work	$d'W = F_x\, dx$	$d'W = \tau\, d\phi$	Work
Power	$\mathcal{P} = F_x v_x$	$\mathcal{P} = \tau\, \omega$	Power

D. Problem solving

Available mechanics laws

We have now available three basic mechanics laws, each of which specifies a relation between a system's motion and its interactions. These laws are applicable to any system and can be used jointly to solve a wide variety of mechanics problems. These laws (summarized in greater detail in Sec. F at the end of the chapter) are the following:

* **Momentum law ($d\vec{P}/dt = \vec{F}_{ext}$).** This law also describes the motion of the center of mass, and is identical to Newton's law in the case of a single particle.
* **Angular-momentum law ($dL/dt = \tau_{ext}$).** This law is especially useful to deal with rotational motion.
* **Energy law ($\Delta E = W_{oth}$).** This law is especially useful if one is interested in relating speeds and positions, without being explicitly interested in the time.

Problem-solving method. Our increased knowledge of mechanics allows us now to deal with a large range of more complex and interesting problems. The general problem-solving method discussed in Chapter 11 is still applicable. As usual, the solution of such a problem requires appropriate choices to find useful relations and then to combine them. (See Fig. D-1.) However, a useful relation about motion and interaction can now be obtained by applying any of the preceding *three* mechanics laws. One may thus need to choose among these three available options. But this choice is ordinarily easy, especially if one keeps in mind the conditions (mentioned above) under which these laws are especially useful.

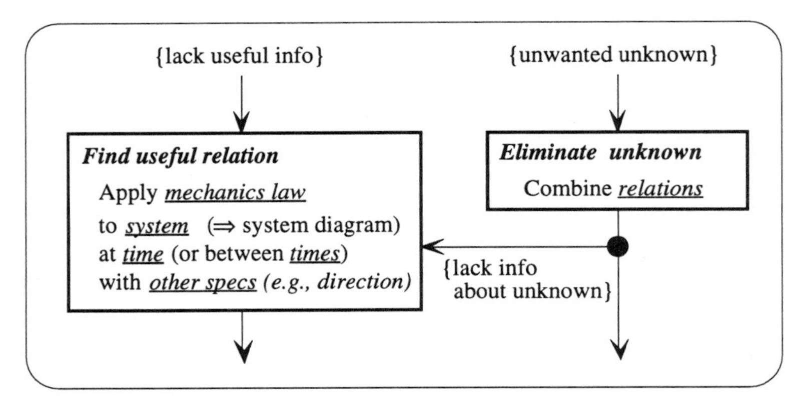

Fig. D-1. Choosing subproblems for solving a mechanics problem.

A system must always be adequately described before a mechanics law can be applied to it. As usual, it is thus important to draw a system diagram which summarizes all pertinent information about the motion and interactions of the system.

Prototype problem: Weight descending from a pulley

The following problem (similar to the one considered in Sec. 17C) illustrates the application of several mechanics laws to solve a problem involving both rotational and linear motion.

Problem statement. A block, of mass m, is attached to a thin string which is wrapped around a pulley free to rotate around a fixed horizontal axle. This pulley has a radius R and a moment of inertia I about its axle. All friction forces are negligible.

(a) What is the acceleration of the descending block?

(b) What is this acceleration if the pulley is a solid wheel of mass M?

Solution. The following illustrates the solution of this problem.

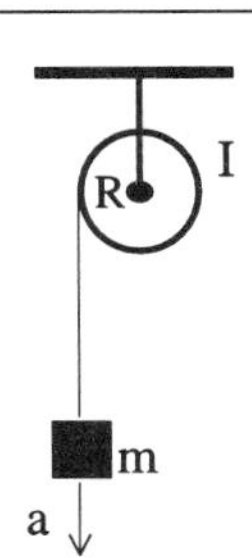

Known: m, R, I.

Goals:
(a) a = ?
(b) a = ? (for solid pulley of mass M)

Analysis of problem

Construction of solution

Motion of block.

Apply Newton's law, to block, downward

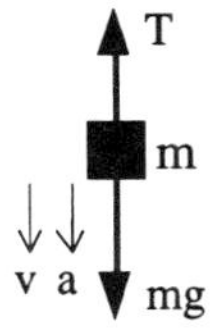

$$m\vec{a} = \vec{F}_{tot}$$

$$m\underline{\underline{a}} = mg - \underline{T} \quad (1)$$

Motion of pulley.

(Find T.) Apply ang.-mom. law, to pulley, about center, counter-clockwise

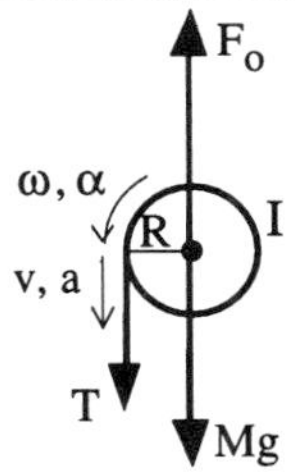

$$I\alpha = \tau_{ext}$$

$$I\underline{\alpha} = R\underline{T} \quad (2)$$

The force F_o on the pulley axle by the ceiling, and the gravitational force acting on the pulley center, both have zero torques since they act at the rotation axis.

Also: $a = R\alpha$ (3)

The tangential acceleration of the pulley at the contact point of the string is the same as the acceleration of the block.

By (2) and (3):

$$Ia = R^2T \quad (4)$$

Eliminate T. By (4) and (1):

$$T = \left(\frac{I}{R^2}\right) a$$

$$ma = mg - \left(\frac{I}{R^2}\right) a$$

$$\left(1 + \frac{I}{mR^2}\right) a = g$$

$$\boxed{a = \frac{g}{1 + \dfrac{I}{mR^2}}} \quad (5)$$

For solid pulley of mass M, $I = \frac{1}{2}MR^2$

$$\therefore \quad \boxed{a = \frac{g}{1 + \dfrac{M}{2m}}} \quad (6)$$

Checks (done in writing or mentally)

Checks

Units of (5): Same as g. OK

Check extreme cases.

If I or M large: Expect small acceleration of pulley particles, hence small a. Eq. (5) OK.

➔ ***Go to Sec. 18D of the Workbook.***

E. Conservation of angular momentum

Rotation with no external torques

The angular-momentum law (B-1)

$$\frac{dL}{dt} = \tau_{ext} \tag{E-1}$$

relates the rate of change of a system's angular momentum to the total external torque on it. If the total external torque on a system is zero, the angular momentum of the system remains, therefore, unchanged, i.e.,

$$\text{if } \tau_{ext} = 0, \qquad L = \text{constant}. \tag{E-2}$$

This conclusion is analogous to the one discussed in Chapter 15 (i.e., that the linear momentum $\vec{P}$ of a system is conserved if the total external force on it is zero).

Thus one says that the system's angular momentum is conserved.

Rotation with constant moment of inertia. The angular momentum is proportional to the angular velocity, i.e.,

$$L = I\omega\,. \tag{E-3}$$

If an object's angular momentum remains unchanged, then its angular velocity must remain unchanged *if* its moment of inertia is constant (as would be the case for a rigid object).

For example, consider a grinding wheel rotating around its axle after the driving motor has been switched off. If friction forces on the axle are negligibly small, the external torque on the wheel is zero. In this case the wheel would then continue rotating with constant angular velocity.

Rotation with changing moment of inertia. As pointed out in Sec. A, the angular-momentum law is also valid even if the moment of inertia is not constant. If the total external torque is zero, the constancy of the angular momentum in (E-3) then implies only that the *product* $I\omega$ must remain constant. Hence the angular velocity can change if the moment of inertia changes.

For example, consider a ballet dancer pivoting on her toes (or ice-skater pivoting on her skates). The external torques exerted on the dancer's feet by friction forces are then quite small, so that the dancer's angular momentum remains nearly constant. Consider the dancer initially spinning with her arms outstretched (as indicated in Fig. E-1a) so that her moment of inertia is large. Suppose that the dancer now pulls her arms closer to her body, thus decreasing her moment of inertia. Then her angular velocity must correspondingly increase to keep her angular momentum the same. In this way dancers can achieve dizzying rates of spin.

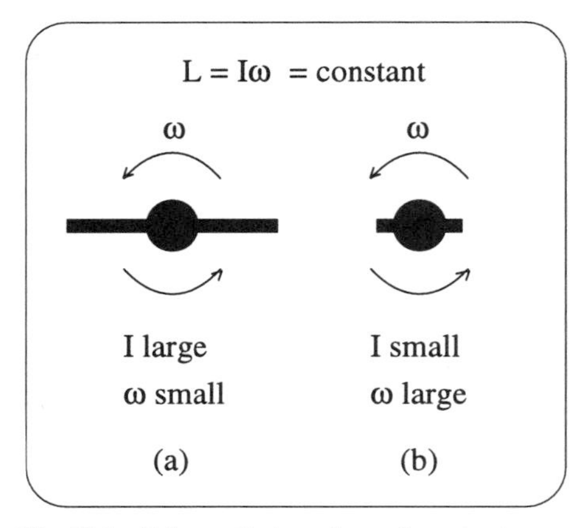

Fig. E-1. Schematic top view of a spinning dancer moving her arms toward her body.

> ***Example: Platform rotation produced by a walking person***
>
> A small merry-go-round consists of a circular horizontal platform, of mass M and radius R, which is free to rotate with negligible friction about a vertical axis through its center. (See Fig. E-2.) A woman, of mass m, stands initially near the edge of the stationary platform. What happens to the platform when the woman starts walking along the rim of the platform with a speed u relative to it?

Since friction forces are negligible, there is no external torque on the system consisting of the platform and woman. Hence the total angular momentum of this system must remain constant, equal to its initial zero value when everything was at rest relative to the ground. When the woman starts walking along the rim, she acquires some angular momentum L_W. To keep the total angular momentum L constant, the platform must thus start rotating so as to acquire an angular momentum L_P of equal magnitude in the opposite sense.

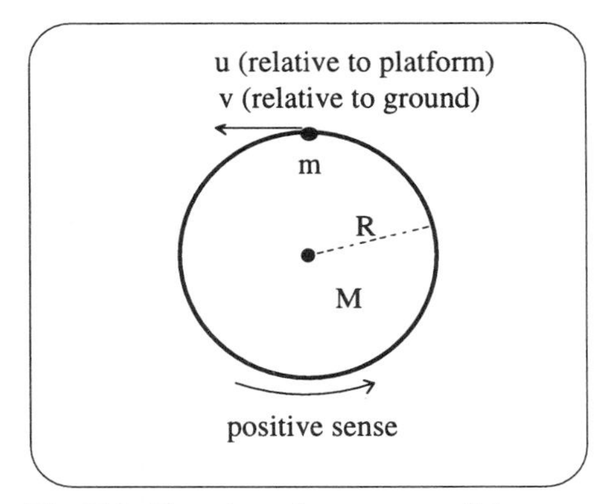

Fig. E-2. Top view of a woman walking along the edge of a platform.

The final angular velocity ω of the platform can be found by noting that the sum of the final angular momenta must be equal to the zero initial angular momentum, i.e., that

$$L_P + L_W = 0. \qquad \text{(E-4)}$$

But the final angular momentum of the platform is

$$L_P = I\omega = \frac{1}{2} MR^2 \omega \qquad \text{(E–5)}$$

where we have used the result (17D-7) for the moment of inertia of a solid disk. Similarly, if the woman walks along the edge of the platform with a speed v *relative to the ground*, her angular velocity relative to the ground is v/R and her angular momentum is

$$L_W = (mR^2)\frac{v}{R} = mRv = mR(u + R\omega)\,. \qquad \text{(E-6)}$$

Here we have used the fact that the woman's speed v relative to the ground is equal to ⟨her speed u relative to the edge of the platform⟩ plus ⟨the speed Rω of the edge of the platform relative to the ground⟩.

Hence (E-4) implies that

$$\frac{1}{2}MR^2\omega + mR(u + R\omega) = 0$$

or $$\left(\frac{1}{2}M + m\right) R^2\omega + mRu = 0\,.$$

Thus $$\omega = -\left(\frac{m}{\frac{1}{2}M + m}\right)\left(\frac{u}{R}\right) \qquad \text{(E-7)}$$

Here the minus sign indicates that the angular velocity of the platform has a sense opposite to the sense in which the woman is walking. As expected, (E-7) indicates that the angular velocity acquired by the platform will be larger if the woman's mass m is larger.

Note that the preceding angular-momentum argument needed to consider only the initial and final situations, and required no knowledge whatever of how the woman walked (e.g., whether she came up to speed slowly or rapidly, or what forces she exerted on the platform while walking).

Short collisions

Total angular momentum change. The angular-momentum law (E-1) allows one to find how the angular momentum of a system changes during any infinitesimal time interval dt, i.e.,

$$dL = \tau_{ext}\, dt\,. \qquad \text{(E-8)}$$

The entire change ΔL of the system's angular momentum during any longer time interval Δt can then be found by adding all the successive infinitesimal changes of angular momentum during this time Δt. Thus

$$\Delta L = \int \tau_{ext}\, dt = \tau_{ext,av}\, \Delta t \tag{E-9}$$

where we have defined the *average* external torque by

$$\tau_{ext,av} = \frac{1}{\Delta t} \int \tau_{ext}\, dt\,. \tag{E-10}$$

A knowledge of a system's angular-momentum change during any time interval thus provides information about the average external torque acting on the system during this time interval.

Total angular momentum change in a short collision. Irrespective of the detailed way in which the external torque on a system varies with time, the right side of (E-9) becomes negligibly small if the time interval Δt is sufficiently short and the external torque is not unduly large. For example, this is true for a system consisting of two objects colliding with each other during a very short time. Thus

$$\boxed{\text{in a short collision,} \qquad \Delta L \approx 0\,.} \tag{E-11}$$

During a very short time, the angular momentum of the system remains thus constant (to very good approximation) even if the external torque is not zero.

During a short collision, the mutual torques exerted on one object by another may, of course, be quite large so that the angular momentum of each changes appreciably. But such internal torques do not affect the system's *total* angular momentum which remains essentially unchanged.

All the preceding comments are completely analogous to those in Sec. 15C which pointed out that the *linear* momentum of a system remains essentially unchanged in a short collision.

Example: Person jumping onto a rotating platform.

A circular horizontal platform, of mass M and radius R, is rotating about its vertical central axis while gradually slowing down (because of frictional torques exerted on the axle by the bearings). At a particular instant when the platform rotates with an angular velocity ω, a person of mass m jumps vertically down onto the platform near its outer edge. What then is the angular velocity of the platform just after the person has landed on it?

The angular-momentum law can be applied to the system consisting of the platform and the person. Thus

$$\frac{dL}{dt} = \tau_{ext}$$

where the total external torque is that due to friction forces in the bearing. But, during the short time Δt of the person's impact with the platform,

$$\Delta L = L' - L = 0 \quad \text{or} \quad L' = L \tag{E-12}$$

to excellent approximation.

The angular momentum L of the system about the axis just *before* the impact is merely the angular momentum of the platform, i.e.,

$$L = I\omega = \frac{1}{2}MR^2\,\omega\,.$$

The angular momentum L' of the system just *after* the impact is that due to the platform and the person moving jointly with a final angular velocity ω'. Thus

$$L' = (I + mR^2)\,\omega' = (\frac{1}{2}MR^2 + mR^2)\,\omega' = (\frac{1}{2}M + m)R^2\omega'$$

since the person then contributes a moment of inertia mR^2 to the system. Thus (E-12) implies that

$$(\frac{1}{2}M + m)R^2\omega' = \frac{1}{2}MR^2\,\omega$$

or

$$\omega' = \left(\frac{M}{M + 2m}\right)\omega. \qquad \text{(E-13)}$$

➜ ***Go to Sec. 18E of the Workbook.***

F. Summary

Angular momentum of a particle: $\ell = mr^2\omega = mrv_{\perp r} = \pm\, rp_{\perp r}$

Angular momentum of a system: $L = \sum \ell_n = I\omega$

Angular-momentum law: $\boxed{\dfrac{dL}{dt} = \tau_{ext}}$

If I constant, $I\alpha = \tau_{ext}$

New abilities

You should now be able to do the following:

(1) Solve mechanics problems by using jointly the momentum law (which includes Newton's law as a special case), the angular-momentum law, and the energy law.

(2) In particular, apply the angular-momentum law in special situations where the angular momentum is conserved.

Grand summary: Mechanics laws for any system

The following chart summarizes all the mechanics laws discussed in this book. (See also Appendices E and F.) They are sufficient to deal with a wide variety of mechanics problems.

Each law relates the motion of the system to its interactions. (All laws are derivable from Newton's law for a particle.)

Momentum law

* *Relates motion and external interactions.*
* *Relates velocities at any two instants if total external force is zero.*

$$\frac{d\vec{P}}{dt} = \vec{F}_{ext}$$

$m\,\vec{a} = \vec{F}_{tot}$ (for a particle) {*Newton's law*}

$M\,\vec{A}_c = \vec{F}_{ext}$ (for center of mass)

Angular-momentum law

(for fixed-direction axis)

* *Relates rotational motion and external interactions.*
* *Relates ang. velocities at any two instants if total external torque is zero.*

$$\frac{dL}{dt} = \tau_{ext}$$

$I\,\alpha = \tau_{ext}$ (if I is constant)

Energy law

* *Relates speeds and positions at any two instants (without mention of time).*

$$\Delta E = W_{oth}$$

➜ ***Go to Sec. 18F of the Workbook.***

19 Rolling and Equilibrium

Our previous knowledge about linear and rotational motions is sufficient to deal with a very broad range of practical applications. The present chapter discusses briefly two such kinds of applications. The first of these deals with objects that rotate about *moving* axes (e.g., with the rolling wheels of moving vehicles). The second deals with systems that remain at rest (e.g., with bridges, houses, and other structures designed so that they do *not* move).

A. General motion of a rigid object

Many objects are approximately rigid, i.e., they consist of parts that remain at fixed distances from each other. We now review briefly how to describe some simple motions of such objects, and then show how this descriptive knowledge can be extended to describe more complex motions.

As usual, these motions must be described relative to some chosen reference frame (e.g., some inertial frame).

Translation. The simplest motion of a rigid object is a *translation* (i.e., a motion where all parts of the object move along parallel paths, as illustrated in Fig. A-1). All particles in the object move then, at any instant, with the same velocity and with the same acceleration. Accordingly, the motion of the object can be completely described by the motion of any one of its points (e.g., by the motion of its center of mass).

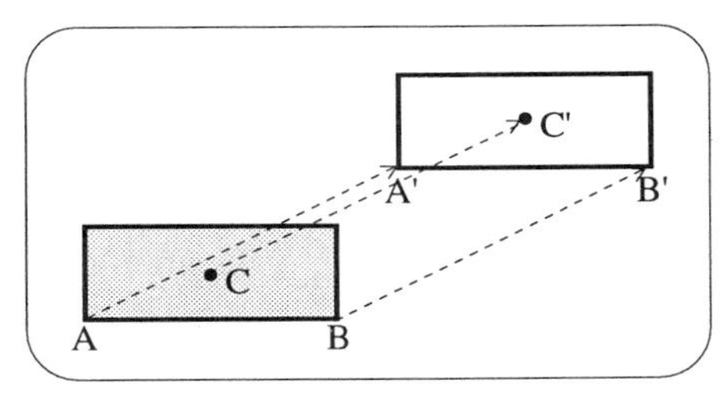

Fig. A-1. Translational motion of a rectangle. (The points A, B, C move to A', B', C'.)

Rotation about a fixed axis. Another simple motion of a rigid object is a *rotation* about a fixed axis (the kind of motion discussed in the preceding two chapters). As illustrated in Fig. A-2, all the particles located on the rotation axis remain then at rest, while all other particles move along circular paths centered around this axis.

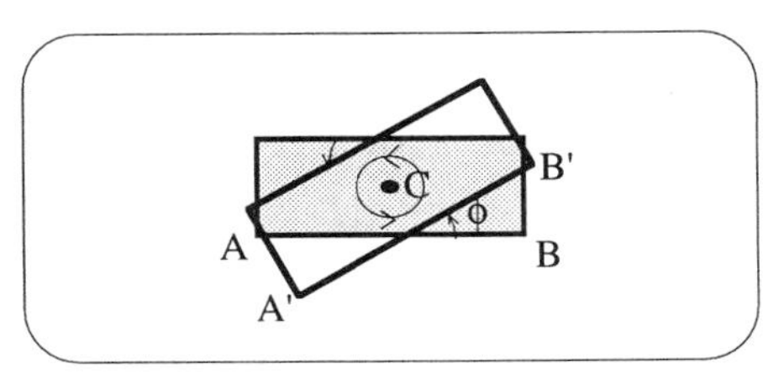

Fig. A-2. Rotational motion of a rectangle about a fixed axis through its center. (The points A and B move to A' and B'.)

General motion as rotation about a moving axis. Consider now the *general* motion of a rigid object in a plane. (For example, consider the motion of the rectangle in Fig. A-3 from its initial position where its lower side is the line AB, to its final position where its lower side is the line A'B'.) Such a motion can always be imagined to be produced by the following two successive operations:

(1) Translating the object, so that its orientation remains the same, but so that one of its points is in the desired final position.
(2) Rotating the object about this point, by the required angle, so that it attains its final desired orientation.

For example, in Fig. A-3a the rectangle is brought to its final position by first translating the rectangle so that its left corner A is brought to the desired final position A', and then rotating the rectangle through the angle ϕ so that it attains the final desired orientation. Alternatively, in Fig. A-3b the rectangle is brought to its final position by first translating the rectangle so that its center C is brought to the desired final position C', and then rotating the rectangle through the angle ϕ so that it attains the desired final orientation.

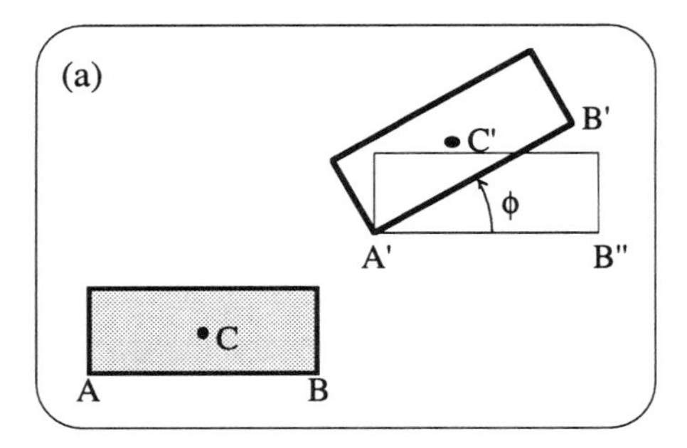

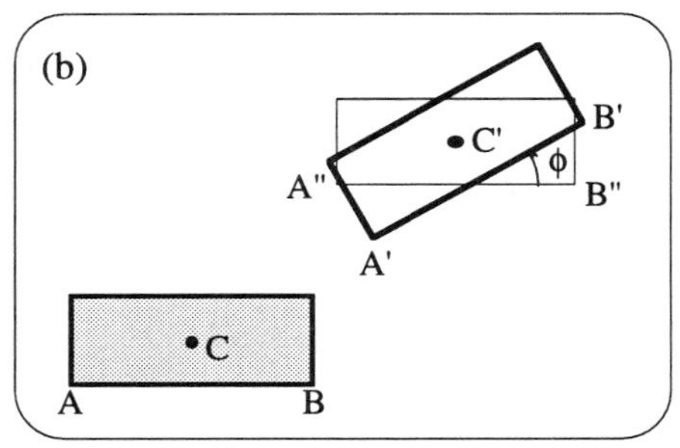

Fig. A-3. General motion of a rectangle.

Although the rotation axis may be moving, its *direction* remains unchanged if the object moves in a plane.

Note that it does not matter which point is first brought to its desired final position by a translation. Since this translation does not change the object's orientation, the angle by which the object needs to be rotated is always the same. This leads to the following conclusion:

The angular displacement of an object is the same, irrespective of the choice of axis of rotation.	(A-1)

Rate of motion. It is then quite easy to describe how fast any rigid object moves. One needs merely to describe the motion of any point in the object (e.g., by specifying this point's velocity and acceleration), and to describe the rotation of the object about an axis through this point (e.g., by specifying the object's angular velocity and angular acceleration). Furthermore, the conclusion (A-1) implies that the angular velocity or angular acceleration of an object does not depend on the chosen location of the axis of rotation.

Description in terms of center of mass. It is ordinarily most convenient to describe the general motion of a rigid object by specifying the motion of its center of mass, and its rotation about an axis passing through the center of mass. The reasons are the following:

(1) As discussed in Sec. 15D, the motion of the center of mass is specified by the momentum law which implies that its acceleration $\vec{A}_c$ is such that

$$M\vec{A}_c = \vec{F}_{ext}\,. \qquad \text{(A-2)}$$

(2) The angular-momentum law is true when applied about any axis fixed in an inertial frame. More generally, one can show that it is also true when applied about an axis passing through the center of mass, *no matter how the center of mass might be moving*, if the rotational motion is described relative to the center of mass. In other words,

$$\text{relative to center of mass, always,} \quad \frac{dL}{dt} = \tau_{ext}\,. \qquad \text{(A-3)}$$

The mechanics laws (A-2) and (A-3) can thus be applied to deal with *any* motion of a rigid object.

→ ***Go to Sec. 19A of the Workbook.***

B. Rolling motion

Rolling wheels are very common and practically important. (Indeed, the invention of vehicles with rolling wheels marked a major advance of our civilization.) Let us then examine more closely the kind of motion involved in rolling.

Characteristics of rolling motion

Fig. B-1a illustrates a circular wheel rolling on the ground (or any other surface). As the wheel rolls, successive points on its rim come into contact with the ground. For instance, at the instant shown in Fig. B-1a, the point A on the wheel's rim touches the point A' on the ground; at a slightly later time, the point B on the rim touches the point B' on the ground; at a slightly later time after that, the point C on the rim touches the point C' on the ground; and so forth.

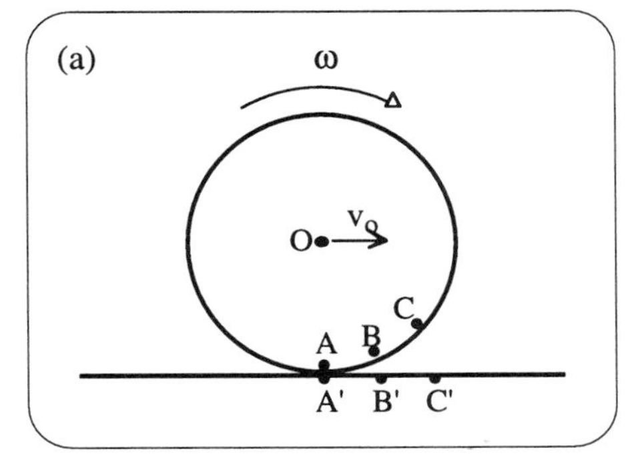

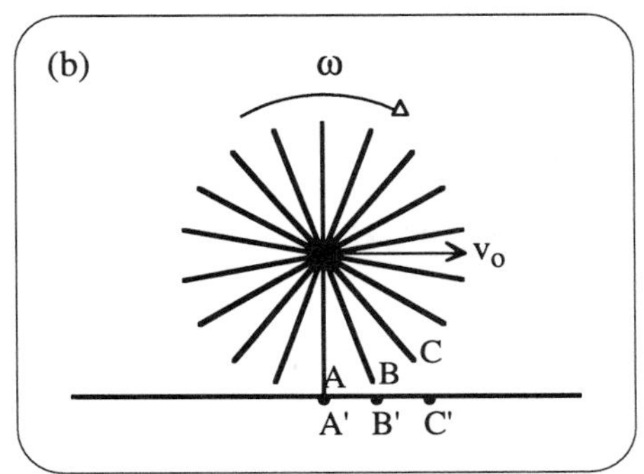

Fig. B-1. A rolling wheel. (a) Circular wheel. (b) Wheel consisting of spokes.

As a result of this kind of motion, the center O of the wheel moves relative to the ground with some speed v_O to the right. At the same time, the wheel rotates in a clockwise sense about an axis through this center with some angular velocity of magnitude ω. (The speed v_O of the wheel's center is, of course, the same as the speed of the vehicle to which the wheel's axle is attached.)

The rolling motion may perhaps be visualized more easily by considering a special wheel, like that illustrated in Fig. B-1b, which is not circular, but consists merely of many identical radial spokes which successively touch the ground. (As the number of such spokes becomes very large, this wheel becomes identical to the circular wheel in Fig. B-1a.)

Rolling without slipping

Zero relative velocity of contact point. In the simplest and most common case, rolling occurs without slipping between the wheel and the surface on which it rolls. In this case, the point on the wheel, in contact with the surface at any instant, does not move *relative to the surface*. Hence the velocity of the wheel's contact point is always the *same* as that of the touching surface. In particular, if the surface is that of the ground, the wheel's point of contact with the ground at any instant must have zero velocity relative to the ground. In other words, the wheel's contact point is always momentarily at rest relative to the ground.

Relation between central and angular speeds. When a wheel rolls without slipping, there must be a close relation between the speed v_O of its center and its angular speed ω (i.e., the magnitude of its angular velocity).

Fig. B-2a shows such a wheel, of radius R, rolling along the horizontal ground without slipping. The following statements describe this situation:

(1) The center O of the wheel has, relative to the ground, some velocity of magnitude v_O to the right.

(2) The wheel rotates about O clockwise with some angular speed ω. Hence the lowest point A of the wheel has, relative to its center O, a velocity $R\omega$ to the left.

(3) The lowest point A of the wheel is in contact with the ground. Hence this point must have zero velocity relative to the ground if the wheel rolls without slipping.

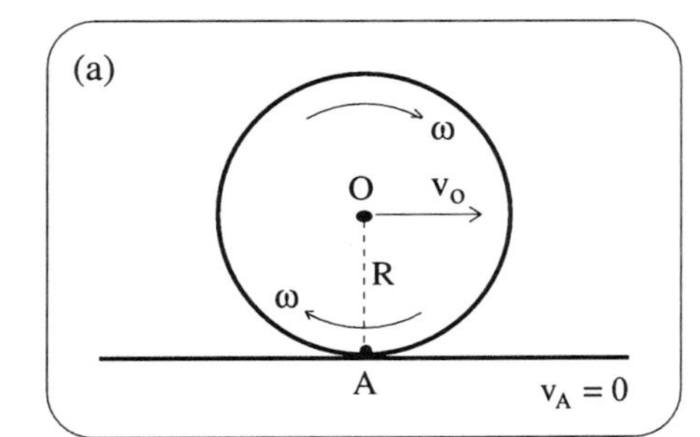

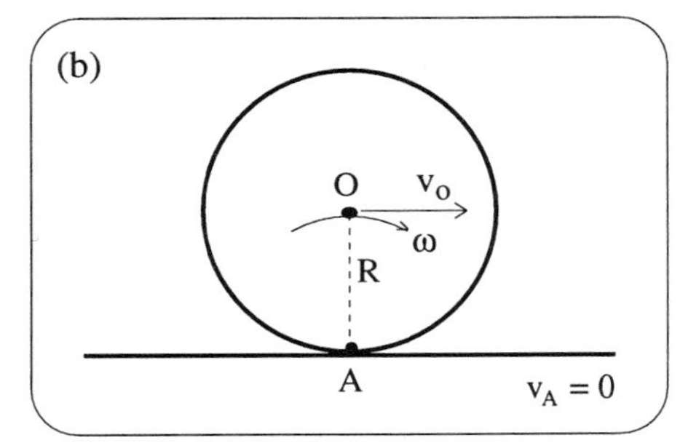

Fig. B-2. A wheel rolling without slipping. (a) Viewed as rotating about its center O. (b) Viewed as rotating about its contact point A.

But the general relation (8E-4) between relative velocities implies that ⟨the velocity of the lowest point A of the wheel relative to the ground⟩ must be equal to ⟨the velocity of A relative to the center O of the wheel⟩ plus ⟨the velocity of O relative to the ground⟩. Considering velocity components to the right, this statement implies the following relation

$$0 = v_O - R\omega$$

or

$$\boxed{v_O = R\omega\,.} \tag{B-1}$$

Alternative description leading to the preceding relation. Alternatively, one may describe the motion of the wheel by considering that it rotates, at any instant, about its stationary contact point with the ground. By (A-1), its angular velocity about this contact point is also ω clockwise, as indicated in Fig. B-2b. As result of this rotation, the center O of the wheel should then be moving to the right (relative to the ground) with a speed

$$v_O = R\omega\,. \tag{B-2}$$

The result obtained from this alternate point of view thus agrees properly with the previous result (B-1).

Corresponding relation for accelerations. The relation (B-1) implies a corresponding relation for the rates of change of these quantities. Thus the magnitude a_O of the acceleration of the wheel's center of mass is related to its angular acceleration α so that

$$a_O = R\alpha\,. \tag{B-3}$$

Zero work done by the friction force. To prevent a wheel from slipping along the surface on which it is rolling, a friction force must ordinarily be exerted on the wheel by this surface. This friction force acts on the wheel at the point where it touches the surface. But, if the wheel does not slip, this contact point does not move. Hence *this friction force does no work on the wheel.*

All the preceding arguments are based upon the assumption that the wheel and touching surface are rigid. However, rigidity is always only an approximation since real objects tend to get deformed. Indeed, small frictional effects do arise if deformations of the wheel and the touching surface are taken into account.

Rolling (unlike sliding or the repeated impacts involved in walking) thus does not lead to a transformation of macroscopic energy into random thermal energy. This is why rolling wheels provide very efficient means of transportation.

➜ ***Go to Sec. 19B of the Workbook.***

C. Problems involving rolling motion

The following typical problems illustrate how the mechanics laws can be used to solve problems involving rolling objects.

Acceleration of a rolling cylinder: Method I

Problem statement. A uniform solid cylinder rolls, without slipping, down along an inclined plane making an angle θ with the horizontal. With what acceleration does the cylinder's center move?

Solution. The following illustrates one method of solving this problem. (This method describes the cylinder's motion by considering the linear motion of its center, and its rotational motion about this center.)

Analysis of problem

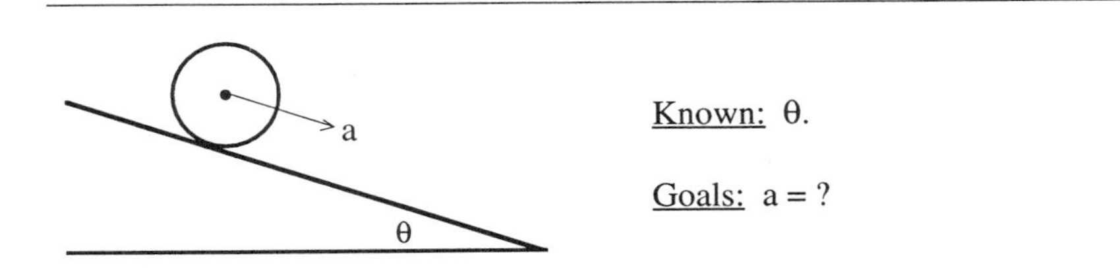

Known: θ.

Goals: a = ?

Construction of solution

M = mass of cylinder
R = radius of cylinder

Apply momentum law, to center of mass, down along plane

$$M\vec{A} = \vec{F}_{ext}$$

$$M\underline{\underline{a}} = Mg\sin\theta - \underline{f} \qquad (1)$$

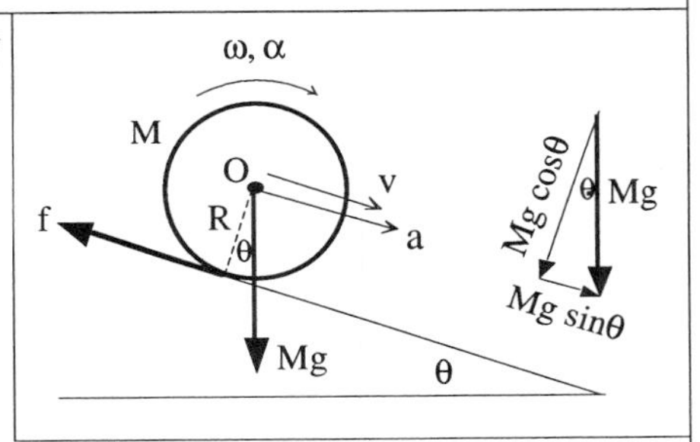

Friction force must exist to prevent slipping of the rolling cylinder.

Underline wanted unknown twice, unwanted unknown once.

(Find f.) Apply angular-momentum law, about O, clockwise

Since the cylinder rolls without slipping, $v = R\omega$ and $a = R\alpha$.

$$I\alpha = \tau_{ext}$$

I is the moment of inertia about the center of mass O of the cylinder.

$$\left(\frac{1}{2}MR^2\right)\left(\frac{a}{R}\right) = Rf$$

Since the gravitational force acts at the center of mass O, its torque about O is zero. Hence only the friction force contributes an external torque about O.

$$\frac{1}{2}M\underline{\underline{a}} = \underline{f} \qquad (2)$$

Eliminate f. Put (2) into (1):

$$Ma = Mg\sin\theta - \frac{1}{2}Ma$$

$$\frac{3}{2}Ma = Mg\sin\theta$$

$$\boxed{a = \frac{2}{3}g\sin\theta} \qquad (3)$$

Note that this result for the acceleration does not depend on the cylinder's mass or radius.

Checks (done in writing or mentally)

Checks

Units of (3): Same as g, i.e., m/s^2. OK

Check extreme cases.

If g or θ large: Expect large acceleration. Eq. (3) OK.

Dependence on M and R:

Possible dependence of a on the following: g, θ, M, R.

Units of these: a [m/s^2]; g [m/s^2], θ [none], M [kg], R [m].

Can get consistent units for a only if it does not depend on M or R.

Self-consistency check: There is no combination of g, θ, M, and R which would yield the correct units for the acceleration a if this acceleration would also depend on M and R.

Acceleration of a rolling cylinder: Method II

Alternative solution. The following illustrates an alternative method of solving the same problem. (This method describes the cylinder's motion by considering it to be rotating about its momentarily stationary contact point with the inclined plane.)

Analysis of problem

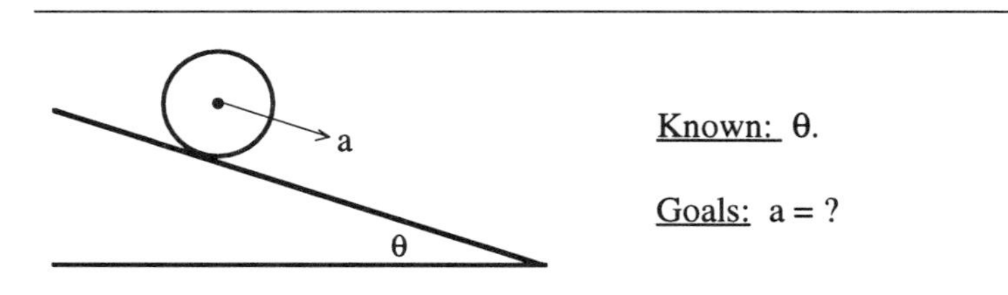

Known: θ.

Goals: a = ?

Construction of solution

Describe the cylinder's motion as rotation about the momentarily stationary contact point P. Apply the angular-momentum law about P.

Apply ang.- mom. law, to cylinder, about contact point P, clockwise:

$$I\alpha = \tau_{ext}$$

$$I_P\, \alpha = R\, Mg \sin\theta \qquad (1)$$

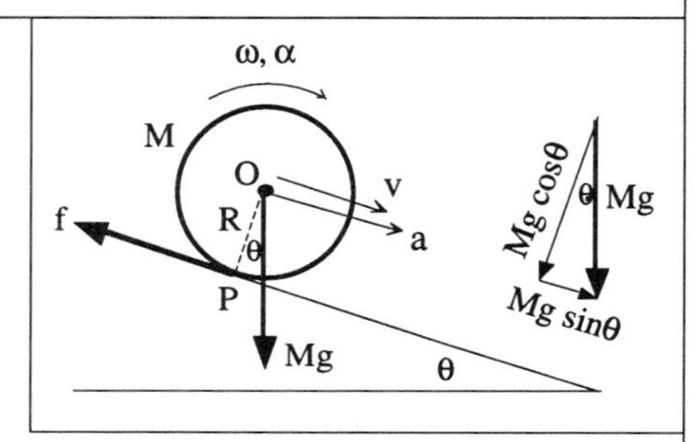

I_P is the moment of inertia about P.

The friction force exerts no torque about the point P, only the gravity force does.

By parallel-axes relation

$$I_P = MR^2 + \tfrac{1}{2}MR^2 = \tfrac{3}{2}MR^2 \qquad (2)$$

Relate I_P to the known moment of inertia $MR^2/2$ about the center of the cylinder.

Put (2) into (1):

$$\left(\tfrac{3}{2}MR^2\right)\left(\tfrac{a}{R}\right) = R\, Mg \sin\theta \qquad (3)$$

Use the fact that a = Rα if the cylinder rolls without slipping.

$$\tfrac{3}{2}MR\, a = R\, Mg \sin\theta$$

$$\boxed{a = \tfrac{2}{3}\, g \sin\theta} \qquad (4)$$

This result is the same as that obtained by the first method.

Speed of a rolling cylinder

Problem statement. After starting from rest, a uniform solid cylinder rolls, without slipping, down along an inclined plane making an angle θ with the horizontal. What is the speed of the cylinder's center after it has descended a distance D along the plane?

Solution. This problem asks about the *speed* attained by the cylinder, but is otherwise similar to the previous problem which asked about the *acceleration* of the cylinder. Indeed, using the previous result found for the acceleration, we could find the time required by the cylinder to descend the distance D, and then find the final speed attained by the cylinder after this time. However, since this problem asks us merely to relate speeds and positions, it is easiest to apply the energy law. Furthermore, it affords a good opportunity to illustrate how the energy law can be applied to a problem involving rolling motion.

Analysis of problem

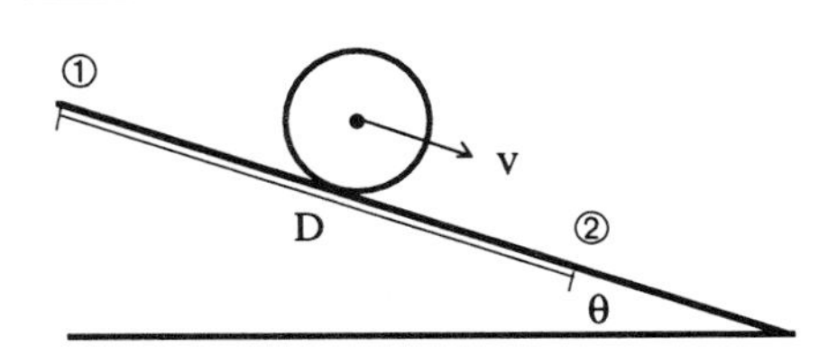

①: $v = 0$

②: $v = v'$

Known: θ, D.

Goals: $v' = ?$

Construction of solution

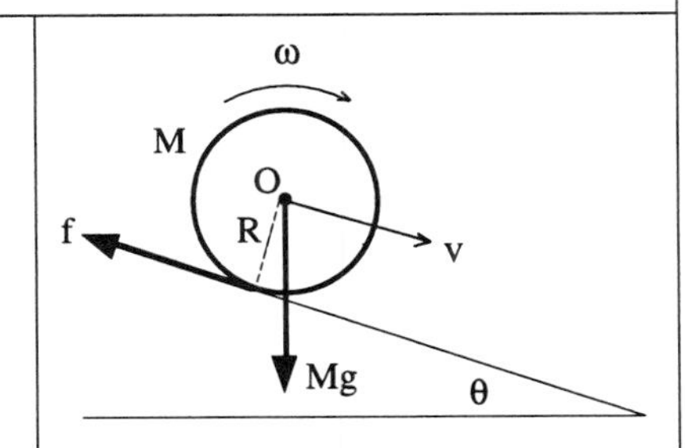

Apply energy law, to cylinder, ① → ②:

$$\Delta E = W_{oth}$$

$$\Delta K + \Delta U_g = 0 \qquad (1)$$

U_g is the potential energy due to the gravity force. Only other force is the friction force which does no work on the cylinder since the contact point does not move.

$$\Delta K = [\tfrac{1}{2}Mv'^2 + \tfrac{1}{2}I_o\omega'^2] - 0$$

$$= \tfrac{1}{2}Mv'^2 + \tfrac{1}{2}(\tfrac{1}{2}MR^2)\left(\frac{v'}{R}\right)^2 = \tfrac{3}{4}Mv'^2 \qquad (2)$$

The kinetic energy is that of the center of mass plus that of rotation about the center of mass.

v' and ω' are the cylinder's final linear and angular speeds. I_o is its moment of inertia about O.

$$\Delta U_g = Mg\,\Delta h = Mg\,(-D\sin\theta) = -MgD\sin\theta \qquad (3)$$

The center of mass of the cylinder descends a vertical distance D sinθ.

Put (2) & (3) into (1):

$$\tfrac{3}{4}Mv'^2 - MgD\sin\theta = 0$$

$$\tfrac{3}{4}v'^2 = gD\sin\theta$$

$$\boxed{v' = \sqrt{\tfrac{4}{3}gD\sin\theta}} \qquad (4)$$

This result for the cylinder's speed again does not depend on its mass or radius.

Checks (done in writing or mentally)

Checks

Units of (4): $\frac{m}{s} = \sqrt{\left(\frac{m}{s^2}\right)m} = \frac{m}{s}$ OK

If g or θ large: Expect large speed. Eq. (4) OK.

Check extreme cases.

→ Go to Sec. 19C of the Workbook.

D. Static equilibrium

Equilibrium conditions

Definition of equilibrium. A system is said to be in *equilibrium* (or *static equilibrium*) relative to some reference frame if it remains at rest relative to this frame. Many systems (like houses, bridges, ...) must be appropriately designed so that they don't fall down but remain in equilibrium. Thus it is important for civil engineers, and many people in everyday life, to know what conditions must prevail so that a system does remain in equilibrium.

Equilibrium conditions. Consider any system which is in equilibrium relative to the earth or some other inertial frame. The conditions needed to ensure that this system remains in equilibrium are then immediately apparent from the familiar mechanics laws.

The momentum law states that $d\vec{P}/dt = \vec{F}_{ext}$. But, if the system remains at rest, its momentum $\vec{P}$ does not change (e.g., its center of mass does not move). Hence,

$$\boxed{\text{by momentum law,} \qquad \vec{F}_{ext} = 0\,,} \qquad \text{(D-1)}$$

i.e., the total external force on the system must be zero.

Similarly, the angular momentum law states that $dL/dt = \tau_{ext}$. But, if the system remains at rest, its angular momentum L does not change (e.g., the system does not rotate). Hence,

$$\boxed{\text{by angular-momentum law,} \quad \tau_{ext} = 0\,,} \qquad \text{(D-2)}$$

i.e., the total external torque on the system must be zero about any axis.

In the case of a rigid object, the conditions (D-1) and (D-2) are sufficient to determine all the conditions that must prevail to ensure that the object remains in equilibrium (i.e., to determine what forces must be exerted at what points so that the object remains at rest).

> ***Application of the equilibrium conditions***
>
> The angular-momentum condition (D-2) can be applied about any convenient axis. (Application of this condition to any other axis does not yield any new results.) Hence it is often simplest to apply it about an axis located at a point where complex forces act (since these forces don't produce any torques about this axis).

Example: Ladder leaning against a wall

The following is a prototypical problem illustrating how the momentum and angular-momentum laws [i.e., the equilibrium conditions (D-1) and (D-2)] can be used to solve problems involving the equilibrium of rigid objects.

Problem statement. A ladder leans against a vertical wall. The coefficient of static friction between the ladder and the horizontal ground is μ, but friction forces between the ladder and the wall are negligible. What then is the minimum value of the angle of the ladder from the horizontal so that the ladder remains stationary without slipping relative to the ground?

Solution. The following illustrates the solution of this problem.

Analysis of problem

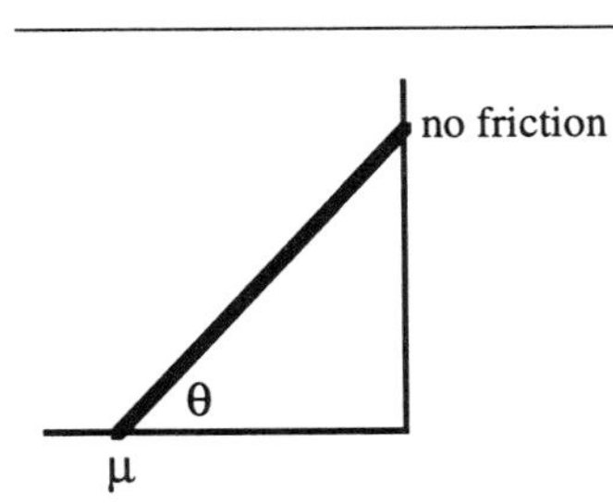

Ladder at rest.

Known: μ.

Goals: Minimum $\theta = ?$

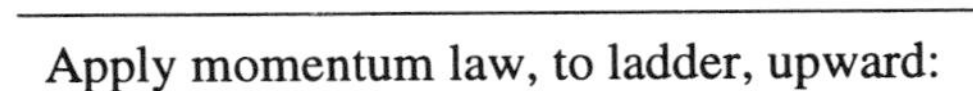

Construction of solution

Apply momentum law, to ladder, upward:

$$\vec{F}_{ext} = 0$$

$$N - Mg = 0$$

$$N = Mg \qquad (1)$$

Apply same, to right:

$$f - N' = 0$$

$$f = N' \qquad (2)$$

M = mass of ladder
L = length of ladder

The system diagram indicates all the forces on the ladder. The gravity force is shown acting at the center of mass at the midpoint of the ladder (since this represents properly the total extenal force and torque due to the gravitty forces on all particles of the ladder).

Friction force

$$f \leq \mu N \qquad (3)$$

Put (1) & (2) into (3):

$$\underline{N'} \leq \mu\, Mg \qquad (4)$$

(Find N'.) Apply angular-momentum law, to ladder, about A, clockwise

$$\tau_{ext} = 0$$

$$\left(\frac{L}{2}\right) \cos\theta\, Mg - L \sin\theta\, N' = 0$$

$$N' = \frac{1}{2} Mg \frac{\cos\theta}{\sin\theta} = \frac{Mg}{2 \tan\theta} \qquad (5)$$

It is convenient to apply the law about the point A since the two unknown forces acting there do not contribute to the torques about this point.

Torques about A are exerted by the gravity force and by the normal force N'.

The lever arm (i.e., the distance from the axis, perpendicular to the force) is (L/2) cosθ for the gravity force Mg, and L sinθ for the force N'.

Eliminate N'. Put (5) into (4):

$$\frac{Mg}{2 \tan\theta} \leq \mu\, Mg$$

$$\boxed{\tan\theta \geq \frac{1}{2\mu}} \qquad (6)$$

or

$$\boxed{\theta_{min} = \arctan\left(\frac{1}{2\mu}\right)} \qquad (7)$$

Note that this result does not depend on M, L, or g.

Checks (done in writing or mentally)

<u>Checks</u>

Units of (6): OK (no units on either side).

If μ small: Expect that ladder must be nearly vertical (i.e., θ must be large) if ladder is not to slip. Eq. (6) is consistent. OK

Check extreme case.

Dependence on g, mass M, and length L:

Possible dependence of θ on following: μ, g, M, L.

Units of these: θ [none]; μ [none], g [m/s^2], M [kg], L [m].

Can get consistent units for θ only if it does not depend on g, M, or L.

Self-consistency check: There is no combination of g, M, and L which would yield an angle (which has no units).

→ ***Go to Sec. 19D of the Workbook.***

E. Summary

Description of general motion: Motion of center of mass & rotation about center of mass.

Rolling without slipping:

Central and angular speeds: $v_0 = R\omega$

Friction force at contact point: Does no work.

Equilibrium conditions:

By momentum law: $\vec{F}_{ext} = 0$

By angular-momentum law: $\tau_{ext} = 0$ (about any axis)

New abilities

You should now be able to do the following:

(1) Solve mechanics problems involving rolling objects.

(2) Solve problems involving simple systems in static equilibrium.

→ ***Go to Sec. 19E of the Workbook.***

Appendices

A. Basic mathematics

Exponents

$$x^{-a} = \frac{1}{x^a}, \qquad x^{a+b} = x^a x^b, \qquad x^{a-b} = \frac{x^a}{x^b}$$

$$\sqrt[n]{x} = x^{1/n}$$

Quadratic equation

If $ax^2 + bx + c = 0$,

$$x = \frac{-b \pm \sqrt{b^2 - 4ac}}{2a}$$ (See Fig. 1.)

Derivation:

$$x^2 + \frac{b}{a}x = -\frac{c}{a}$$

$$(x + \frac{b}{2a})^2 = \frac{b^2}{4a^2} - \frac{c}{a} = \frac{b^2 - 4ac}{4a^2}$$

$$x + \frac{b}{2a} = \pm\frac{\sqrt{b^2 - 4ac}}{2a}$$

Fig. 1.

Angles

Angles with mutually perpendicular sides are equal. (See Fig. 2.)

Angle (in radians): $\theta = \frac{s}{r} = \frac{\text{arc length}}{\text{radius}}$ (See Fig. 3.)

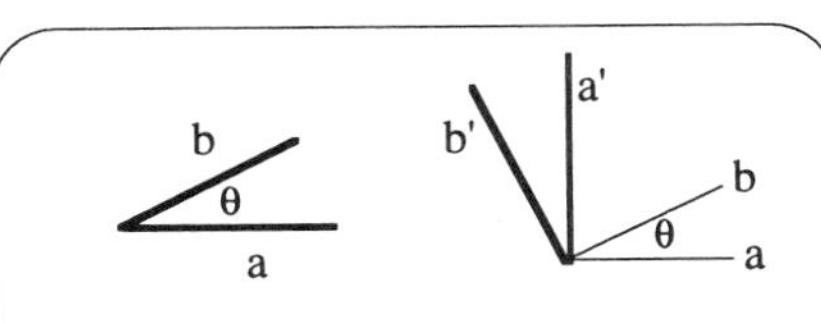

$\angle(b, a) = \theta.$ $a' \perp a,$ $b' \perp b$

$\therefore \angle(a', a) = 90°,$ $\angle(b', a) = 90° + \theta$

$\therefore \angle(b', a') = \angle(b', a) - \angle(a', a) = \theta$

Fig. 2.

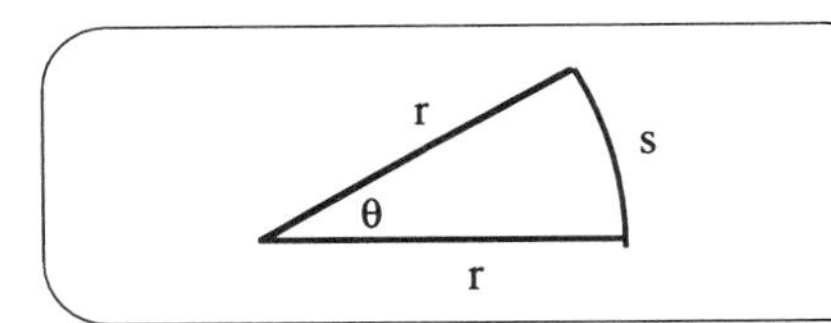

Fig. 3.

Trigonometry definitions

(See Fig. 4.)

$$\sin\theta = \frac{b}{c} = \frac{\text{opposite side}}{\text{hypotenuse}}$$

$$\cos\theta = \frac{a}{c} = \frac{\text{adjacent side}}{\text{hypotenuse}}$$

$$\tan\theta = \frac{b}{a} = \frac{\text{opposite side}}{\text{adjacent side}} = \frac{\sin\theta}{\cos\theta}$$

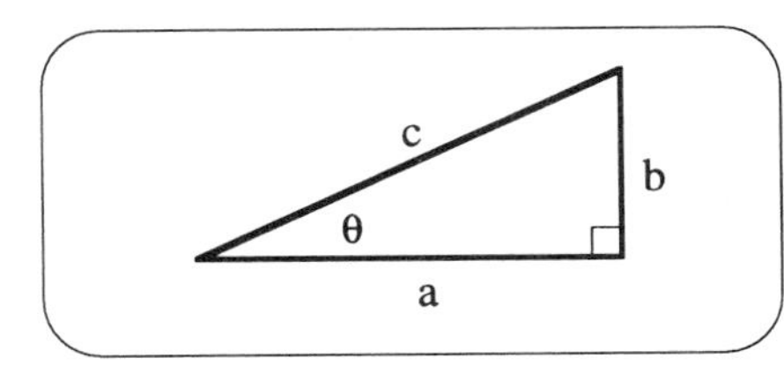

Fig. 4.

If $\theta \approx 0$, $\sin\theta = \tan\theta \approx \theta$, $\cos\theta \approx 1$, (See Fig. 5.)

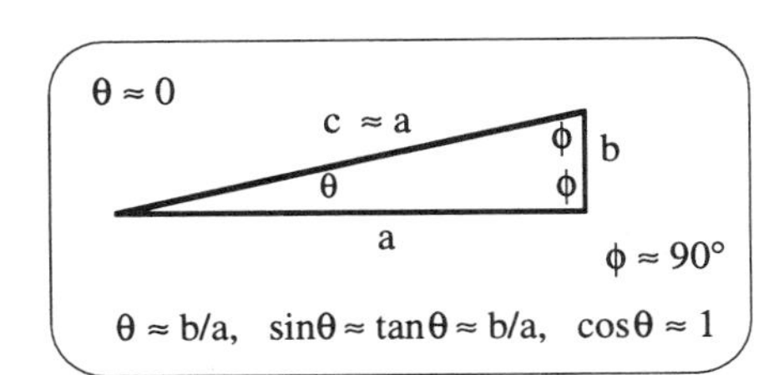

Fig. 5.

Right triangle

Pythagorean theorem: $a^2 + b^2 = c^2$ (See Fig. 6.)

Implication: $\cos^2\theta + \sin^2\theta = 1$

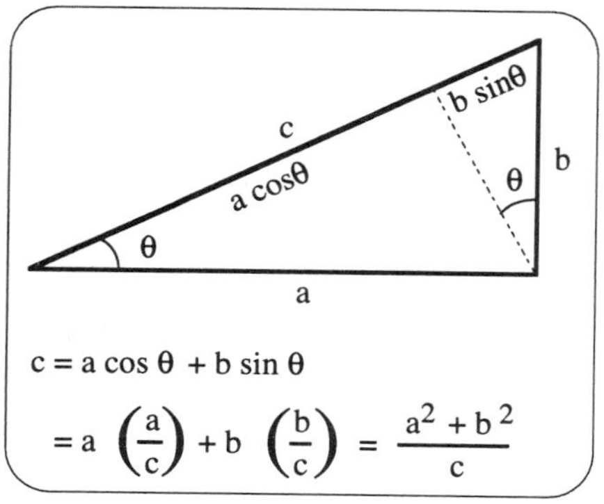

Any triangle

Law of sines: $a \sin\theta = c \sin\phi$ (See Fig. 7.)

Area: $A = \frac{1}{2} b\, h$ (b = base, h = height)

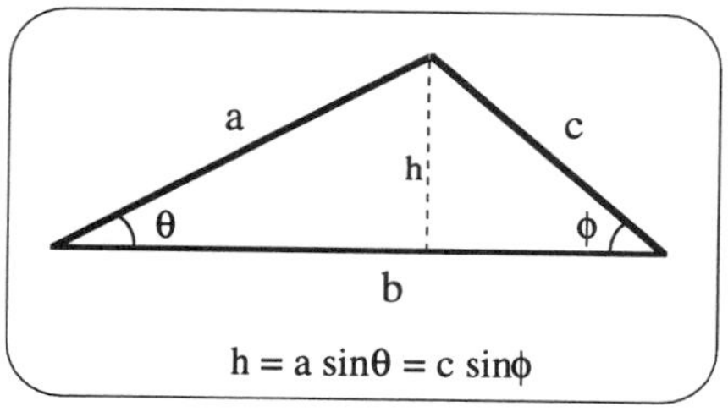

Circle

Circumference: $C = 2\pi r$ (r = radius)

Definition of π: $\pi = \frac{C}{D}$ ($\pi \approx 3.14$) (D = 2r = diameter)

Area: $A = \pi r^2$ (See Fig. 8.)

Vectors

Vector sum: $\vec{S} = \vec{A} + \vec{B}$ (See Fig. 9.)

Vector difference: $\vec{D} = \vec{A} - \vec{B}$ (See Fig. 10)

Component: $A_i = A \cos\theta$ (See Fig. 11.)

Dot product: $\vec{A} \cdot \vec{B}$ (See Fig. 12.)

$$\vec{A} \cdot \vec{B} = A\, B \cos\theta = A\, B_A = B\, A_B$$

$$\vec{A} \cdot \vec{B} = \vec{B} \cdot \vec{A}$$

$$\vec{A} \cdot (\vec{B} + \vec{C}) = \vec{A} \cdot \vec{B} + \vec{A} \cdot \vec{C}$$

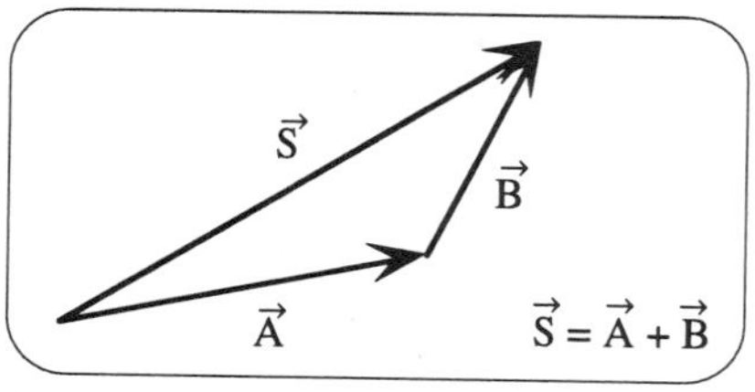

$c = a \cos\theta + b \sin\theta$

$= a\left(\frac{a}{c}\right) + b\left(\frac{b}{c}\right) = \frac{a^2 + b^2}{c}$

Fig. 6.

$h = a \sin\theta = c \sin\phi$

Fig. 7.

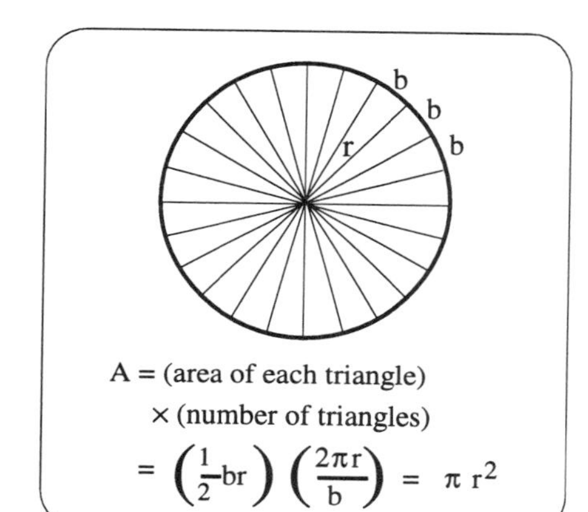

A = (area of each triangle)
× (number of triangles)
$= \left(\frac{1}{2} br\right)\left(\frac{2\pi r}{b}\right) = \pi r^2$

Fig. 8.

Fig. 9.

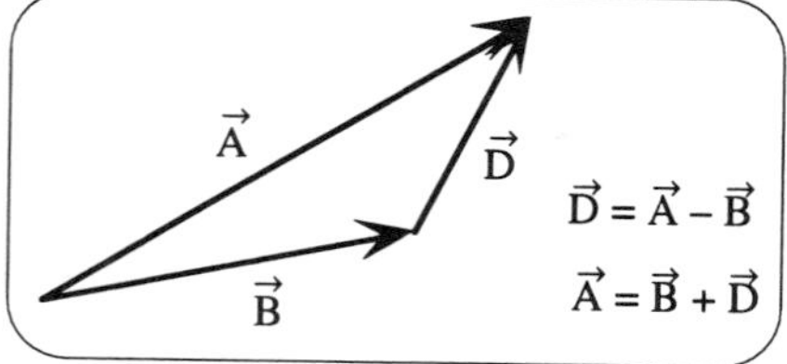

Fig. 10.

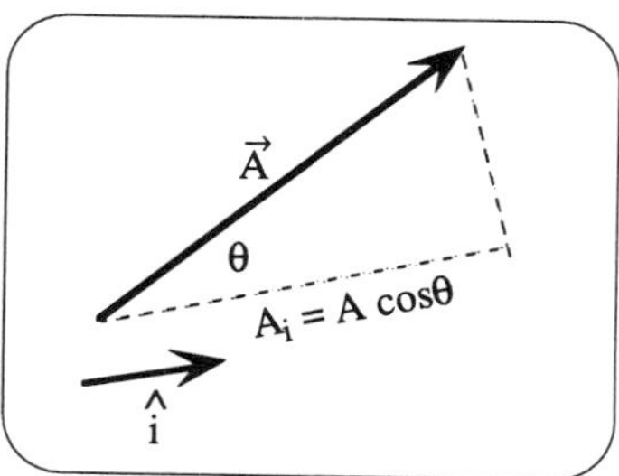

Fig. 11.

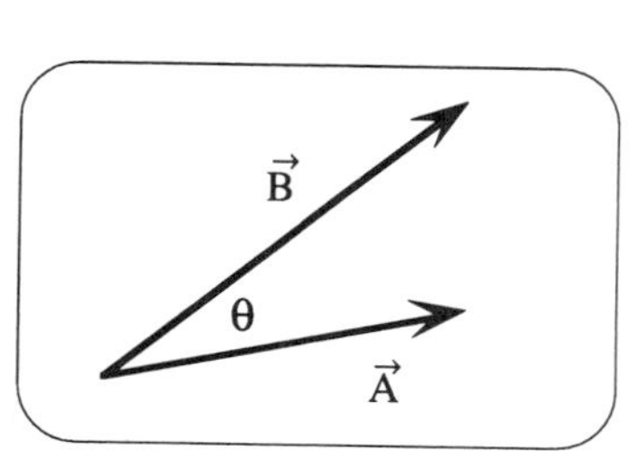

Fig. 12.

Rates of change (derivatives)

$$\frac{dc}{dt} = 0 \ \{\text{if c is a constant}\}$$

$$\frac{d(t)}{dt} = 1, \quad \frac{d(t^2)}{dt} = 2t$$

$$\frac{d(t^n)}{dt} = n\, t^{n-1}$$

$$\Delta(cQ) = c\, \Delta Q\,, \qquad \frac{d(cQ)}{dt} = c\,\frac{dQ}{dt}$$

$$\Delta(Q_1 + Q_2) = \Delta Q_1 + \Delta Q_2\,, \qquad \frac{d(Q_1+Q_2)}{dt} = \frac{dQ_1}{dt} + \frac{dQ_2}{dt}$$

B. Overview of mechanics

Goal: The goal of the science of mechanics is to discover concepts and principles enabling one to predict or explain motions.

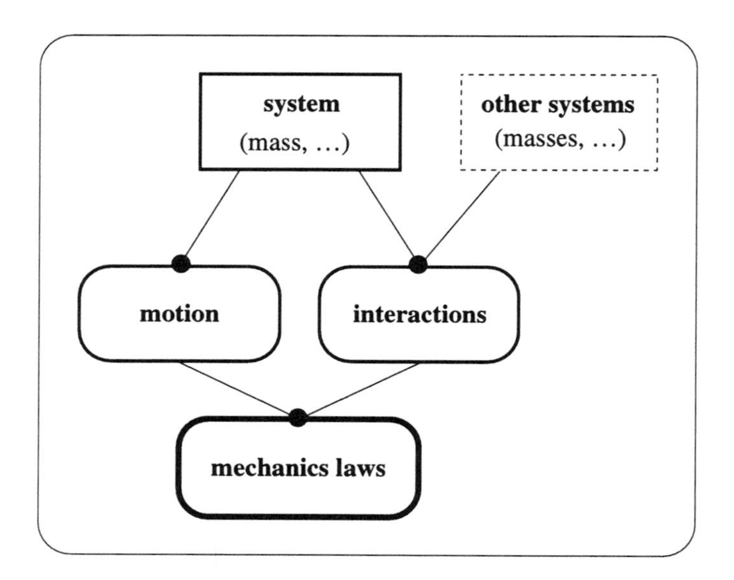

System. A single particle or any specified set of particles.
Characteristics: Mass, charge, moment of inertia, ...

Motion.
Descriptive concepts:
Basic: Position, velocity, acceleration.
Angular position, angular velocity, angular acceleration.
Complex: Momentum, angular momentum, kinetic energy, center of mass.

Interactions.
Descriptive concepts: Force, torque, work, potential energy.
Kinds of interactions and their properties:
Long-range interactions: Gravitational, electric, magnetic, ...
Contact interactions: With spring, with string, with touching solid, ...

Mechanics laws. (Relations between a system's motion and its interactions.)
Basic law for a particle: Newton's law.
Laws for a system: Momentum law, angular-momentum law, energy law.

C. Motion of a particle

Descriptive concepts

Basic concepts

Position vector .... $\vec{r}$

Velocity .... $\vec{v} = d\vec{r}/dt$

Acceleration .... $\vec{a} = d\vec{v}/dt$

Rotational motion

Angular position.................... ϕ

Angular velocity.................... $\omega = d\phi/dt$

Angular acceleration .... $\alpha = d\omega/dt$

Distance .... $s = r\phi$ $(r = \text{radius})$

Speed .... $v = r\omega$

Accel. components .... $a_v = r\alpha$, $a_c = r\omega^2$

Special motions

Straight-line motion (acceleration a_x = constant)

Rel (vel, t) ....$v_x - v_{xo} = a_x t$

Rel (disp, t).............$D_x = x - x_o = v_{xo}t + \frac{1}{2}a_x t^2$

Rel (vel, disp).........$v_x{}^2 - v_{xo}{}^2 = 2\,a_x D_x$

Circular motion (speed v, radius r)

Acceleration components

Along $\vec{v}$ .... $a_v = dv/dt$

$\perp$ to $\vec{v}$ (to center)....... $a_c = v^2/r$

Period.............................. $T = 2\pi r/v$

Frequency........................ $f = 1/T$

Motions relative to different frames

Positions .... $\vec{r}_{PA} = \vec{r}_{PB} + \vec{r}_{BA}$

Velocities .... $\vec{v}_{PA} = \vec{v}_{PB} + \vec{v}_{BA}$

Accelerations............... $\vec{a}_{PA} = \vec{a}_{PB} + \vec{a}_{BA}$

(Particle P, frames A and B.)

($\vec{r}_{PA}$ = position of P relative to A, ...)

D. Interactions

(Interactions of a particle with another particle or object)

Relation between mutual forces: $\vec{F}_{12} = -\vec{F}_{21}$ {Forces have equal magnitudes and opposite directions}

Interaction	Force: Direction	Force: Magnitude	Potential energy
Long-range interactions			
Gravitational			
Due to earth, near surface	$\vec{F}$ vertically down (to center of earth)	$F = mg$ {$g = 9.80$ m/s^2}	$U = mgh$
Due to any particle	$\vec{F}$ attractive	$F = Gm_1m_2/R^2$ {$G = 6.67 \times 10^{-11}$ N m^2/kg^2}	$U = -Gm_1m_2/R$
Electric			
Due to charged particle	$\vec{F}$ repulsive for charges of same sign, attractive for opposite signs	$F = k_e \lvert q_1q_2 \rvert /R^2$ {$k_e = 8.99 \times 10^9$ N m^2/C^2}	$U = k_eq_1q_2/R$
Contact interactions			
Due to spring	$\vec{F}$ along spring, to undeformed position (opposes deformation)	F increases with deformation x. ($F = 0$ if $x = 0$.)	
For small deformation	....	$F_x = -kx$	$U = \frac{1}{2}kx^2$
Due to string	$\vec{F}$ attractive {tension force} (opposes elongation)	$F \neq 0$ if string is taut, $= 0$ if slack.	
Due to touching object	$\vec{F} = \vec{N} + \vec{f}$		
Normal force	$\vec{N}$ perpendicular to surface, repulsive (opposes compression)	$N \neq 0$ if touching, $= 0$ otherwise.	
Friction force	$\vec{f}$ parallel to surface (opposes *relative* sliding)		
Kinetic (relative motion)	Direction opposite to relative velocity	$f = \mu_k N$	
Static (no relative motion)	Direction so that relative accel = 0	f so that relative accel = 0; $f \leq \mu_s N$	

E. Mechanics laws

Relations (motion ↔ interactions) for any system

(All derivable from Newton's law $m\vec{a} = \vec{F}_{tot}$ for a particle.)

Momentum law

* *Relates motion and external interactions.*
* *Relates velocities at any two instants if total external force is zero.*

$$\frac{d\vec{P}}{dt} = \vec{F}_{ext}$$

$m\,\vec{a} = \vec{F}_{tot}$ (for a particle) {*Newton's law*}

$M\,\vec{A}_c = \vec{F}_{ext}$ (for center of mass)

Angular-momentum law
(for fixed-direction axis)

* *Relates rotational motion and external interactions.*
* *Relates ang. velocities at any two instants if total external torque is zero.*

$$\frac{dL}{dt} = \tau_{ext}$$

$I\,\alpha = \tau_{ext}$ (if I is constant)

Energy law

* *Relates speeds and positions at any two instants (without mention of time).*

$$\Delta E = W_{oth}$$

(In the above, each law has been accompanied by an indication of the useful information which it can provide.)

F. Mechanics laws and definitions

(Each law relates motion and interactions)

Momentum law

Validity: * *For inertial frame.*
Utility: * *Relates motion & ext. interactions.*
* *Relates velocities at any two instants if total external force is zero.*

$$\frac{d\vec{P}}{dt} = \vec{F}_{ext}$$

$m\vec{a} = \vec{F}_{tot}$ (for a particle) *{Newton's law}*

$M\vec{A}_c = \vec{F}_{ext}$ (for center of mass)

Motion		Interactions	
$\vec{P} = \Sigma\, \vec{p}_n$	{momentum}	$\vec{F}_{ext} = \Sigma\, \vec{F}_{ns'}$	{total external force = sum of forces on all internal particles n by all external particles s')
$\vec{p} = m\, \vec{v}$	{momentum of particle}	$\vec{F}_{int} = 0$	{total internal force = sum of forces on all internal particles by all other internal particles = 0}
$\vec{P} = M\, \vec{V}_c$			
$M = \Sigma\, m_n$	{mass of system}		
$\vec{R}_c = (\Sigma\, m_n \vec{r}_n)/M$	{position of center of mass}		
$\vec{V}_c = d\vec{R}_c/dt,\ \vec{A}_c = d\vec{V}_c/dt$	{velocity & accel of CM}		

Angular-momentum law

(for fixed-direction axis)

Validity: * *For inertial frame.*
* *Relative to CM, always.*
Utility: * *Relates rotation & ext. interactions.*
* *Relates ang. velocities at any two instants if total ext. torque is zero.*

$$\frac{dL}{dt} = \tau_{ext}$$

$I\alpha = \tau_{ext}$ (if I is constant)

Motion		Interactions	
$L = \Sigma\, \ell_n$	{angular momentum}	$\tau_{ext} = \Sigma \tau_{ns'}$	{total external torque}
$\ell = mrv \sin\theta = rp \sin\theta$	{ang. momentum of particle}	$\tau = rF \sin\theta = rF_{\perp r} = Fr_{\perp F}$	{torque on particle}
$L = I\, \omega$	{ω = ang. velocity, $\alpha = d\omega/dt$}		
$I = \Sigma\, m_n r_n^2$	{moment of inertia}		
$I = MR_c^2 + I'$	{I' = moment about center of mass}		

Energy law

Validity: * *For inertial frame.*
Utility: * *Relates speeds & positions at any two instants (without mention of time).*

$$\Delta E = W_{oth}$$

$$E = K + U$$

Motion		Interactions	
$K = \Sigma\, K_n$	{kinetic energy}	W_{oth}	{other work, by all forces not included in U}
$K = \frac{1}{2} mv^2$	{kinetic energy of particle}	$W = W_{int} + W_{ext}$	{work on system = internal work + external work}
$K = K_c + K'$		$W = \int d'W = \int \vec{F} \cdot d\vec{r}$	{work on particle}
$K_c = \frac{1}{2} MV_c^2$	{kinetic energy of CM}	$U_A = W_{AS}$	{potential energy in state A} {S = standard state} *(for conservative forces only)*
K'	{kin. energy relative to CM}	$\Delta U = -W$	{W = work by conservative forces}
$K_{rot} = \frac{1}{2} I\omega^2$	{rotational kinetic energy}	$U = U_{int} + U_{ext}$	{potential energy of system = internal pot. energy of all particle pairs plus external potential energy}

G. Problem-solving method

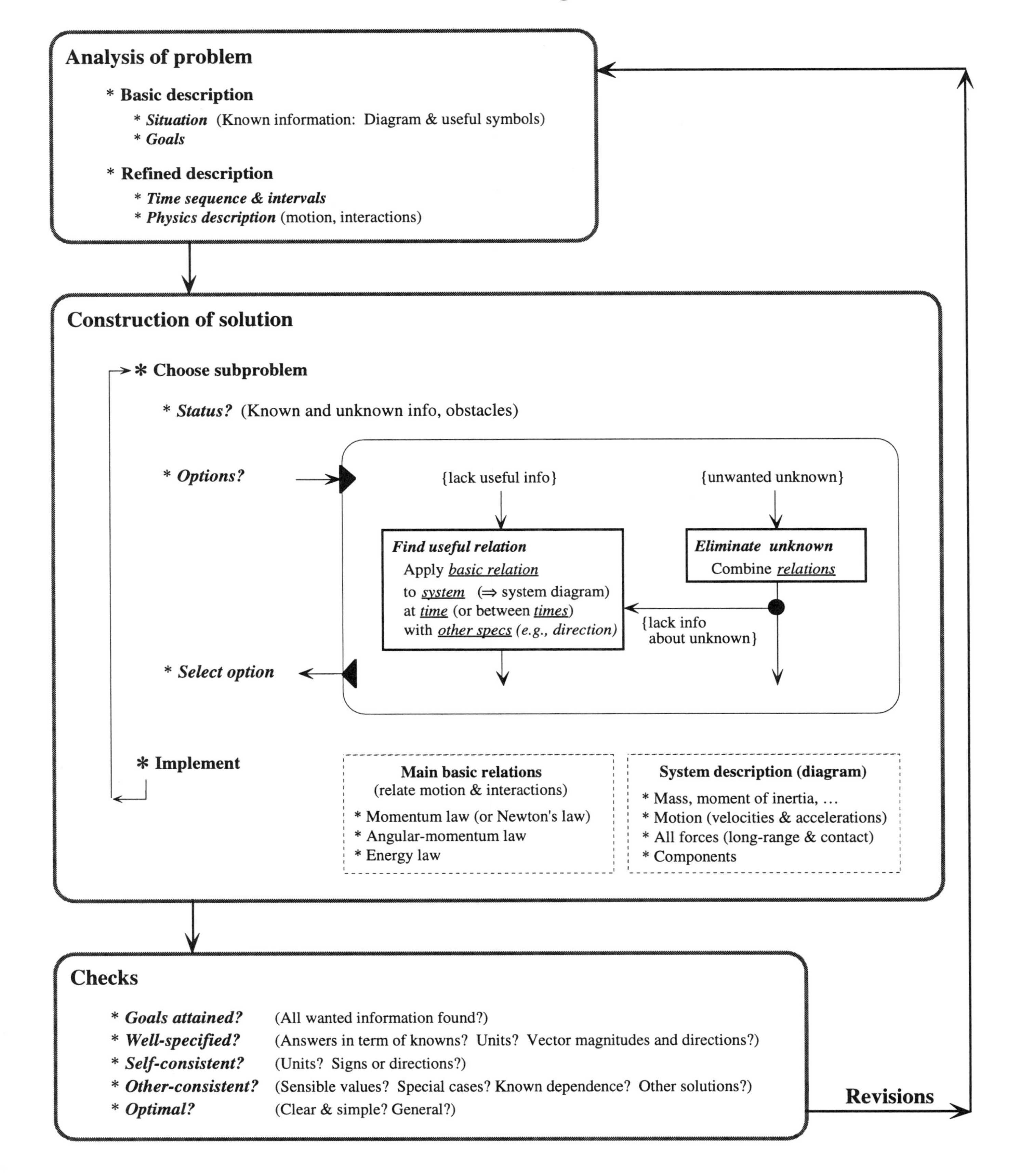

Index

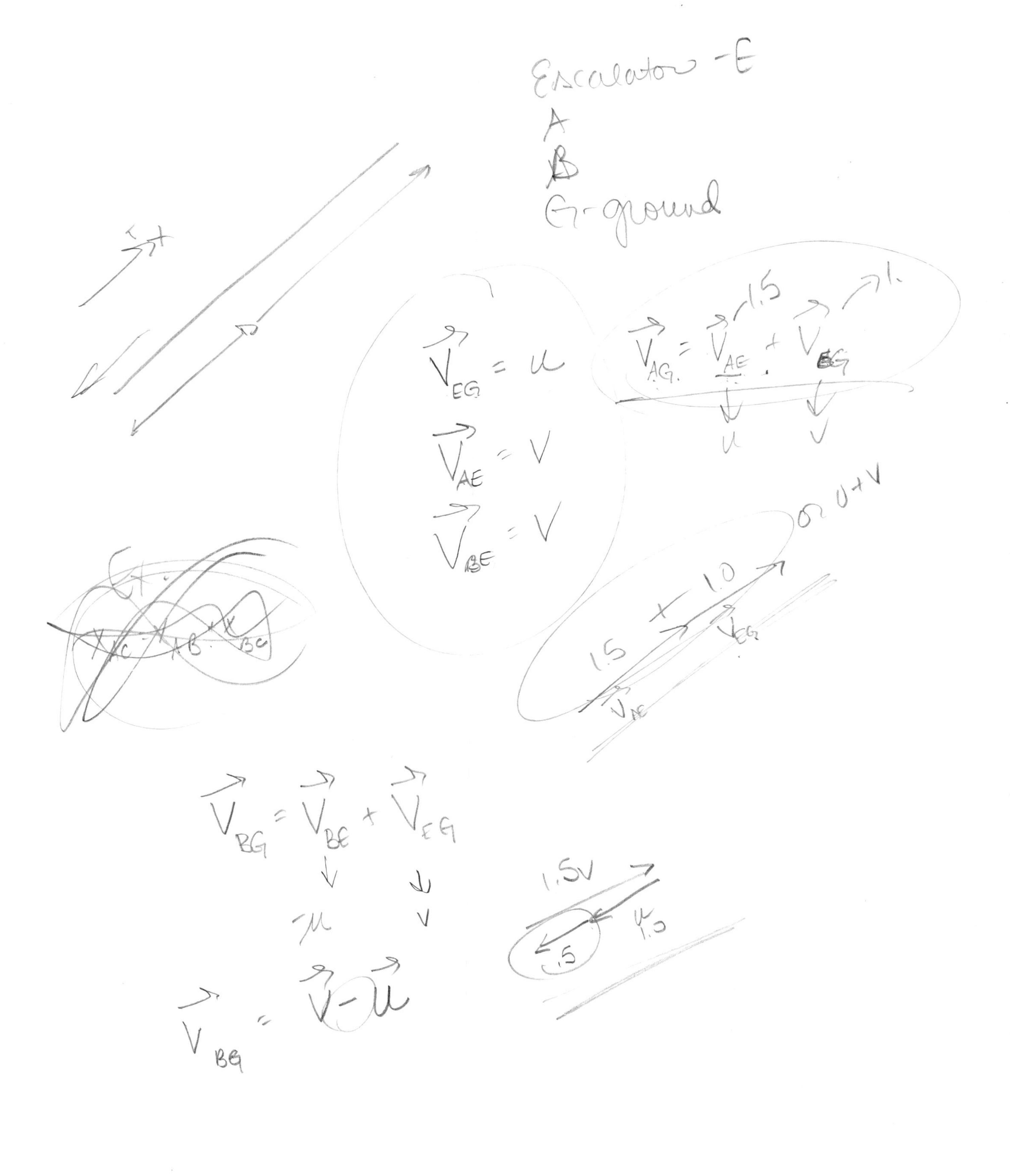

Escalator - E
A
B
G - ground
V EG = u
V AE = V
V BE = V
V AG = V AE + V EG
1.5
1.
u
V
or u+V
1.5
+
1.0
V EG
V AE
V BG = V BE + V EG
-u
V
V BG = V - u
1.5V
u
1.0
.5

mvpeach@yahoo.com.

87

348

97
95
92
284

348